Tribology of Textured Surfaces

Tribology of Textured Surfaces

Guest Editors

Min Zou
Mahyar Afshar-Mohajer
Robert A. Fleming

MDPI

Basel • Beijing • Wuhan • Barcelona • Belgrade • Novi Sad • Cluj • Manchester

Guest Editors

Min Zou
Department of Mechanical Engineering
University of Arkansas
Fayetteville
USA

Mahyar Afshar-Mohajer
Department of Mechanical Engineering
University of Arkansas
Fayetteville
USA

Robert A. Fleming
College of Engineering & Computer Science
Arkansas State University
Jonesboro
USA

Editorial Office
MDPI AG
Grosspeteranlage 5
4052 Basel, Switzerland

This is a reprint of the Special Issue, published open access by the journal *Lubricants* (ISSN 2075-4442), freely accessible at: https://www.mdpi.com/journal/lubricants/special_issues/BAAFUG04RN.

For citation purposes, cite each article independently as indicated on the article page online and as indicated below:

Lastname, A.A.; Lastname, B.B. Article Title. *Journal Name* **Year**, *Volume Number*, Page Range.

ISBN 978-3-7258-5103-4 (Hbk)
ISBN 978-3-7258-5104-1 (PDF)
https://doi.org/10.3390/books978-3-7258-5104-1

Contents

About the Editors

Min Zou

Min Zou is a Distinguished Professor of Mechanical Engineering at the University of Arkansas (UA), where she holds the Twenty-First Century Endowed Chair in Materials, Manufacturing, and Integrated Systems. She is also affiliated with the Institute for Nanoscale Science and Engineering and the Materials Science and Engineering Graduate Program at UA. She earned her Ph.D. in Mechanical Engineering from the Georgia Institute of Technology and previously worked at the Shanghai Aircraft Research Institute and Seagate Technology. Her research focuses on advanced three-dimensional (3D) nanoprinting, nanoscale surface engineering, nanomechanics, and tribology, with applications in manufacturing, coatings, and biomedical fields. Her work has been widely recognized for both its fundamental contributions and technology transfer, including the commercialization of innovative surface technologies.

She is an elected Fellow of the National Academy of Inventors, the American Society of Mechanical Engineers, the Society of Tribologists and Lubrication Engineers (STLE), and the American Institute for Medical and Biological Engineering. Among her honors are the STLE Al Sonntag Award, the Edmond E. Bisson Award, the Southeastern Conference Faculty Achievement Award, and the US National Science Foundation CAREER Award. She has also received the Faculty Distinguished Achievement Award for Research from the Arkansas Alumni Association and the Science and Technology Advancement Award from the Ministry of Aeronautics and Aerospace of China. Beyond her research, she has served on editorial boards of leading journals, chaired the STLE Annual Meeting Program Committee, and served as a member of the STLE Board of Directors.

Mahyar Afshar-Mohajer

Mahyar Afshar-Mohajer graduated with a Ph.D. in Mechanical Engineering from the University of Arkansas in 2022 where he subsequently spent a year as a postdoctoral fellow. During his Ph.D. and postdoctoral work, his research focused on the 3D printing of micro/nano-textured surfaces and in situ SEM observation of their friction and deformation.

Robert A. Fleming

Robert "Drew" Fleming is an Associate Professor of Mechanical Engineering in the College of Engineering & Computer Science at Arkansas State University (A-State). He earned a Ph.D. in Engineering from the University of Arkansas in 2017, where he was a National Science Foundation Graduate Research Fellow. At A-State, he directs both the Advanced Functional Materials Laboratory and the Experimental Nanomechanics Facility, which jointly conduct research on a broad portfolio of projects in the areas of surface science/engineering, computational materials science, and experimental nanomechanics. His research has been funded by the DOE, NSF, NIH, and NASA, and he is the recipient of the Ralph E. Powe Junior Faculty Enhancement Award from ORAU and the Edmond E. Bisson Award from the STLE. He currently serves as an Associate Editor for the ASME *Journal of Tribology*.

lubricants

Editorial

Tribology of Textured Surfaces

Min Zou [1,2,*], Mahyar Afshar-Mohajer [1,2] and Robert A. Fleming [3]

[1] Department of Mechanical Engineering, University of Arkansas, Fayetteville, AR 72701, USA; mahyar.am@hotmail.com
[2] Center for Advanced Surface Engineering, University of Arkansas, Fayetteville, AR 72701, USA
[3] Department of Mechanical Engineering, Arkansas State University, Jonesboro, AR 72467, USA; rofleming@astate.edu
* Correspondence: mzou@uark.edu

1. Introduction

Surface texturing has emerged as a transformative approach to improving friction, lubrication, and wear across a wide range of mechanical components. From bio-inspired micro-dimples to engineered macro-grooves, texture designs are increasingly applied in biomedical implants, bearings, seals, and automotive systems. This Special Issue, "Tribology of Textured Surfaces," compiles 14 papers that collectively cover fabrication techniques, multi-scale optimization, predictive modeling, and diverse application areas.

The following sections summarize the major themes of this Special Issue. Section 2 reviews the fabrication techniques and texture scales addressed by the authors. Section 3 examines modeling and numerical studies, while Section 4 categorizes the work by application domains. Section 5 highlights the balance between experimental and computational methods, and Section 6 outlines emerging trends and future directions.

2. Fabrication Techniques and Texture Scales

Several papers in this Special Issue focus on advanced fabrication methods for surface texturing. Laser surface texturing (LST) is a recurring theme due to its precision and versatility. Bennett and Zou (contribution 1) produced 5 μm micro-dimples on CoCrMo surfaces to enhance the tribological performance of artificial hip joints. Makowski et al. (contribution 2) used LST to prepare steel substrates for ta-C/MoSx coatings, achieving improved tribological performance in both air and vacuum. Similarly, Liu et al. (contribution 3) employed laser texturing to create millimeter-scale warhead-type grooves on rubber stators of screw pumps, optimizing their geometry to improve pressure generation and wear resistance. Yue et al. (contribution 4) conducted a detailed study of the combined effects of plateau roughness and laser-textured dimples under oil lubrication. By systematically varying dimple density and roughness parameters, they revealed how micro-textures interact with background roughness to optimize oil film formation and reduce boundary friction.

Hybrid approaches that combine coatings and surface textures are also prominently featured. Wos et al. (contribution 5) explored the synergy between diamond-like carbon (DLC) coatings and textured dimples, achieving lower friction and improved seizure resistance in reciprocating sliding. Makowski et al. (contribution 2) extended this concept by integrating micro-textures with coatings to meet the demands of different operating environments. Mechanical grinding and multi-scale texturing were also studied. Parameswaran et al. (contribution 6) demonstrated that controlled grinding of steel counterfaces can significantly reduce sliding friction against aluminum. Tewelde et al.

(contribution 7) optimized multi-scale textures on prosthetic hip implant surfaces, improving lubricant film thickness and reducing wear.

The studies in this section highlight the wide range of scales that textured surfaces can cover, from micro-dimples (5–100 μm) (contributions 1 and 2) to macro-scale grooves approaching 3 mm (contribution 3). This diversity underscores the adaptability of surface texturing for applications requiring tailored tribological properties.

3. Modeling and Numerical Innovations

A number of contributions focus on modeling as a tool for understanding and optimizing textured surfaces. Ekşioğlu and Zou (contribution 8) developed a non-linear optimization framework for micro-electromechanical systems and nano-electromechanical systems, designed to minimize the real contact area between surfaces. Li et al. (contribution 9) employed computational fluid dynamics to study lubrication and frictional behavior in textured angular contact ball bearings. Shi et al. (contribution 10) presented a fluid dynamic model of rhombus-like textures in screw pumps, while Liu et al. (contribution 3) applied finite element modeling to evaluate the effects of warhead-type grooves.

Scharf et al. (contribution 11) conducted a comprehensive numerical study of wedge-shaped convergent oil film gaps, demonstrating how surface geometry can influence lubrication efficiency. Luz et al. (contribution 12) used deterministic simulations to examine the performance of piston ring/cylinder liner systems in free-piston engines. Ma et al. (contribution 13) combined multi-field coupling and experiments to evaluate the sealing performance of textured metal components used in roller cone bits. Cohen et al. (contribution 14) carried out a stability analysis of secondary piston motions, showing how surface texturing influences lateral and rotational dynamics. Their work highlights how micro-scale texturing can mitigate unwanted secondary motions, reducing energy losses and potential scuffing in reciprocating piston-cylinder systems.

4. Application Categories

The applications of textured surfaces span multiple industries and engineering challenges. In the biomedical field, Bennett and Zou (contribution 1) demonstrated that laser-textured micro-dimples on CoCrMo can reduce wear in artificial hip joints, while Tewelde et al. (contribution 7) optimized multi-scale textures to enhance lubrication and extend the longevity of prosthetic hip implants. For bearings and seals, Li et al. (contribution 9) showed how textured topographies improve lubrication performance in angular contact ball bearings, and Ma et al. (contribution 13) examined how surface textures enhance the performance of metal seals used in roller cone bits.

In energy and automotive systems, Luz et al. (contribution 12) evaluated the effects of surface texturing on the frictional behavior of piston ring/cylinder liner systems in free-piston engines. Cohen et al. (contribution 14) studied textured pistons, providing insights into secondary motion dynamics and frictional losses. Industrial machinery applications were addressed by Shi et al. (contribution 10) and Liu et al. (contribution 3), who investigated screw pumps and demonstrated how textured surfaces can improve wear resistance and efficiency. Coating and sliding surface studies, including those by Wos et al. (contribution 5), Makowski et al. (contribution 2), Yue et al. (contribution 4), and Parameswaran et al. (contribution 6), explored strategies to reduce friction and improve surface durability under diverse lubrication conditions.

5. Experimental vs. Computational Balance

This Special Issue achieves a balance between experimental and modeling studies. Experimental investigations, such as those by Bennett (contribution 1), Parameswaran (contribution 6), and Yue (contribution 4), provide direct validation of tribological performance improvements. Modeling-heavy studies, including those by Ekşioğlu (contribution 8), Scharf (contribution 11), and Luz (contribution 12), offer predictive insights into surface behavior. Hybrid studies, such as those by Liu at al. (contribution 3) and Ma et al. (contribution 13), combine computational simulations with experimental validation, reinforcing the reliability of their findings.

6. Emerging Trends and Outlook

Several important trends emerge from this Special Issue. There is a clear emphasis on integrating computational modeling with experiments, enabling better optimization of texture parameters. Scale-aware design is another recurring theme, with micro-dimples (contribution 1) tailored for biomedical wear reduction and macro-grooves (contribution 3) engineered for fluid transport in screw pumps. The combination of coatings and surface textures (contributions 2 and 5) reflects an interest in multi-functional surfaces that can perform well under demanding conditions. Looking ahead, AI-driven optimization, additive manufacturing for customized textures, and the development of multi-functional coatings offer promising avenues for future research.

7. Conclusions

The 14 papers in this Special Issue highlight the diversity and impact of textured surface research. By advancing fabrication techniques, optimizing texture scales, and integrating computational modeling with experiments, these contributions provide effective strategies to reduce friction, enhance lubrication, and improve wear resistance across biomedical, automotive, and industrial applications.

Acknowledgments: We would like to thank all contributing authors for sharing their high-quality research and the reviewers for their valuable feedback, which significantly strengthened the papers in this Special Issue. Finally, we appreciate the engagement of the broader tribology community, whose efforts continue to drive advances in textured surface research.

Conflicts of Interest: M.Z. is a co-author of contributions 1 and 8 in this Special Issue. The other authors declare no conflicts of interest.

List of Contributions:

1. Bennett, W.B.; Zou, M. Small Laser-Textured Dimples for Improved Tribological Performance of CoCrMo in Artificial Hip Joints. *Lubricants* **2025**, *13*, 158.
2. Makowski, S.; Härtwig, F.; Soldera, M.; Ojeil, M.; Lorenz, L.; Kaulfuß, F.; Lasagni, A.F. Improved Tribological Performance of ta-C/MoSx Coatings Deposited on Laser Micro-Structured Steel Substrates. *Lubricants* **2024**, *12*, 200.
3. Liu, X.; Sun, Y.; Liu, C.; Shi, X.; Niu, X.; Zheng, G.; Wei, W.; Wei, S.; Huang, S. Lubrication Characteristics of a Warhead-Type Irregular Symmetric Texture on the Stator Rubber Surfaces of Screw Pumps. *Lubricants* **2024**, *12*, 397.
4. Yue, H.; Schneider, J.; Deng, J. Laser Surface Texturing for Ground Surface: Frictional Effect of Plateau Roughness and Surface Textures under Oil Lubrication. *Lubricants* **2024**, *12*, 22.
5. Wos, S.; Koszela, W.; Dzierwa, A.; Pawlus, P. Combined Effects of DLC Coating and Surface Texturing on Seizure and Friction in Reciprocating Sliding. *Lubricants* **2025**, *13*, 244.
6. Parameswaran, G.; Jayaram, V.; Kailas, S.V. Predicting Low Sliding Friction in Al–Steel Reciprocating Sliding Experiment after a Controlled Grinding of the Steel Counterface. *Lubricants* **2024**, *12*, 292.
7. Tewelde, F.B.; Allen, Q.; Zhou, T. Multiscale Texture Features to Enhance Lubricant Film Thickness for Prosthetic Hip Implant Bearing Surfaces. *Lubricants* **2024**, *12*, 187.
8. Ekşioğlu, S.D.; Zou, M. A Non-Linear Optimization Model for Controlling the Real Area of Contact in Surface Texture Design. *Lubricants* **2025**, *13*, 163.

9. Li, Z.; Yin, S.; Zhang, Q.; Zhang, X.; Zhang, H. Analysis of Lubrication Characteristics and Friction Test of Texture Topography of Angular Contact Ball Bearing Based on CFD. *Lubricants* **2025**, *13*, 41.
10. Shi, X.; Liu, X.; Liu, C.; Hao, Z.; Huang, S.; Sun, Y.; Niu, X. Numerical Simulation of the Lubrication Performance of the Stator and Rotor Friction Pair Surface Rhombus-Like Texture in Screw Pumps. *Lubricants* **2025**, *13*, 157.
11. Scharf, R.; Maier, M.; Pusterhofer, M.; Grün, F. A Comprehensive Numerical Study of a Wedge-Shaped Textured Convergent Oil Film Gap. *Lubricants* **2024**, *12*, 121.
12. Luz, F.K.C.; Profito, F.J.; dos Santos, M.B.; Silva, S.A.N.; Costa, H.L. Deterministic Simulation of Surface Textures for the Piston Ring/Cylinder Liner System in a Free Piston Linear Engine. *Lubricants* **2024**, *12*, 12.
13. Ma, Y.; Li, Z.; Yuan, Z.; Meng, X.; Peng, X.; Jiang, J. Multi-Field Coupling Numerical Analysis and Experimental Validation of Surface-Textured Metal Seals in Roller Cone Bits. *Lubricants* **2024**, *12*, 15.
14. Cohen, I.; Tresser, S.; Goltsberg, R. Stability Analysis of the Secondary Motion of a Textured Piston. *Lubricants* **2024**, *12*, 274.

lubricants

MDPI

Article

Small Laser-Textured Dimples for Improved Tribological Performance of CoCrMo in Artificial Hip Joints

William B. Bennett [1,2] and Min Zou [1,2,*]

[1] Department of Mechanical Engineering, University of Arkansas, Fayetteville, AR 72701, USA; wbennett@uark.edu
[2] Center for Advanced Surface Engineering, University of Arkansas, Fayetteville, AR 72701, USA
* Correspondence: mzou@uark.edu; Tel.: +1-479-575-6671

Abstract: This study investigates the impact of small dimples on the tribological properties of CoCrMo (CCM) surfaces. Laser-ablated textures with 5 μm diameter dimples were fabricated at varying aspect ratios (0.1, 0.2, 0.3) and surface densities (5%, 15%, 25%) to evaluate their effects on friction and wear when paired with ultra-high molecular weight polyethylene (UHMWPE) counterfaces. The results showed that small dimples significantly reduced and stabilized the coefficient of friction (CoF) and wear compared to untextured CCM and larger dimples as reported in the literature. The texture configuration with a 5% surface density and 0.1 aspect ratio achieved the best combination of friction and wear performance by facilitating the formation of a stable and uniform lubricant film during sliding. These findings underscore the potential of small, precisely engineered surface textures to improve the tribological performance of CCM, offering a promising approach for reducing friction and wear in artificial hip joints.

Keywords: small dimples; laser surface texturing; CoCrMo; artificial hip joint; surface engineering; friction and wear performance

1. Introduction

Total hip replacement (THR) surgery has significantly improved quality of life for individuals suffering from severe arthritis or joint injuries. In 2021 alone, over 450,000 THR procedures were performed in the United States, and the number is expected to rise with the aging population [1]. Despite their success, the longevity of artificial hip joints remains a critical concern, as wear-induced implant failure is one of the primary reasons for revision surgeries [2]. The articulation between the femoral head and acetabular liner generates wear particles that may lead to osteolysis, ultimately necessitating revision surgery [3,4].

To address wear-related challenges, extensive research has explored strategies to reduce friction and wear in metal-on-polyethylene (MoP) hip implants. Approaches such as low-friction surface coatings [5,6], cross-linked polyethylene liners [7,8], and surface texturing of CoCrMo (CCM) femoral heads have been investigated. Among these, surface texturing has emerged as a promising technique due to its ability to enhance lubrication, reduce contact stress, and improve load distribution. Textured dimples serve as lubricant reservoirs, reducing direct asperity contact and improving tribological performance [9–11].

Artificial hip joints operate under dynamic loading conditions, where lubrication transitions among boundary, mixed, and elasto-hydrodynamic (EHL) regimes depending on motion, contact pressure, and lubricant properties. Textured surfaces influence these lubrication transitions by promoting fluid retention and pressure redistribution, improving load-carrying capacity in orthopedic applications [12,13]. Optimized textures can extend the

https://doi.org/10.3390/lubricants13040158

duration of EHL lubrication, preventing lubricant starvation and reducing wear-induced damage [14,15].

Laser surface texturing (LST) offers a precise, repeatable method to fabricate micro- and nano-scale textures on CCM surfaces [16–18]. Prior studies have shown that textured CoCrMo surfaces with dimples (~100 μm) improve wear resistance and reduce friction in hip implants [19]. Chyr et al. [20] demonstrated that a 5% surface density of 100 μm diameter dimples lowered the coefficient of friction (CoF) from 0.26 to 0.12 under a maximum contact pressure of 0.57 MPa. However, at higher contact pressures (1.13 MPa), untextured surfaces outperformed all textured samples, as the microtextures failed to generate sufficient load-carrying capacity to support the increased load and reduce friction. These findings emphasize the importance of tailoring texture designs for specific operating conditions.

The performance of dimpled textures depends on parameters such as diameter, depth, and surface density [21]. Langhorn et al. [22] reported a 68% wear reduction using 100 μm diameter dimples with an optimal aspect ratio of 0.01 (1 μm depth) and 10% surface density on CCM. Despite these findings, most existing studies focus on large dimples (≥20 μm diameter) and exhibit significant wear after extended testing, often without detailed friction data to clarify wear mechanisms [23].

Small dimples (~5 μm in diameter) offer a biomimetic advantage, closely resembling the smallest natural features of human cartilage [24]. The higher dimple density associated with smaller dimples allows for improved lubricant retention, better hydrodynamic lift, and enhanced load distribution. For example, at 25% surface density, the 300 mm^2 contact area in a hip joint can accommodate 15.3 million 5 μm diameter dimples—44 times more than the 38,000 dimples possible with a 100 μm diameter (Figure 1) [25]. This increased density reduces the spacing between dimples, ensuring consistent lubricant reservoirs and superior tribological performance.

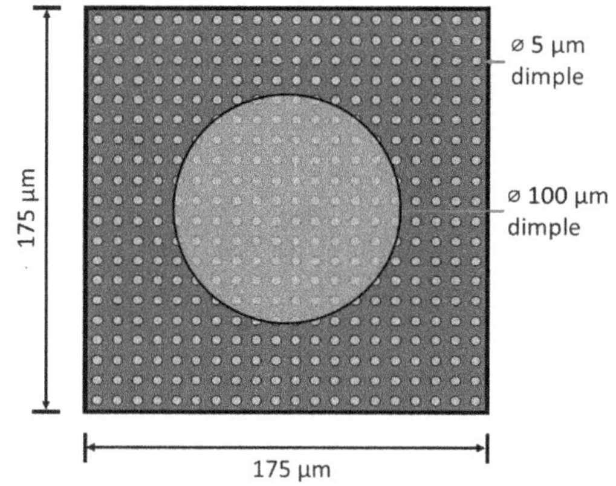

Figure 1. At a 25% surface texture density, a 175 μm × 175 μm area can accommodate only a single 100 μm diameter dimple, but can fit up to 400 dimples with a diameter of 5 μm.

Despite their potential, small dimples fabricated via laser ablation have rarely been studied in tribological research for artificial hip joints. This study investigates how varying aspect ratios and surface densities of 5 μm diameter dimples affect the friction and wear performance of CCM femoral heads. The tribological properties of textured samples are compared against those of untextured CCM to quantify the advantages of small dimple textures. By demonstrating that small dimples outperform the untextured surfaces and larger dimples reported in the literature, this study aims to extend the lifespan of MoP artificial hip joints and improve the quality of life for THR recipients.

2. Experimental Methods

2.1. Sample Fabrication

In this study, ASTM F1537 CoCrMo (CCM) discs, supplied by United Performance Metals (Cincinnati, OH, USA), were tested against ultra-high molecular weight polyethylene (UHMWPE) pins made from GUR1020-E, a highly cross-linked material treated with 100 kGy of radiation and stabilized with vitamin E (VEHXL) from Orthoplastics (Bacup, UK). CCM was selected due to its high mechanical strength, exceptional corrosion resistance, and biocompatibility, making it a preferred material for artificial hip joints [1]. UHMWPE was chosen as the counterface material for its proven performance in metal-on-polyethylene (MoP) configurations, offering excellent biocompatibility, durability, and chemical inertness [26].

The CCM discs were fabricated from a 12.7 mm diameter rod and machined into discs approximately 1.5 mm thick. These discs were polished in several stages to achieve a mirror finish and flatness on both sides. The polishing process began with coarse polishing under a 10 lb load using 1200-grit sandpaper for two 50 s intervals. This was followed by intermediate polishing for three minutes using a 6 μm polycrystalline diamond suspension, and a final polishing step using a 0.06 μm amorphous colloidal silica suspension for 90 s. During polishing, the specimen holder rotated at 60 rpm, while the polishing wheels operated at 150 rpm. After polishing, the discs were cleaned by submerging them in acetone and placing them in a 10 min sonication bath (Branson CPX2800, Branson Ultrasonics, Brookfield, CT, USA). They were then rinsed with isopropyl alcohol (IPA) and deionized (DI) water and dried with nitrogen gas.

Laser surface texturing (LST) was used to enhance the tribological performance of the CCM discs. Textures were created using a femtosecond laser machining system (Oxford Laser A5 Femtosecond, Oxford Lasers Ltd., Didcot, UK), which employed an Ytterbium-doped Potassium Gadolinium Tungstate (Yb:KGW) femtosecond laser with a wavelength of 532 nm, a pulse duration of less than 290 fs, and a maximum power output of 2.6 W. Grid patterns of circular dimples (Table 1) were fabricated using varying laser parameters, resulting in dimple diameters of 5 μm and depths of 0.5, 1.0, and 1.5 μm. These laser parameters, shown in Table 2, were chosen because they yielded the highest quality and most consistent dimples across textures. Surface densities of 5%, 15%, and 25% were achieved by adjusting the spacing between laser pulses. A dwell time of 0.1 s was applied between each pulse to ensure uniformity in dimple shape and depth. The texture density and depth ranges were selected to encompass the optimal values identified in the literature [20,22].

Table 1. Design characteristics of various laser-textured samples.

Texture Type	Diameter (μm)	Depth (μm)	Aspect Ratio	Surface Density (%)
A1	5	0.5	0.1	5
A2	5	1.0	0.2	5
A3	5	1.5	0.3	5
B1	5	0.5	0.1	15
B2	5	1.0	0.2	15
B3	5	1.5	0.3	15
C1	5	0.5	0.1	25
C2	5	1.0	0.2	25
C3	5	1.5	0.3	25

Table 2. Laser parameters for fabricating dimples with different aspect ratios.

Aspect Ratio	Power (%)	Drill Time (s)	RA Divider	Frequency (Hz)
0.1	15	0.08	800	700
0.2	20	0.10	800	700
0.3	22	0.15	800	700

The VEHXL UHMWPE pins were machined from a 40 mm diameter, 500 mm long rod into individual pins measuring 4 mm in diameter and 12 mm in length using a CNC machine (PCNC 440, Tormach, Madison, WI, USA). The machining process included an initial leveling step to flatten the bulk material, followed by precise fabrication of the pins to their target dimensions. A small notch was drilled into each pin edge to serve as a reference point. Prior to testing, the pins were cleaned thoroughly with IPA and DI water and dried using nitrogen gas to ensure surface cleanliness.

2.2. Surface Topography Characterization

The surface topography of the fabricated CCM discs and VEHXL UHMWPE pins was characterized using a 3D laser scanning confocal microscope (LSCM, VK-X260, Keyence Corporation, Itasca, IL, USA). To ensure reliable and reproducible results, three scans were conducted for each sample. Key parameters, including dimple depth profiles, average surface roughness (Sa), and root mean square surface roughness (Sq), were measured and analyzed. Standard deviations were calculated to provide a comprehensive assessment of the surface texture and its consistency.

2.3. Water Contact Angle Measurements

Water contact angle (WCA) measurements were conducted using a goniometer (OCA 15, DataPhysics Instruments, Charlotte, NC, USA) to assess the wettability of both textured and untextured samples. Prior to measurement, the samples were cleaned by submerging them in acetone, followed by a 10 min sonication bath. After rinsing with DI water, the samples were dried using nitrogen gas. A 3 μL droplet of DI water was carefully deposited onto the sample surface using a syringe, and a backlit image of the droplet was captured. The left and right contact angles were measured and averaged to determine the WCA. For consistency, three measurements were performed for each sample.

2.4. Pin-on-Disc Tribological Testing

Tribological testing was conducted using a pin-on-disc setup with a tribometer (UMT-2, Bruker, San Jose, CA, USA), as shown in Figure 2. To simulate the lubricating environment of a natural hip joint, a mixture of 25% bovine calf serum (BCS, Sigma-Aldrich, St. Louis, MO, USA) and 75% DI water was used as the lubricant. The lubricant temperature was maintained at 37 °C to replicate physiological conditions. A VEHXL UHMWPE pin, measuring 4 mm in diameter, was positioned to slide against a CCM disc in an oscillating motion along a circular arc. The tribometer applied a normal load of 15 N, with the pin oscillating at a frequency of 1 Hz within a 30-degree arc, positioned 4 mm from the disc center. The sliding speed was set at 1 mm/s, and the test was conducted for a total duration of eight hours. These parameters were chosen to best mimic the dynamic motion and contact mechanics of a hip joint during walking, where the femoral head undergoes an arcuate motion against the acetabular cup, rather than a purely linear reciprocating motion. The CoF was calculated as the ratio of the measured friction force to the applied normal load, providing reliable data on the tribological performance of the samples.

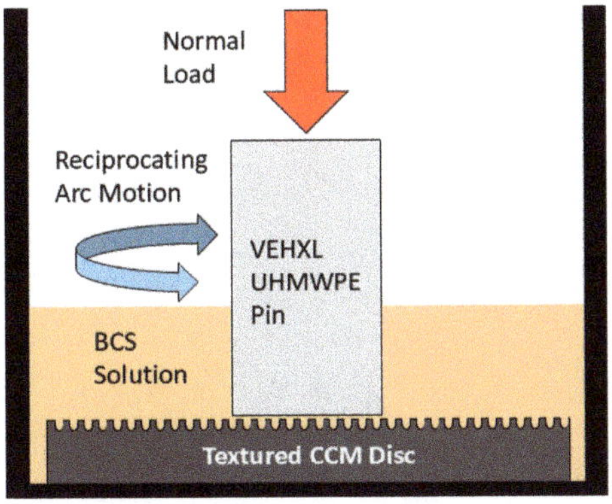

Figure 2. Pin-on-disc experimental setup, showing a VEHXL UHMWPE pin pressed against a textured CoCrMo (CCM) sample submerged in bovine calf serum (BCS) and kept at 37 °C (not to scale).

2.5. Wear Analysis

Wear analysis was performed based on the wear track on the CCM discs and changes in the surface roughness of the VEHXL UHMWPE. High-resolution imaging of the wear tracks was conducted using scanning electron microscopy (SEM) and 3D LSCM to visualize the wear patterns and transfer films. Changes in the surface roughness of the VEHXL UHMWPE pins were analyzed by generating 3D surface maps using LSCM. The pre- and post-test roughness values (Sa) were calculated, and the differences were used to assess the extent of pin wear.

3. Results

3.1. Surface Topography of Fabricated CCM Disc and UHMWPE Pin

The sample surfaces were analyzed at 20× magnification using 3D LSCM. The arrays of dimples for all textured samples are shown in Figure 3. Different surface densities result in varying numbers of dimples within the same area by altering their pitch. Specifically, at 5% surface density, 12 dimples are observed, while 48 and 88 dimples are present at 15% and 25% surface densities, respectively. Textures with lower aspect ratios display a lighter interior surrounded by a darker heat-affected zone around the edges of the dimples.

Figure 4 illustrates the cross-sectional profiles of dimples with three different aspect ratios at a 5% surface density. The femtosecond laser used for fabrication achieved precise dimple depths with excellent dimensional accuracy, even at microscale feature sizes. However, slight protrusions, rising above the surface, sporadically appear on the perimeters of some dimples. These protrusions may contribute to the generation of wear debris during extended rubbing.

The average surface roughness (Sa) and root mean square roughness (Sq) of the textured CCM discs are plotted in Figure 5. As expected, all textured samples exhibit higher roughness values compared to the untextured sample. In general, textures with higher surface densities show greater roughness, while aspect ratio has a much smaller impact on the roughness of the samples. However, as surface density increases, and more dimples contribute to the surface topography, aspect ratio has a greater effect on surface roughness.

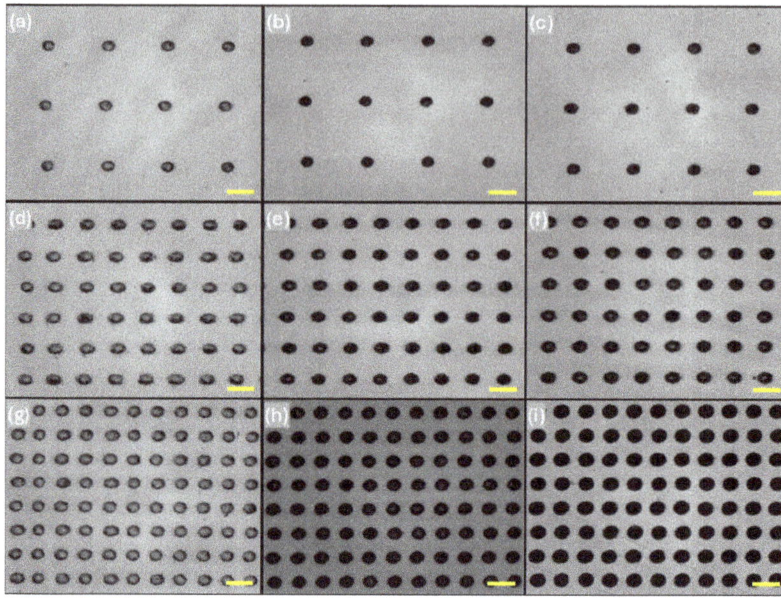

Figure 3. LSCM images of all surface textures: (**a**) A1, (**b**) A2, (**c**) A3, (**d**) B1, (**e**) B2, (**f**) B3, (**g**) C1, (**h**) C2, and (**i**) C3 (all scale bars are 10 μm).

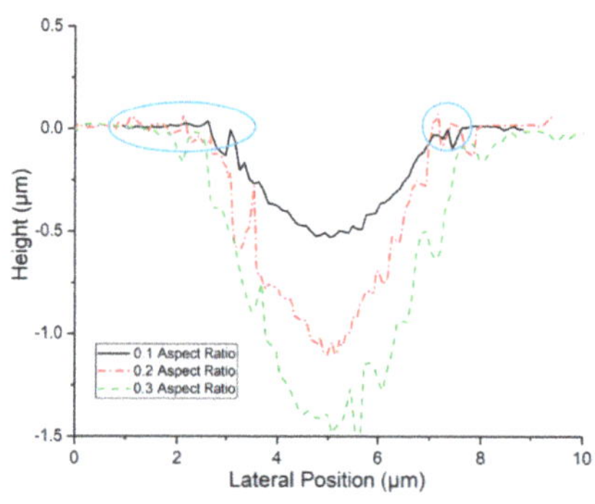

Figure 4. Cross-sectional profiles of dimples at different aspect ratios of 5% surface density textures. Protrusions around the edge have been circled.

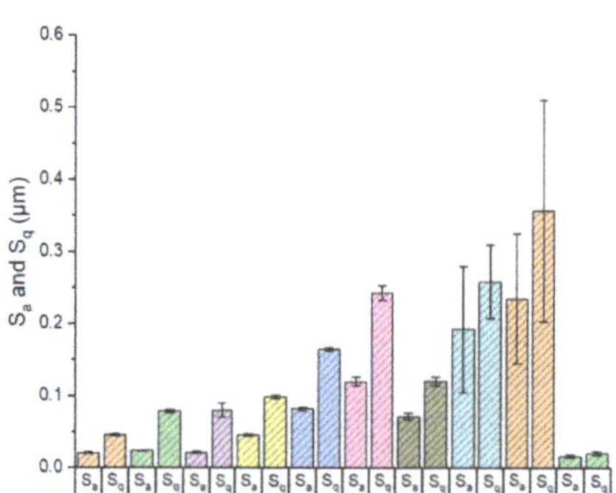

Figure 5. The average surface roughness, S_a, and root mean square roughness, S_q, of different CCM textures.

For the VEHXL UHMWPE pins, the CNC machining process produced a grooved surface texture, as shown in Figure 6. The R_a and R_q of the pins were measured at 2.168 μm and 2.612 μm, respectively. These grooves closely resemble the texture and roughness of new UHMWPE cups used in artificial hip replacements. For instance, Choudhury et al. reported an R_a value of 1.96 μm for a brand new UHMWPE cup [27].

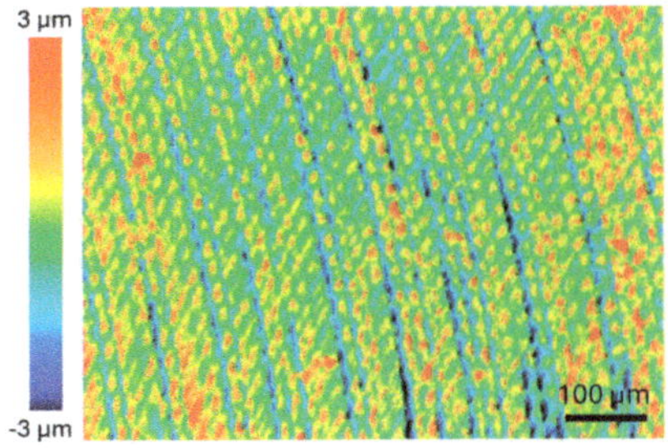

Figure 6. Surface finish of CNC-fabricated VEHXL UHMWPE pin showing machining grooves like new artificial cups.

3.2. *Surface Wettability*

The WCAs for the nine laser-textured surfaces and the untextured CCM sample, as shown in Figure 7, reveal significant differences in surface wettability. The untextured CCM sample exhibited slight hydrophobicity, with a mean WCA of 91.8°. In contrast, all laser-textured surfaces consistently displayed hydrophilic behavior. Among the textured samples, Texture C3 was the least hydrophilic, with a mean WCA of 84.9°, while Texture A2 demonstrated the highest hydrophilicity, achieving a mean WCA of 62.4°. Surface density was identified as the most influential factor, with textures at 5% density exhibiting the lowest WCAs and strongest hydrophilicity. Aspect ratio, on the other hand, had no significant impact on surface wettability.

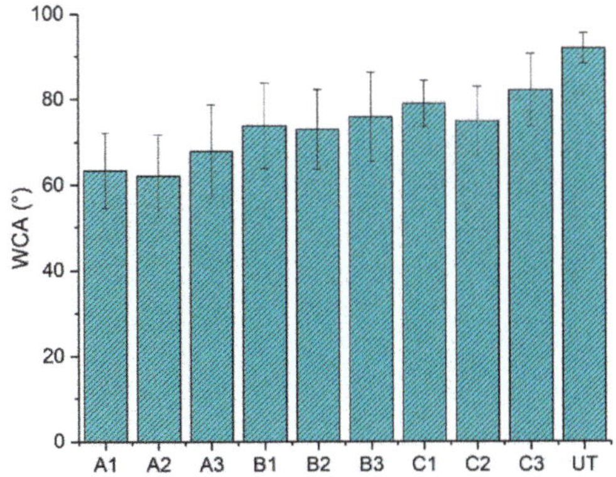

Figure 7. Water contact angles of different textures on CCM.

The reduction in WCA due to laser texturing can be attributed to changes in both surface chemistry and geometry. Laser texturing modifies the oxide layer, leading to changes in surface energy and wettability. While metals generally have higher surface free energy than their oxides, the passive oxide layer on CoCrMo plays a critical role in wetting behavior. Laser processing promotes the growth and restructuring of chromium oxide, which can enhance hydrophilicity depending on its thickness, composition, and nanostructuring. Additionally, the geometric features of the surface influence wetting behavior through the Wenzel [28] and Cassie–Baxter models [29]. For example, textures such as A1 and A2 (5% density, aspect ratios 0.1 and 0.2) achieved the lowest WCAs (62–70°), consistent with the Wenzel model, where water fully penetrates the surface features, increasing the effective solid–liquid contact area. In contrast, denser textures, such as C3 (25% density, aspect ratio 0.3), exhibited higher WCAs (~85°) due to the Cassie–Baxter model, where air pockets trapped within the texture reduce the effective contact area. Despite this, Texture C3 still showed a lower WCA than the untextured surface, underscoring the effects of laser-induced oxide modification and topographical changes on wettability.

Overall, the data in Figure 7 highlight that both oxidation-induced hydrophilicity and texture geometry work synergistically to enhance surface wettability. Sparse, shallow textures like A1 and A2 maximize water spreading due to high hydrophilicity, while denser, deeper textures like C3 retain moderate hydrophilicity. This enhanced hydrophilicity is particularly advantageous for self-lubrication in joint applications, as it helps retain lubricants on the textured surfaces, reducing friction and improving tribological performance.

3.3. Friction and Wear Performance

3.3.1. Coefficient of Friction

Tribological tests were conducted on CCM with VEHXL UHMWPE as the counterface material, and the average CoF results are shown in Figure 8. All laser-textured samples exhibited lower CoF values compared to untextured CCM, with the lowest CoF reaching just 60% of the untextured sample. This underscores the effectiveness of laser texturing in enhancing tribological performance.

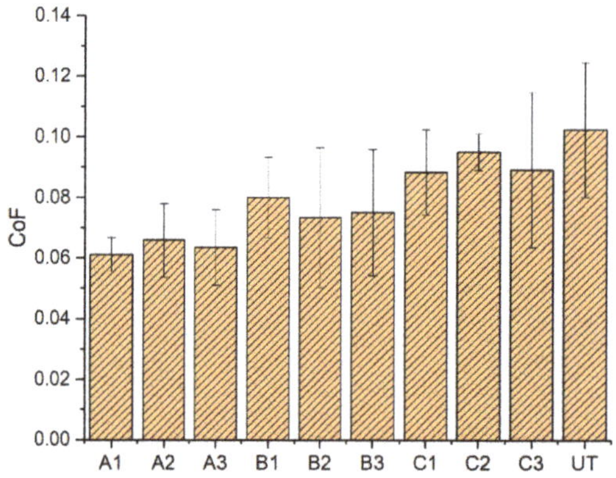

Figure 8. Coefficient of friction values of different dimple textures.

Among the factors studied, surface density emerged as the most influential in reducing CoF. Textures with a 5% surface density achieved the lowest CoF values, outperforming those with higher surface densities. A clear trend was observed: as surface density increased, CoF values rose, which aligns with the trend observed in WCA data. These findings highlight a strong correlation between surface hydrophilicity and reduced CoF.

In contrast, the aspect ratio exhibited a much weaker and inconsistent effect on CoF, which suggests that the aspect ratio may play a less significant role in the performance of small diameter dimples, where the dominant factors are likely the overall density and ability to promote lubricant retention. Notably, the lowest CoF value recorded in this study demonstrated a friction reduction, from 15% to 50%, compared to the lowest CoF values reported in the literature for textured CCM under similar testing conditions [30,31], further highlighting the effectiveness of small-diameter dimples when paired with optimized surface density.

These results further emphasize the potential of laser texturing, particularly through optimizing surface density, to significantly improve the friction performance of CCM in tribological applications.

3.3.2. Frictional Behavior

The representative CoF plots in Figure 9 reveal that all laser-textured samples exhibit lower and more stable friction compared to untextured CCM. While the untextured samples displayed substantial fluctuations in CoF throughout the eight-hour tests, the textured surfaces maintained consistent and steady values. This stability is likely due to the enhanced ability of textured surfaces to retain and support the formation of a stable lubricant film. In contrast, untextured surfaces are less effective at maintaining lubricant coverage, leading to intermittent lubricant loss and sharp increases in CoF.

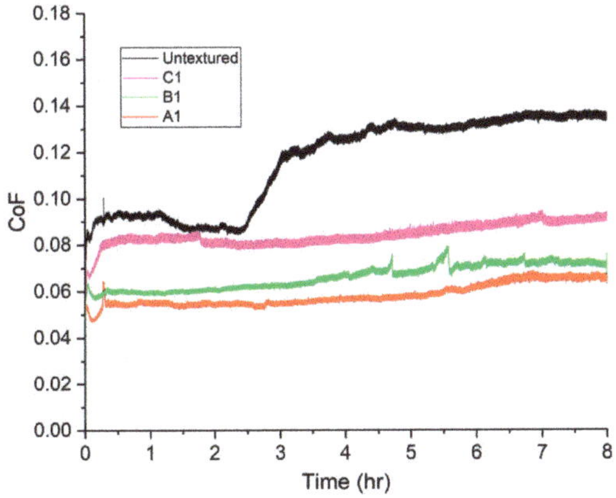

Figure 9. Coefficients of friction for representative textured CCM samples compared to an untextured sample, tested against a VEHXL UHMWPE pin.

Among the textured samples, Texture A1 achieved the lowest CoF, with an average reduction of approximately 40% compared to the untextured samples. This remarkable improvement highlights the effectiveness of laser texturing in not only reducing friction, but also providing a more reliable and consistent tribological performance. These findings emphasize the critical role of laser texturing in optimizing frictional behavior, particularly by promoting stable lubrication and reducing frictional resistance.

3.3.3. Wear Performance and Mechanisms

Figure 10 presents SEM and LSCM images of the CCM wear tracks and dimples within the wear track after testing, along with transfer film thickness measurements. The LSCM images (Figure 10c,d) correlate closely with the SEM images of the same region (Figure 10a,b), showing that the darker regions in the SEM images correspond to areas with thicker transfer films from the VEHXL UHMWPE pin. In these regions, the transfer film thickness reaches up to 109 nm, as measured in Figure 10e. Conversely, the lighter areas on the wear tracks exhibit little to no measurable transfer film, indicating minimal material transfer from the pin and negligible wear on the CCM surface. These findings highlight the localized nature of material transfer during testing, where thicker, more uniform transfer films are associated with increased surface interactions and higher CoFs.

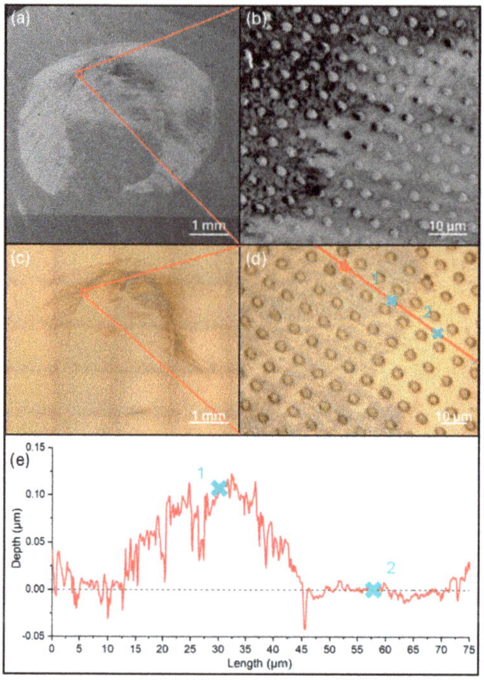

Figure 10. SEM images of a C1 sample after tribological testing: (**a**) wear track on the CCM surface and (**b**) magnified view of the dimples within the boxed region in (**a**). Corresponding optical image show (**c**) the wear track and (**d**) a magnified view of the dimples within the boxed region in (**c**). (**e**) Thickness profile of the transfer film along the red line in (**d**), with positions 1 and 2 labeled in cyan for reference.

Most CCM wear tracks showed no detectable wear. However, some samples exhibited 1–5 minor visible scratches, with depths reaching up to 108 nm (Figure 11). These scratches were observed near the transfer film areas, even when no visible wear was present on other areas of the wear track. These scratches corresponded to more severe wear on the VEHXL UHMWPE counterface pin. Given that VEHXL UHMWPE is significantly softer than the CCM disc, it is unlikely that the pin itself caused these scratches. Instead, the scratches are likely due to abrasive particles dislodged from the protruding rims around the edges of the laser-ablated dimples, as circled in the dimple profiles in Figure 4.

Out of nine tests conducted for each surface density, scratches were observed in two tests with 5% surface density, three tests with 15% surface density, and four tests with 25% surface density. This trend indicates that textures with higher surface densities are more prone to scratches on their wear tracks. The increased susceptibility to scratches is attributed to the greater number of dimples in higher density textures, which leads to more protrusions and, consequently, a higher likelihood of abrasive damage.

These protrusions form during laser ablation when certain regions of the CCM surface are insufficiently heated for complete material removal. Instead, these areas undergo a phase change that alters their surface topography, creating small, irregular protrusions. These protrusions are likely oxidized, making them harder, yet more brittle, than the surrounding material. During sliding, these protrusions can detach and act as abrasive debris, leading to three-body wear and causing the observed scratches on the CCM surface. The presence of these scratches correlated with a higher CoF and more significant transfer film, indicating higher wear.

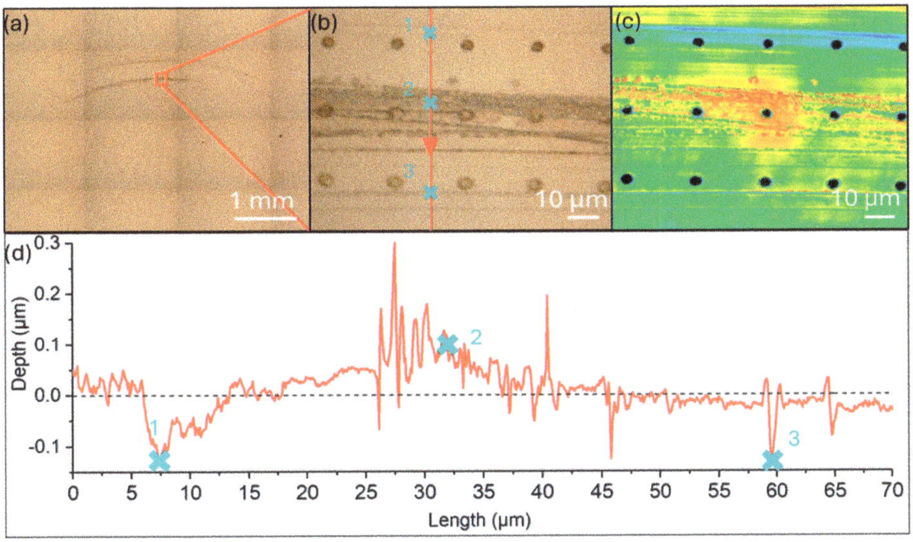

Figure 11. Scratches on an A1 sample after tribological testing: (**a**) optical image of the full wear track; (**b**) magnified optical image of a scratch and transfer film within the boxed region in (**a**); (**c**) height map corresponding to the close-up in (**b**); and (**d**) topographical line profile extracted along the red line in (**b**), showing scratch depths at positions 1 and 3, and the presence of transfer film at position 2.

The findings underscore the detrimental effects of these protrusions on the tribological performance of textured surfaces. The increased frequency of scratches in textures with higher surface densities further highlights the need to optimize laser processing parameters to minimize protrusion formation. By reducing these defects, the durability and tribological performance of laser-textured surfaces can be significantly improved.

The wear of the VEHXL UHMWPE pin was minimal, rendering it impractical to measure wear weight by comparing the pin's weight before and after testing. Instead, an indirect method was used to assess wear by evaluating changes in the pin's surface roughness. The initial surface of the pins displayed grooves resulting from the CNC machining process, as shown in Figure 6. After testing, surface roughness measurements conducted using LSCM revealed a reduction in these grooves, indicating wear (Figure 12). By comparing Figure 12b with Figure 12a, and Figure 12d with Figure 12c, reductions in height around the edges and the centers of the pins can be clearly observed.

Figure 13 illustrates the average surface roughness changes in the VEHXL UHMWPE pins for different textures. Among the 27 textured CCM samples, 19 showed smaller roughness changes compared to the untextured samples, indicating reduced pin wear when paired with laser-textured surfaces; in contrast, the remaining 8 samples exhibited larger roughness changes, primarily due to significant scratches on the VEHXL UHMWPE pin surface after testing. The variations in roughness changes within each sample group were largely driven by differing levels of scratches caused by individual samples within each group. Some samples, however, showed uncharacteristically low changes in roughness when compared to other tests on the same texture. For example, test #2 on texture C2 showed a 98% decrease in friction when compared to the average untextured test. This is due to its lower frequency of protrusions after fabrication, and more localized wear on the pin surface.

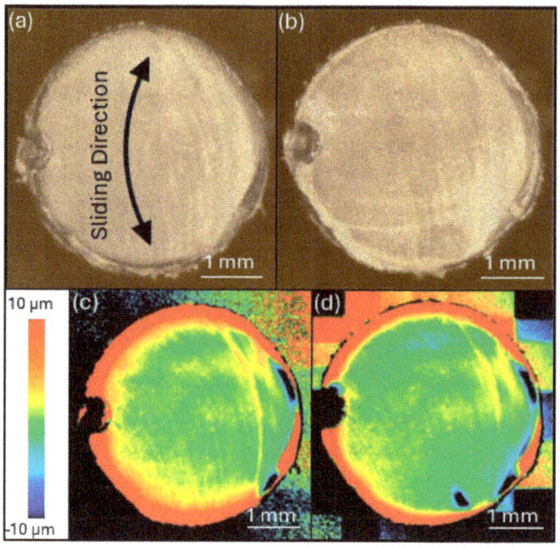

Figure 12. LSCM images of VEHXL UHMWPE pin surface (**a**,**c**) before and (**b**,**d**) after tribological testing.

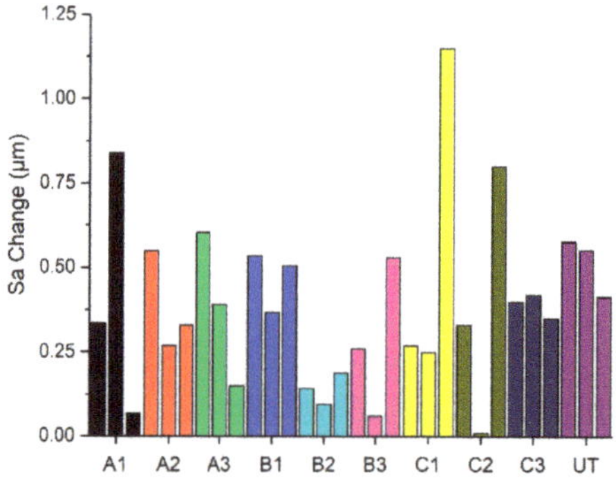

Figure 13. Average surface roughness changes in VEHXL UHMWPE pin after rubbing against different CCM textures.

4. Discussion

4.1. Wear

Quantifying wear in this study was challenging due to the shorter test duration compared to longer studies that use simpler methods, such as measuring pin weight loss [20,22] or wear track dimensions [32]. Despite this limitation, minimal wear was observed on the CCM surface, highlighting the durability and longevity of laser-textured surfaces. In contrast, wear primarily occurred on the VEHXL UHMWPE pins, as evidenced by the smoothing of their surfaces and the presence of transfer films on the CCM. Studies under similar conditions showed measurable wear on the UHMWPE pins despite fewer testing cycles [31], and/or showed lesser differences in wear when compared to an untextured sample [33]. These small-dimpled textures showed overall improvement in wear when compared to an untextured CCM sample.

Larger areas of wear tracks, marked by thicker transfer films on the CCM surface, indicate greater material transfer from the pin to the disc, signifying higher wear rates. Similarly, the significant reduction in the height of grooves on the VEHXL UHMWPE pins after testing reflects increased pin wear. This reduction in grooves on the pin surface, combined with LSCM imaging confirming transfer film on the surface on the CCM, indicates that the transfer film is from the wear on the pin, not from the CCM itself.

Smaller CCM wear tracks, minimal changes in VEHXL UHMWPE pin surface roughness, and thinner transfer films on the CCM wear tracks served as clear indicators of reduced wear rates. These results underscore the critical importance of optimizing surface texture design to minimize wear and improve the tribological performance of CCM-VEHXL UHMWPE pairings.

4.2. Effect of Protrusions on Texture Performance

A side effect of LST was the formation of protruding rings around the dimple edges. These protrusions formed when regions of the CCM surface, insufficiently heated for complete ablation, underwent phase changes that altered the surface topography. Han et al. [34] reported a similar phenomenon, where pronounced and evenly distributed protrusions enhanced friction and wear performance by contributing to consistent surface interactions.

In this study, however, the protrusions were sporadic, shorter, and sharper, likely due to the lower laser power and shorter drill times used to create smaller dimples. Textures with lower aspect ratios were exposed to faster laser pulses, which further reduced the formation of protrusions. Unlike Han et al.'s study, the inconsistent and sharp protrusions in this work may have increased resistance during sliding, negatively impacting both friction and wear performance.

During testing, these protrusions often wore off and, in some cases, became trapped between the surfaces, leading to three-body wear and causing visible scratches on the CCM surface. Figure 11 highlights scratches as deep as 100 nm adjacent to transfer film marks, indicating higher wear in regions with scratches. This underscores the detrimental effects of irregular protrusions on tribological performance, emphasizing the need to optimize laser processing parameters to minimize their formation. Post-fabrication polishing steps were performed in an attempt to remove these protrusions and minimize their impacts on friction and wear performance. However, due to their small size and localized nature, effective removal proved challenging. Gentle polishing was insufficient to eliminate them, while more aggressive polishing unevenly altered the aspect ratio of the dimples across the texture, introducing additional variability in tribological performance.

4.3. Effect of WCA and Roughness on Tribological Performance

The formation of a stable and consistent lubricant film, which separates asperities between the two surfaces under lubricated conditions, is a key advantage of surface texturing. As shown in Figures 7 and 8, the trends in average WCA and CoF values suggest a correlation between increased wettability and improved tribological performance. Textures with lower average CoF values likely benefit from enhanced hydrophilicity, which promotes better lubricant retention. In addition to acting as lubricant reservoirs and enhancing hydrodynamic pressure, small dimples appear to increase the surface's affinity for lubricant, particularly at lower surface densities, where the Wenzel model dominates. This greater hydrophilicity facilitates the formation of a more stable lubricant film, thereby improving tribological properties.

In contrast, surface roughness played a less significant role in the tribological performance of the textures compared to WCA. At lower surface densities, the presence of small dimples did not substantially alter roughness compared to untextured CCM samples. However, at higher surface densities, an increased number of dimples led to a greater presence of protrusions, which had a stronger impact on tribological behavior than roughness alone. Comparing the CoF trends in Figure 8 with the roughness values in Figure 6, no clear correlation emerges, especially when accounting for variations in aspect ratio. These findings suggest that wettability, rather than roughness, is the dominant factor influencing friction and wear in our study.

4.4. Mechanisms Governing Tribological Performance

The tribological improvements observed in this study are attributed to several interconnected mechanisms. Dimples function as lubricant reservoirs, enhancing the retention and distribution of bovine calf serum lubricant. This effect is particularly crucial in artificial hip joints, where lubrication shifts among boundary, mixed, and EHL regimes. In mixed lubrication, dimples provide micro-reservoirs that sustain lubrication film thickness, preventing dry contact and reducing wear. Under EHL, the localized pressure differentials around dimples generate hydrodynamic lift, facilitating stable lubricant film formation and minimizing direct asperity interactions.

Furthermore, surface texturing redistributes contact stresses, reducing peak contact pressures and mitigating material loss. This is particularly important under high-load conditions, where lubricant depletion can lead to increased friction and wear on untextured surfaces. Dimples also act as debris traps, capturing wear particles that would otherwise contribute to third-body abrasion, thereby extending implant longevity.

The effectiveness of these mechanisms is highly dependent on dimple geometry. The optimal texture identified in this study (5% surface density, 0.1 aspect ratio) demonstrated the lowest friction and wear due to a balance among effective lubricant retention, load distribution, and hydrodynamic enhancement. The findings underscore the critical role of surface texturing in regulating tribological behavior and enhancing the performance of metal-on-polyethylene hip implants.

5. Conclusions

This study demonstrates that LST is a highly effective method for fabricating precise sub-10 micron surface textures on CCM surfaces, significantly improving tribological performance. Small dimples with a 5 μm diameter were shown to reduce friction and wear effectively, making them well-suited for application in artificial hip joints.

Among the configurations tested, the texture design featuring a 5% surface density and a 0.1 aspect ratio was the most effective. This configuration achieved the lowest CoF and favorable wear rates, underscoring the critical role of optimizing texture geometry and density. The presence of dimples facilitated the formation of a stable and uniform lubricant film during sliding, which reduced asperity contact, lowered friction, and enabled consistent tribological performance.

However, this study also identified challenges associated with the laser texturing process. Sporadic protrusions formed around the edges of the dimples due to incomplete ablation, which negatively impacted performance by introducing irregular resistance and contributing to three-body wear. These protrusions caused visible scratches on the CCM surface and localized increases in wear, highlighting the importance of refining laser processing parameters to minimize their formation.

In summary, this work highlights the potential of LST to enhance the performance and longevity of artificial hip joints by improving lubrication and minimizing wear through precisely engineered surface textures. The findings provide valuable insights for optimizing texture designs and laser processing techniques, paving the way for more durable and efficient medical implants.

Future studies on small-dimpled textures should explore the effects of further reducing dimple diameter and adjusting key parameters, such as surface density, to optimize tribological performance. Given that the lowest coefficient of friction (CoF) was observed at the lowest tested surface density (5%), additional experiments should investigate whether even lower surface densities could yield further reductions in CoF. Additionally, research should focus on mitigating the formation of protrusions, either by refining laser processing parameters or introducing a post-processing step before testing, to minimize their impact on wear behavior.

Author Contributions: Conceptualization, W.B.B. and M.Z.; methodology, W.B.B.; software, W.B.B.; validation, W.B.B.; formal analysis, W.B.B. and M.Z.; investigation, W.B.B.; resources, M.Z.; data curation, W.B.B.; writing—original draft preparation, W.B.B.; writing—review and editing, M.Z.; visualization, W.B.B.; supervision, M.Z.; project administration, M.Z.; funding acquisition, M.Z. All authors have read and agreed to the published version of the manuscript.

Funding: This research was funded by the University of Arkansas and the US National Science Foundation under Grant Award #OIA-1457888. Any opinions, findings, and conclusions or recommendations expressed in this material are those of the authors and do not necessarily reflect the views of the US National Science Foundation.

Data Availability Statement: The data used in this study are available upon reasonable request from the corresponding author.

Acknowledgments: We thank Orthoplastics for generously providing the VEHXL UHMWPE pins used in this study.

Conflicts of Interest: The authors declare that they have no known competing financial interests or personal relationships that could have appeared to influence the work reported in this paper.

Abbreviations

The following abbreviations are used in this manuscript:

CCM	CoCrMo
UHMWPE	Ultra-high molecular weight polyethylene
CoF	Coefficient of friction
THR	Total hip replacement
LST	Laser surface texturing
MoP	Metal-on-polyethylene
VEHXL	Vitamin E-stabilized highly cross-linked
IPA	Isopropyl alcohol
DI water	Deionized water
LSCM	Laser scanning confocal microscopy
WCA	Water contact angle
BCS	Bovine calf serum
SEM	Scanning electron microscopy

References

1. El-Desouky, I.I.; Helal, A.H.; Mansour, A.M.R. Ten-Year Survival of Ceramic-on-Ceramic Total Hip Arthroplasty in Patients Younger than 60 Years: A Systematic Review and Meta-Analysis. *J. Orthop. Surg. Res.* **2021**, *16*, 679. [CrossRef] [PubMed]
2. Nebergall, A.K.; Greene, M.E.; Laursen, M.B.; Nielsen, P.T.; Malchau, H.; Troelsen, A. 2017, Vitamin E Diffused Highly Cross-Linked Polyethylene in Total Hip Arthroplasty at Five Years. *Bone Jt. J.* **2017**, *99*, 577–584. [CrossRef]
3. Nakanishi, Y.; Nakashima, Y.; Fujiwara, Y.; Komohara, Y.; Takeya, M.; Miura, H.; Higaki, H. Influence of Surface Profile of Co-28Cr-6Mo Alloy on Wear Behaviour of Ultra-High Molecular Weight Polyethylene Used in Artificial Joint. *Tribol. Int.* **2018**, *118*, 538–546. [CrossRef]
4. Saleh, K.J.; Thongtrangan, I.; Schwarz, E.M. Osteolysis: Medical and surgical approaches. *Clin. Orthop. Relat. Res.* **2004**, *427*, 138–147. [PubMed]
5. Ching, H.A.; Choudhury, D.; Nine, M.J.; Abu Osman, N.A. Effects of Surface Coating on Reducing Friction and Wear of Orthopaedic Implants. *Sci. Technol. Adv. Mater.* **2014**, *15*, 014402. [CrossRef]

6. Kalayarasan, M.; Prabhu Raja, V.; Nithin, V.; Vivinkumar, S.; Vishwaanth, M. Comparative Study of Wear Behavior of Multilayer Coatings for Human Hip Prosthesis. *Appl. Mech. Mater.* **2015**, *787*, 528–533. [CrossRef]
7. Moro, T.; Kawaguchi, H.; Ishihara, K.; Kyomoto, M.; Karita, T.; Ito, H.; Nakamura, K.; Takatori, Y. Wear Resistance of Artificial Hip Joints with Poly(2-Methacryloyloxyethyl Phosphorylcholine) Grafted Polyethylene: Comparisons with the Effect of Polyethylene Cross-Linking and Ceramic Femoral Heads. *Biomaterials* **2009**, *30*, 2995–3001. [CrossRef]
8. Lestari, W.D.; Nugroho, A.; Ismail, R.; Jamari, J.; Bayuseno, A.P. Study of Wear Performance of Crosslinking UHMWPE Acetabular Liner for Artificial Hip Joint Made from CNC Milling. *IOP Conf. Ser. Mater. Sci. Eng.* **2021**, *1078*, 012009. [CrossRef]
9. López-Cervantes, A.; Domínguez-López, I.; Barceinas-Sánchez, J.D.O.; García-García, A.L. Effects of Surface Texturing on the Performance of Biocompatible UHMWPE as a Bearing Material during in Vitro Lubricated Sliding/Rolling Motion. *J. Mech. Behav. Biomed. Mater.* **2013**, *20*, 45–53. [CrossRef]
10. Zhao, C.; Jia, X.; Zhao, Q.; Ma, H.; Yu, Z.; Zhang, Z.; Du, W.; Zhang, H. Research Progress on the Effect of Surface Texture on the Friction Properties of CoCrMo Alloys. *Compos. Interfaces* **2024**, *31*, 1269–1290. [CrossRef]
11. Hidayat, T.; Jamari, J.; Bayuseno, A.P.; Ismail, R.; Tauviqirrahman, M.; Wijaya, P.N. Study of Lubrication Fluid Pressure in Artificial Hip Joint During Bowing (Ruku'). In *Lecture Notes in Mechanical Engineering, Proceedings of the 7th International Conference and Exhibition on Sustainable Energy and Advanced Materials (ICE-SEAM 2021), Melaka, Malaysia, 1 July 2022*; Springer: Singapore, 2022; pp. 303–306. [CrossRef]
12. Gao, L.; Yang, P.; Dymond, I.; Fisher, J.; Jin, Z. Effect of Surface Texturing on the Electrohydrodynamic Lubrication Analysis of Metal-on-Metal Hip Implants. *Tribol. Int.* **2010**, *43*, 1851–1860. [CrossRef]
13. Ranuša, M.; Odehnal, L.; Kučera, O.; Nečas, D.; Hartl, M.; Křupka, I.; Vrbka, M. Effect of Surface Texturing on Friction and Lubrication of Ti6Al4V Biomaterials for Joint Implants. *Tribol. Lett.* **2015**, *73*, 15. [CrossRef]
14. Shinkarenko, A.; Kligerman, Y.; Etsion, I. The Effect of Surface Texturing in Soft Elasto-Hydrodynamic Lubrication. *Tribol. Int.* **2009**, *42*, 284–292. [CrossRef]
15. Uddin, M.S.; Ibatan, T.; Shankar, S. Influence of Surface Texture Shape, Geometry and Orientation on Hydrodynamic Lubrication Performance of Plane-to-Plane Slider Surfaces. *Lubr. Sci.* **2016**, *29*, 153–181. [CrossRef]
16. Drescher, P.; Oldorf, P.; Dreier, T.; Schnell, G.; Peters, R.; Seitz, H. Ring-Shaped Surface Microstructures for Improved Lubrication Performance of Joint Prostheses. *Lubricants* **2020**, *8*, 45. [CrossRef]
17. Tarabolsi, M.; Klassen, T.; Mantwill, F.; Gärtner, F.; Siegel, F.; Schulz, A. Patterned CoCrMo and Al2O3 Surfaces for Reduced Free Wear Debris in Artificial Joint Arthroplasty. *J. Biomed. Mater. Res.* **2013**, *101*, 3447–3456. [CrossRef]
18. Mao, B.; Siddaiah, A.; Liao, Y.; Menezes, P.L. Laser Surface Texturing and Related Techniques for Enhancing Tribological Performance of Engineering Materials: A Review. *J. Manuf. Process.* **2020**, *53*, 153–173. [CrossRef]
19. Liu, Y.; Zhu, Q.; Wang, C.; Li, J. Tribological Behavior of CoCrMo Artificial Knee Joint with Symmetrically Biomimetic Textured Surfaces on PEEK. *Opt. Laser Technol.* **2023**, *157*, 108774. [CrossRef]
20. Chyr, A.; Qiu, M.; Speltz, J.W.; Jacobsen, R.L.; Sanders, A.P.; Raeymaekers, B. A Patterned Microtexture to Reduce Friction and Increase Longevity of Prosthetic Hip Joints. *Wear* **2014**, *315*, 51–57. [PubMed]
21. Chen, K.; Tang, Y. Research Progress on the Design of Surface Texture in Tribological Applications: A Mini-Review. *Symmetry* **2024**, *16*, 1523. [CrossRef]
22. Langhorn, J.; Borjali, A.; Hippensteel, E.; Nelson, W.; Raeymaekers, B. Microtextured CoCrMo Alloy for Use in Metal-on-Polyethylene Prosthetic Joint Bearings: Multi-Directional Wear and Corrosion Measurements. *Tribol. Int.* **2018**, *124*, 178–183. [CrossRef] [PubMed]
23. Allen, Q.; Raeymaekers, B. Surface Texturing of Prosthetic Hip Implant Bearing Surfaces: A Review. *J. Tribol.* **2020**, *143*, 040801. [CrossRef]
24. Longmore, R.B.; Gardner, D.L. The surface structure of ageing human articular cartilage: A study by reflected light interference microscopy (RLIM). *J. Anat.* **1978**, *126 Pt 2*, 353–365. [PubMed] [PubMed Central]
25. Hidayat, T.; Ammarullah, M.I.; Saputra, E.; Lamura, M.D.P.; KN, C.; Ismail, R.; Bayuseno, A.P.; Jamari, J. A Method for Estimating the Contact Area of a Dual-Mobility Total Hip Prosthesis. *AIP Adv.* **2024**, *14*, 015317. [CrossRef]
26. Puccio, F.D. Biotribology of Artificial Hip Joints. *World J. Orthop.* **2015**, *6*, 77–94. [CrossRef]
27. Choudhury, D.; Ranuša, M.; Fleming, R.A.; Vrbka, M.; Křupka, I.; Teeter, M.G.; Goss, J.; Zou, M. Mechanical Wear and Oxidative Degradation Analysis of Retrieved Ultra High Molecular Weight Polyethylene Acetabular Cups. *J. Mech. Behav. Biomed. Mater.* **2018**, *79*, 314–323. [CrossRef]
28. Wenzel, R.N. Resistance of Solid Surfaces to Wetting by Water. *Ind. Eng. Chem.* **1936**, *28*, 988–994. [CrossRef]
29. Cassie, A.B.D.; Baxter, S. Wettability of Porous Surfaces. *Trans. Faraday Soc.* **1944**, *40*, 546–551. [CrossRef]
30. Shen, G.; Zhang, J.; Fang, F. Study on the Effect of Hydrodynamic Pressure on the Tribological Performance of Textured Bioimplants. *Proc. Inst. Mech. Eng. Part C J. Mech. Eng. Sci.* **2021**, *236*, 3135–3145. [CrossRef]

31. Alvarez-Vera, M.; Ortega, J.A.; Ortega-Ramos, I.A.; Hdz-García, H.M.; Muñoz-Arroyo, R.; Díaz-Guillén, J.C.; Acevedo-Dávila, J.L.; Hernández-Rodriguez, M.A.L. Tribological and Microstructural Characterization of Laser Microtextured CoCr Alloy Tested against UHMWPE for Biomedical Applications. *Wear* **2021**, *477*, 203819. [CrossRef]
32. Yuan, S.; Lin, N.; Wang, W.; Zhang, H.; Liu, Z.; Yu, Y.; Zeng, Q.; Wu, Y. Correlation between surface textural parameter and tribological behaviour of four metal materials with laser surface texturing (LST). *Appl. Surf. Sci.* **2022**, *583*, 152410. [CrossRef]
33. Borjali, A.; Langhorn, J.; Monson, K.; Raeymaekers, B. Using a Patterned Microtexture to Reduce Polyethylene Wear in Metal-on-Polyethylene Prosthetic Bearing Couples. *Wear* **2017**, *392–393*, 77–83. [CrossRef]
34. Han, Y.; Liu, F.; Zhang, K.; Huang, Q.; Guo, X.; Wang, C. A Study on Tribological Properties of Textured Co-Cr-Mo Alloy for Artificial Hip Joints. *Int. J. Refract. Met. Hard Mater.* **2021**, *95*, 105463. [CrossRef]

lubricants

MDPI

Article

Improved Tribological Performance of ta-C/MoSx Coatings Deposited on Laser Micro-Structured Steel Substrates in Both Vacuum and Air

Stefan Makowski [1,*], Fabian Härtwig [1,2], Marcos Soldera [2], Mahmoud Ojeil [1], Lars Lorenz [2], Frank Kaulfuß [1] and Andrés Fabián Lasagni [1,2]

1 Fraunhofer Institute for Material and Beam Technology IWS, 01277 Dresden, Germany
2 Institut für Fertigungstechnik, Technische Universität Dresden, 01069 Dresden, Germany
* Correspondence: stefan.makowski@iws.fraunhofer.de

Abstract: Vacuum and air atmospheres impose very different requirements on tribological-loaded contacts, which usually require different surface materials. While hard tetrahedral amorphous carbon (ta-C) coatings provide good tribological properties in air, soft coatings such as molybdenum disulfide (MoS2) work well in a vacuum. Tribological performance in the respective other environment, however, is poor. In this work, the combination of laser microstructured (direct laser interference patterning) steel substrates and the deposition of ta-C and MoSx coatings with vacuum arc evaporation (LaserArc™) was studied, resulting in steel/DLIP, steel/DLIP/ta-C, steel/DLIP/MoSx, steel/DLIP/ta-C/MoSx, and steel/MoSx surface combinations. The tribological properties were studied using a ball-on-disk tribometer with a steel ball counter body in air and in a vacuum ($p < 5 \times 10^{-7}$ mbar). The type of the topmost coating governed their tribological properties in the respective atmosphere, and no general beneficial influence of the microstructure was found. However, steel/DLIP/ta-C/MoSx performed best in both conditions and endured the highest contact pressure, which is attributed to the mechanical support of the ta-C coating and MoSx reservoir in the remaining structure, as evidenced by Raman spectroscopy. The findings suggest that such combination allows for surfaces bearing a high load capacity that can be applied in both a vacuum and in air, for example, in multi-use space applications.

Keywords: ta-C; DLC; MoS2; direct laser interference patterning; vacuum; tribology

1. Introduction

Vacuum conditions present several challenges to tribological contacts. The absence of gases like oxygen, nitrogen, and water prevents passivation of nascent surfaces. A low pressure promotes fast evaporation of conventional liquid lubricants and furthermore would contaminate technical vacua. Therefore, surfaces that need to endure friction in vacuum conditions must have self-lubricating and self-passivating properties.

Even more challenging are space conditions, which, in addition to low atmospheric pressure, require tribological performance during exposure to higher temperatures and atomic oxygen [1]. Before being sent to space, most technical systems are assembled and rigorously tested within Earth's atmosphere to ensure their reliable performance in both environments. Moreover, these systems must be designed to withstand the challenges of the space vacuum, where they experience significant changes in pressure and atmosphere.

Although coatings made from molybdenum disulfide (MoS_2), which belongs to the group of transition metal dichalcogenides (TMD), are the state of the art for space applications, they have several drawbacks, such as their soft nature resulting in high wear in air as well as their high sensitivity to moisture [1–3]. However, in vacuum conditions, MoS_2 can achieve very low friction by sliding on crystalline atomic smooth layers, and it can even achieve superlubricity in appropriate conditions [4].

Gradt and Schneider studied sputtered MoS_2 coatings in moist air at room temperature and a vacuum at 77 K. They found that the wear performance in low-pressure water vapor depends on the deposition conditions of the coating and can be improved by adding chromium or inducing internal stress [2]. The influence of industrial deposition conditions, such as threefold rotation and a large target–substrate distance, was found to influence the porosity of the coatings, which also affected their tribological properties [5].

In space applications, tribological systems are often susceptible to elevated temperatures due to the absence of convective cooling in a vacuum environment. In low Earth orbit, these systems experience substantial temperature cycling resulting from alternating exposure to direct sunlight and complete shadow periods. Any coatings therefore must be able to withstand a wide temperature range. Friction and wear of sputtered MoS_2 coatings improve up to approximately 100 °C by desorbing water from the surface but deteriorate for higher temperatures as oxidation is favored [1].

Carbon-based amorphous coatings have been considered for vacuum conditions as well. Their tribological behavior strongly depends on whether they contain hydrogen or not and on the availability of water vapor in the atmosphere. In the absence of water vapor, hydrogen-containing a-C:H-coatings have remarkable tribological properties and may achieve superlubricity in a vacuum or hydrogen atmosphere [6]. Similar behavior is found for CNx coatings in a vacuum and nitrogen [7]. In contrast, hydrogen-free amorphous carbon coatings perform well in humid conditions but have high friction in dry air and vacuums [8–10].

In addition to carbon-based coatings, titanium nitride (TiN) coatings have been studied in a vacuum. It was found that they exhibited moderate friction coefficients when operating in a vacuum, whereas their friction coefficient was higher in atmospheric conditions [11,12].

The idea of combining materials that can adapt to different environments has been investigated using various approaches. For instance, a multi-layer coating consisting of a thin film a-C:H/MoS_2/a-C structure was found to achieve superlubricity in 20% rF air via nanoscroll formation [13]. Hybrid coatings, consisting of carbon alloyed TMDs such as MoS_2, WS, and $MoSe_2$, were studied and showed low friction conditions in several terrestrial and space conditions [14–16]. The wear properties of pure MoS_2 coatings could be improved by adding a supramolecular gel [17].

Laser-cladded nickel-based coatings with MoS_2 incorporation showed good high-temperature vacuum performance up to 300 °C [18]. Silver-impregnated thermal-sprayed YSZ ceramic showed moderate friction in a vacuum and an improved wear life [19]. Hot-pressed nickel-based alloys were found to perform well in a vacuum up to 800 °C [20].

Although pure tetrahedral amorphous carbon (ta-C) coatings commonly have high friction and wear in a vacuum and cannot passivate as in air due to the absence of water, some notable exceptions have been reported. Härtwig et al. studied the influence of different counter body materials when sliding against ta-C in a vacuum and found that a brass counter body resulted in unexpectedly low friction and wear [21]. A simple bilayer coating approach of MoS_2 on top of a ta-C coating deposited via vacuum arc deposition was found to improve the wear life in humid and vacuum conditions compared to a pure MoS_2 coating due to the underlaying hard ta-C coating [22].

Patterning a surface with controlled microstructures is another strategy for improving the material's tribological properties by reducing the contact area and promoting traps for wear debris to prevent third-body interactions [23]. In this way, frictional and wear losses can be minimized [24,25]. Among the plethora of texturing methods, direct laser interference patterning (DLIP) is an industrial-scalable laser microfabrication method that is able to process large areas at throughputs already exceeding 1 m^2/min [26]. The method is based on the overlap of multiple laser beams on the sample surface so that an interference pattern can be produced. At the maxima positions of the interference pattern, the material can be locally melted or ablated, whereas at the minima, the material remains unmodified, leading to a periodic surface texture. The spatial period of the texture can be easily controlled by adjusting the overlapping angle between the beams and can be as short as half of the used laser wavelength, yielding finer resolutions compared to other laser-based methods [27]. This technique has already been used to optimize surfaces for tribology. For instance, in the works of Gachot and Rosenkranz [28], the influence of the orientation of line-like textures produced using DLIP on friction (dry) and wear on different metals was investigated. In addition, they found a dependence of the coefficient of friction on the orientation of the tested line-like structure. Furthermore, DLIP has already been applied to ta-C layers, obtaining a reduction in the coefficient of friction by 30% under dry lubrication conditions [29]. Also, DLIP was used to modify the topography of substrates prior to the deposition of carbon nanoparticle coatings, resulting in an enhanced coefficient of friction and wear resistance [30,31].

Motivated by the emergence of new industrial-scale processes for high-throughput surface structuring and advanced coating technologies capable of depositing ta-C and MoS_2 films in one process, this study focuses on the tribological performance of ta-C/MoS_2 bilayer coatings on structured steel surfaces. The friction and wear behavior of these coatings in a vacuum and atmosphere are investigated.

2. Materials and Methods

Steel disks (Rocholl GmbH, Eschelbronn, Germany) and steel balls (TIS Wälzkörpertechnologie, Gauting, Germany) were used as tribological test specimens. Both were made from 100Cr6/EN 1.3505 steel and hardened to 60 HRC. The disks had a diameter of 24 mm and were polished to a roughness of Ra < 0.03 µm. The steel balls had a diameter of 10 mm and had a polished surface as well.

Dot-like patterns were applied to the steel disks using a four-beam DLIP setup. The system was equipped with a 70 ps pulsed laser source (neoLASE GmbH, Hanover, Germany) with an output power of 2.7 W at a wavelength of 532 nm and a repetition rate of 10 kHz. Details on the method and used setup can be found elsewhere [32]. In this study, the spatial period was targeted to 5 µm. The experiments were performed at a peak fluence (energy density) per pulse of 1.32 J/cm^2 with a laser spot size of 50 µm at the focal plane. The structuring strategy consisted of firing 6 pulses per position to reach the targeted structure depth of 500 nm. The sample was translated with an XY stage (travel distance 200 mm, max. speed 300 m/s, resolution 0.1 µm, Aerotech ABL-1500-E318187, Pittsburgh, PA, USA) and a hatch distance of 40 µm between adjacent laser-treated positions.

The coatings on the disks were deposited using a LaserArc™ (Fraunhofer IWS, Dresden, Germany) coating system, which is based on vacuum arc discharge. The general preparation of ta-C reference films has been described previously elsewhere [33]. In this study, a pulse current of 1600 A at 300 Hz and a 100 V bias voltage with a pulse overlap of 175 µs was used.

Additionally, MoSx coatings were fabricated using an unfiltered LaserArc module as described by Krülle et al. [34]. Similarly to graphite evaporation for ta-C coatings, a pure MoS_2 cathode was used. The discharge current was reduced to 250 A, and no bias voltage was applied. Furthermore, the process temperature was set to 150 °C via a radiation heater to avoid the formation of solid sulfur. These process conditions result in a sulfur to molybdenum ratio in the growing film similar to that in the cathode [35]. However, since the element ratio is not perfectly stoichiometric and the coating is mostly amorphous and contains some impurities, the coating type is referred to as MoSx to distinguish it from crystalline MoS_2.

To ensure good adhesion of all the coatings, an approximately 100 nm thick chromium adhesion layer was applied to the substrate using magnetron sputtering. In the case of ta-C coating deposition, the sample surfaces were lapped after ta-C deposition to remove unwanted growth defects from the contacting surface which typically result from the arc process. In the case of ta-C/MoSx bi-layer films, the MoSx top layer was added onto the ta-C coating in an additional batch without another adhesion layer.

The properties of each single-layer coating were measured for reference samples placed in the corresponding coating batch. The coating thickness was obtained using the ball crater method. The coating hardness was measured using nanoindentation using a ZHN nanomechanical test system (Zwick/Roell, Ulm, Germany) with a Berkovich indenter (Synton-MDP Ltd, Nidau, Switzerland), and the quasi-continuous stiffness method (QCSM).

An overview of the substrate and coating materials with their respective properties is given in Table 1.

Table 1. Overview of substrate and coating types with their basic properties used in this work.

Material	Material	Properties
Substrate	100Cr6, hardened	Hardened, 60 HRC
ta-C coating	100% carbon	Thickness: 2.5 µm Hardness: ~63 GPa
MoSx coating	54 at% S 34 at% Mo 12 at% C	Thickness: 0.8 µm Hardness: 2.6 GPa

DLIP structuring and deposition of ta-C and MoSx was applied to the disks in several combinations to obtain different surfaces that allowed for an assessment of single and combined processing steps. The corresponding process steps and their order used to specify each surface are given in Table 2.

The steel balls remained unstructured and uncoated.

Table 2. Overview of tested disk surfaces with assigned names and structuring and coatings steps indicated in their order of processing based on the polished steel disk.

Surface Name	DLIP	ta-C Coating + Lapping	MoS_2 Coating
Steel/DLIP	1st	-	-
Steel/DLIP/ta-C	1st	2nd	
Steel/DLIP/MoSx	1st	-	2nd
Steel/DLIP/ta-C/MoSx	1st	2nd	3rd
Steel/MoSx	-	-	1st

Tribological testing was performed using a UHVT universal vacuum tribometer (Tetra, Ilmenau, Germany) using a ball-on-disk geometry with a rotating motion. The pressure was kept between 1 and 5×10^{-7} mbar for ultrahigh vacuum conditions and at approximately 1013 mbar and 40 to 50% rH for atmospheric conditions. The ball was loaded with a normal force of 5 N. By rotating the disk, the sliding speed was kept constant at 3 mm/s for all experiments by varying the angular speed. The sliding duration for standard experiments was 60 to 100 min, resulting in 12 and 18 m sliding lengths, respectively.

The normal force and friction forces were recorded to calculate the friction coefficient over time. The steady-state friction coefficient was obtained by averaging over the last 2 m of the experiments.

Wear scars on balls and wear tracks on the disks were imaged using light microscopy. A differential interference contrast (DIC) filter was used to visualize the topographic features in the nanometer range.

Additionally, 3D white light interferometry (Keyence VK-X3000, Osaka, Japan) was used to measure the profile of the wear tracks and resulting wear area. Images of the disk wear tracks were always taken at four different locations with an equal 90° distribution. For visualization of the changes to the steel substrate caused by DLIP, the same microscope was used in laser confocal mode.

The wear rate on the disk was calculated based on the averaged cross-section area of the wear track. The wear rate on the counter body ball was based on the apparent contact area obtained using light microscopy, assuming that the volume of the corresponding spherical calotte had been removed. This approach is an easy estimate to overcome the obstacle of measuring the removed volume from the curved surface of the ball; however, as a drawback, it will show systematic failure when the worn surface is not flat or is not representative due to material transfer.

Raman spectra were recorded using an "inVia" Raman microscope from Renishaw plc., Wotton-under-Edge, UK (100× objective, wavelength 514 nm, laser power 8 mW, CCD line detector).

3. Results

3.1. *Topography of Structured Steel Surface*

Figure 1 shows the topography of the DLIP-treated steel surface taken using optical confocal microscopy. The surface shows a square pattern of ablated areas separated by a hatch distance of H = 40 µm. Such ablated regions consist of square arrays of microcraters with a fixed periodicity Λ = 5.2 ± 0.1 µm, as calculated via Fourier analysis. The shape of the ablated crater array resembles the interference pattern produced upon overlapping the four laser sub-beams. Furthermore, these ablated spots have a modulated depth given by the Gaussian intensity distribution of the laser pulses, as can be observed from the topographic profile shown in Figure 1b. The craters feature relatively smooth sidewalls, and they are encircled by elevated rims produced by molten and redeposited material that flowed from the interference maxima positions (craters) to the minima (rims). This feature can be clearly seen in the profile of Figure 1b, where the dashed line corresponds to the average height of the untreated surface. At the center of the Gaussian pulse, the DLIP craters have an average structure depth of 502 nm ± 36 nm.

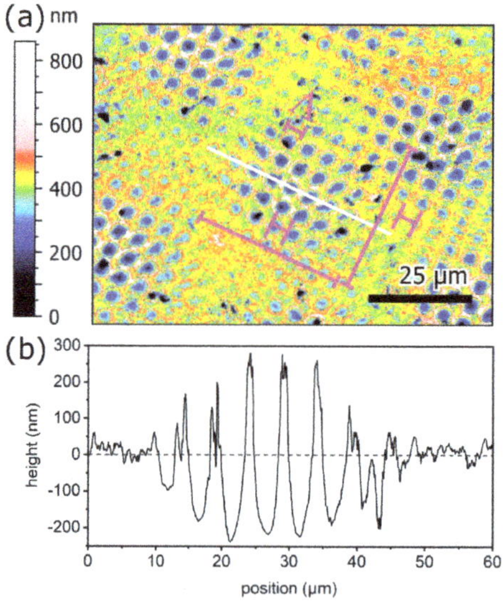

Figure 1. (**a**) Topography image of DLIP-treated stainless steel surface, including the DLIP spatial period Λ and hatch distance H. (**b**) Profile extracted along the white line in (**a**). The dashed line corresponds to the average height of the untreated surface.

3.2. *Tribological Performance in a Vacuum*

All the surfaces described in Table 2 were tested against steel balls in ultrahigh vacuum conditions and studied regarding their wear and friction. The wear scars and wear tracks are shown along the progression of the friction coefficient in Figure 2. Quantitative evaluations of the wear coefficients and steady-state friction coefficient for the same experiments are given in Figure 3.

The surface steel/DLIP (Figure 2a) showed a running-in phenomenon, where COF increases from 0.3 to above 1.1. After the test, well-visible wear marks could be seen on both the ball and disk. The wear coefficient for both bodies was approximately 2×10^{-6} mm^3/(Nm). On the disk, the initial DLIP structure was lost in many parts of the wear track, and signs of scuffing and heavy plastic deformation, including material deposit and removal both up to 0.5 μm, could be found instead.

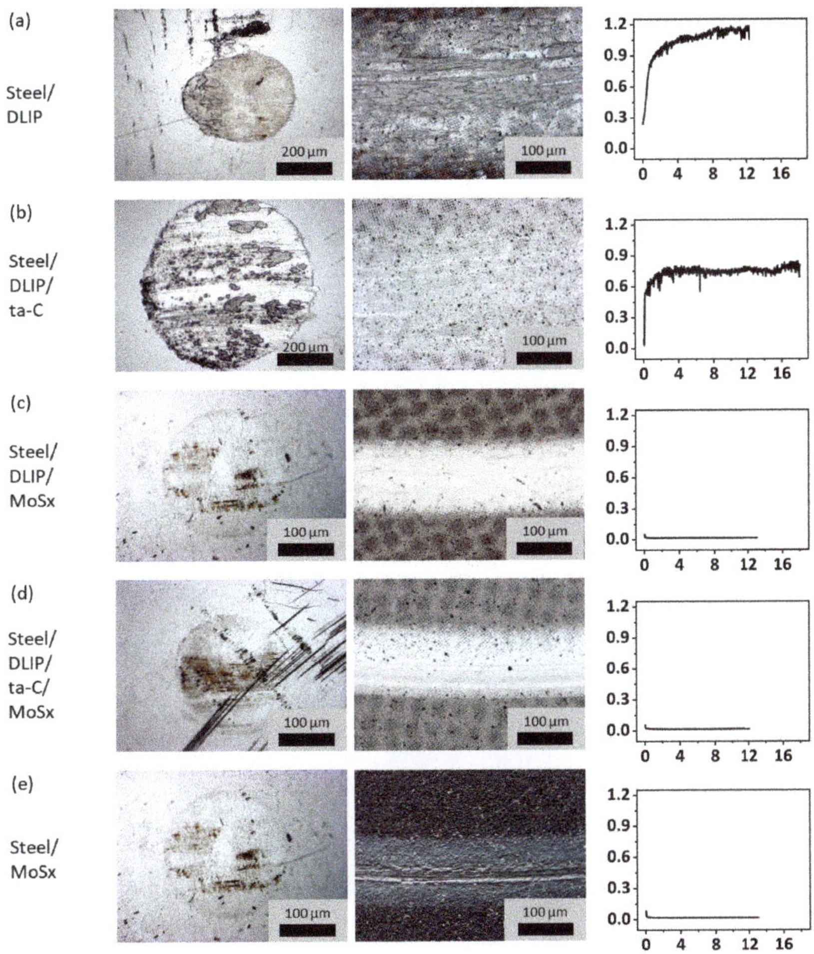

Figure 2. Wear and friction data for tribological short-term testing in an ultrahigh vacuum shown for all surfaces (**a–e**) with light microscope images of the wear scar on the counter body ball (**left**), light microscope images of the wear track on the disk (**center**), and friction (**right**, friction coefficient on Y axis and sliding length in meters on X axis). Please note the different scales for the ball wear scar images. Diagonal black scratches on the ball wear scar in (**d**) resulted from post-test handling and are not a tribological feature.

The surface steel/DLIP/ta-C (Figure 2b) showed an even shorter running in, where the friction coefficient climbed from an initial value of 0.05 to 0.8 in steady state. The wear on the ball wear scar and disk wear track were by far the biggest among all the tested combinations ($k > 10^{-5}$ mm^3/(N·m)). The wear track depth was up to 150 nm, at which point the DLIP structure was not visible anymore. However, in most areas with smaller wear depths, the DLIP structure was still visible. Some metallic scratches parallel to the sliding direction most likely resulted because of material deposition from the steel counter body.

Compared to the steel/DLIP surface, wear is mostly governed by abrasion. Only small patches of plastic behavior can be seen on the ball wear scar, and no plastic behavior can be seen on the disk wear track, which is smooth and even.

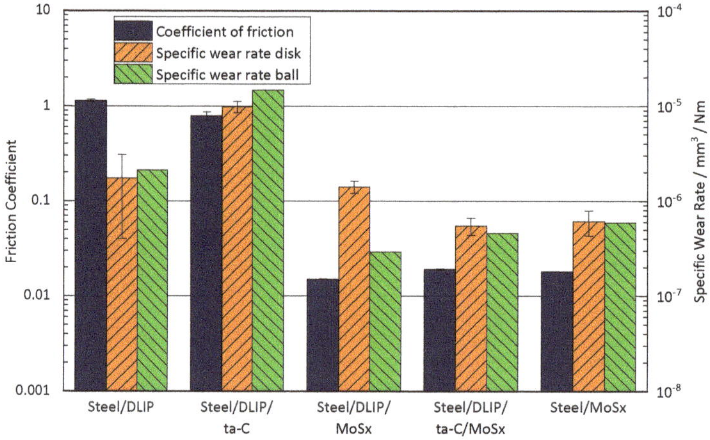

Figure 3. Steady-state friction coefficient, specific wear rate of disk, and specific wear rate of counterbody for all surfaces after tribological testing in an ultrahigh vacuum ($p < 5 \times 10^{-7}$ mbar).

The surface of the steel/MoSx (Figure 2e) had a much lower friction coefficient of 0.018, which remained very stable after nearly no running-in. The ball wear scar showed only mild signs of wear, had an almost circular shape, showed no signs of plasticity, and exhibited some colored patches most likely resulting from the tribo-induced transfer layer. The coated disk showed a wear track with some grooves and shallow voids along the sliding direction and signs of plastic deformation. The specific wear rates were rather low, at $k = 6.1 \times 10^{7}$ $mm^3/(N{\cdot}m)$ for the disk and $k = 5.9 \times 10^{7}$ $mm^3/(N{\cdot}m)$ for the ball.

Two more surfaces with MoSx top layers were studied: steel/DLIP/MoSx (Figure 2c) and steel/DLIP/ta-C/MoSx (Figure 2d). Despite their different designs below the MoSx top layer, their friction and wear behaviors were very similar to the steel/MoSx based on the qualitative assessment.

For the steel/DLIP/ta-C/MoSx, the friction coefficient was 0.019. The wear rates were $k = 5.5 \times 10^{-7}$ $mm^3/(N{\cdot}m)$ for the disk and $k = 4.6 \times 10^{-7}$ $mm^3/(N{\cdot}m)$ for the ball. In contrast to the steel/MoSx, the disk wear track was smooth and did not show signs of plasticity.

For the steel/DLIP/MoSx, the lowest friction of all the combinations was found with $\mu = 0.015$. Furthermore, the ball wear rate was the lowest, with $k = 2.9 \times 10^{-7}$ $mm^3/(N{\cdot}m)$. In return, the disk wear rate was increased to $k = 1.4 \times 10^{-6}$ $mm^3/(N{\cdot}m)$. Again, the disk wear track was smooth and did not show plasticity, as was the case for the steel/MoSx.

In all cases, the material removal was low, in the range of just few hundred nanometers, and the coatings were not destroyed or removed. In most cases, the wear depths were smaller or similar to the DLIP structure depth. For the surfaces with MoSx top coatings, the DLIP structure often could not be seen in the microscopic images, but it can be assumed that the very soft material was smeared into the structure.

3.3. Tribological Performance in Air

Two surfaces from the vacuum tests were tested using the same settings, except in ambient air. Steel/MoSx was chosen to represent the state-of-the-art coating technology, with steel/DLIP/ta-C/MoSx providing an advanced coating structure below the MoSx top layer.

Images of the contact areas and the progression of the friction coefficient are given in Figure 4.

Figure 4. Wear and friction data for testing in ambient air shown for (**a**) steel/MoSx and (**b**) steel/DLIP/ta-C/MoSx with light microscope images of the wear scar on the counter body ball (**left**), the wear track on the disk (**center**), and friction diagram (**right**, friction coefficient on the Y-axis and sliding length in meters on the X-axis).

The steel/MoSx surface (Figure 4a) created a thick and grey-colored tribolayer on the ball. The friction coefficient started at 0.12 and climbed to a steady-state friction of 0.24. The width of the ball contact area was much wider than the visible wear track width on the disk, allowing the conclusion that transfer of the MoSx to the ball had occurred. In order to calculate the ball wear rate, the transferred material was ignored, and just the circular contact area on the ball, corresponding to the disk wear track width, was used. The ball wear rate was found to be $k = 1.9 \times 10^{-6}$ $mm^3/(N{\cdot}m)$.

The disk wear track was smooth, with apparent grooves in the sliding direction in the middle of the contact area. On the side of the contact area, there was partial delamination of the coating. Obviously, the higher friction forces led to higher shearing forces on the interface, resulting in delamination. This finding is unexpected, as the same interface technology was used as for the carbon coatings, which easily withstands such conditions without delamination. It might have been the high mechanical mismatch of the soft MoSx coating with the steel substrate that led to the critical loading conditions. The disk wear rate was calculated to be $k = 1.6 \times 10^{-5}$ $mm^3/(N{\cdot}m)$, omitting the delaminated volume.

The surface steel/DLIP/ta-C/MoSx (Figure 4b) showed a similar transfer layer on the ball. The ball wear rate was slightly lower, at $k = 1.6 \times 10^{-6}$ $mm^3/(N{\cdot}m)$, when compared to the steel/MoSx and only considering the inside contact area again. The disk wear track also showed some grooves in the sliding direction, but the DLIP pattern remained clearly visible. The disk wear rate was $k = 1.7 \times 10^{-5}$ $mm^3/(N{\cdot}m)$. The friction coefficient started at 0.12 and climbed to a steady-state friction coefficient of 0.28.

From a solely quantitative standpoint, the tribological behavior of both surfaces in the atmosphere was very similar, showing the same transfer layer and almost the same friction and wear rates on both the ball and disk. However, the steel/MoSx surface already started to delaminate, which was expected to result in rapid and complete failure of the system.

3.4. Conservation of DLIP Structure after Tribological Loading

To provide a tribological effect, the DLIP structure must withstand tribological loading. In the following, three different outcomes of DLIP structures after the end of the test are found, depending on the types of coatings used.

In the case of the steel/DLIP surface in a vacuum, the structure was lost due to heavy plastic deformation and material build-up where contact occurred on the microscopic scale. Only some areas that were not in contact kept their original structure, as visible in Figure 5.

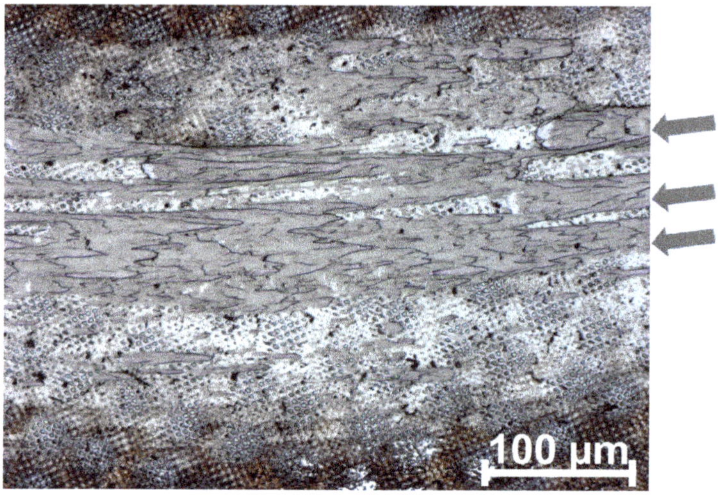

Figure 5. Light microscope image of wear track from steel/DLIP surface after testing in vacuum conditions. Remaining holes from initial DLIP structure can be seen as a periodic pattern. Heavy plastic deformation is seen along some areas along the sliding direction where the DLIP structures were lost (arrows). 3D study of the wear track (not shown here) indicates that both material build-up and material removal occurred.

The surface of the steel/DLIP/ta-C in a vacuum did not show plastic deformation. It was very hard and only showed signs of mild abrasion in the form of polishing of the protruding areas (Figure 6). In the center of the wear track, where the maximum contact pressure was observed, abrasive wear was found, which went deeper into the ta-C coating than the original DLIP structure.

A very interesting case was found for the steel/DLIP/ta-C/MoSx surface (Figure 7), where the very hard ta-C surface was covered with very soft MoSx. At the edge of the wear track, where only low contact pressure was found and only little material was removed, the MoSx coating had only the protruding peaks removed. Towards the middle of the wear track, where a higher contact pressure and higher wear occurred, most of the MoSx volume had been removed. Instead of the MoSx coating, the underlying peaks of the hard ta-C coating provided support and could withstand the locally concentrated contact pressure. No difference in the color or structure could be seen in the optical image, allowing the assumption that the whole area was still covered with MoSx.

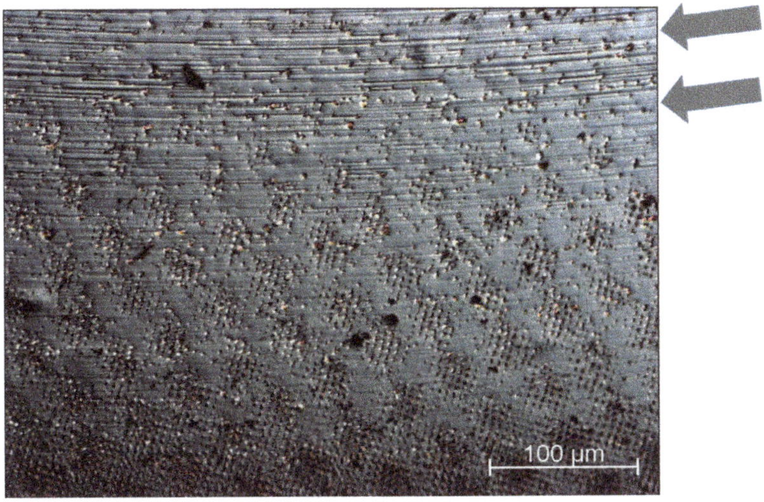

Figure 6. Light microscope image with DIC filter of wear track of steel/DLIP/ta-C surface after testing in vacuum conditions. Sliding direction is from right to left. Original DLIP surface can be seen on the lower border of the image. Towards the middle, the DLIP structure is superimposed with a polishing of the protruding areas of the surface. Abrasive wear with grooves resulting from detached hard particles can be seen in the upper part of the image, as indicated by arrows.

(**a**)

(**b**)

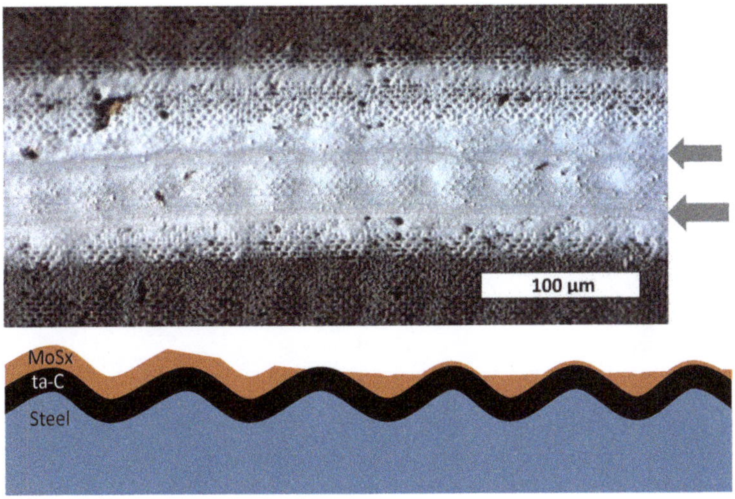

Figure 7. Light microscope image (**a**) with DIC filter of wear track of steel/DLIP/ta-C/MoSx surface after sliding in a vacuum. DLIP reversed structure can be clearly seen when comparing the outer and inner parts of the wear track. Along some circumferential areas, the DLIP structure is fully lost, as indicated by the arrows. Schematic representation (**b**) of pristine and worn areas of steel/DLIP/ta-C/MoSx where all surfaces are still covered with a thin layer of MoSx.

4. Discussion

The tribological performance of the investigated surfaces in a vacuum can be categorized into two groups, whether the top layer was MoSx or not (Figure 3).

Whenever MoSx was the top layer, friction was very low ($\mu < 0.02$). Also, the wear rates were low for both the disk and ball. The DLIP structuring and ta-C sublayer did not change the outcome, except for the steel/DLIP/MoSx, where wear rate was somewhat lower on the ball but was increased for the disk. In general, the tribological properties were mostly governed by the MoSx, which was expected to perform well in a vacuum.

Contrarily, the friction and wear were high when ta-C or steel was the topmost surface. The friction coefficient was $\mu > 0.8$, indicating adhesion of the materials. Both ta-C and steel lost passivation in a vacuum quickly due to the lack of oxygen or water, indicated by an initially low friction coefficient that increased sharply. The steel surfaces in contact reacted with cold welding and heavy plastic deformation. The surface with a ta-C coating was very hard and therefore did not deform plastically but showed heavy abrasion. Consequently, both the disk and ball wear were half a magnitude higher for steel/DLIP and more than one magnitude higher for steel/DLIP/ta-C. Again, no benefit from DLIP structuring was evident based on the analysis of the wear scar and wear track images.

With respect to their performance in air, two of the surfaces were studied in air to investigate whether the steel/DLIP/ta-C/MoSx had any benefit over the steel/MoSx. It was found that friction coefficient and disk and ball wear rates were very similar (Figure 4). In both cases, the MoSx was removed from the disk surface, forming a thick transfer layer on and around the ball contact area.

The disk wear rate for both was more than one order of magnitude higher compared to the vacuum conditions for the same surfaces. Furthermore, the MoSx wear rate in atmosphere was in the same order as the ta-C surface in a vacuum, highlighting the adverse effects of the surrounding conditions on the coating.

Despite its high similarity when evaluated quantitatively, a detailed analysis of the wear mechanism revealed differences. In the case of the steel/MoSx surface, the soft MoSx coating was worn, leading to a large contact area. The grooves in the wear track indicated a localized and uneven wear. Delamination of the coating indicated that the strength of the interface was exceeded. In contrast, on the steel/DLIP/ta-C/MoSx, the soft top coating was removed down to the ta-C coating only. The study of the wear track indicated that the underlying ta-C provided support for the loaded contact, as the DLIP structure was revealed again after most of the MoSx was removed. The MoSx pockets remaining between the ta-C hills acted as a reservoir of solid lubricant.

Structuring also changed the load-bearing capacity of the surface. On the macroscopic scale, the contact pressure was about 1.6 times higher in the case of the steel/DLIP/ta-C/MoSx compared to the steel/MoSx (Table 3).

Table 3. Projected area of contact and resulting macroscopic contact pressure as found after tribological testing in atmosphere.

Surface	Area of Contact/mm^2	Macroscopic Contact Pressure/MPa
Steel/MoSx	0.035	137
Steel/DLIP/ta-C/MoSx	0.021	219

On the microscopic scale, the contact pressure for the steel/DLIP/ta-C/MoSx was much higher, as the load was primarily supported by ta-C-structured surface and not by the soft MoSx. Such very high contact pressures also induced very high shear stresses, which were beneficial to forming shearing planes of MoS2 out of the amorphous MoSx material.

The presence of MoSx on the remaining structure was evidenced using Raman spectroscopy. A 100× objective, leading to a spot size of about 1 µm in diameter, was focused on the peak and valley areas of the remaining structure in the wear track after testing the steel/DLIP/ta-C/MoSx in air. The measurement positions are shown in Figure 8a. For both locations, a single representative spectrum is shown in Figure 8b.

In all locations, dominant spectra corresponding to MoS_2 were found (characteristic peaks at 382 cm^{-1} and 408 cm^{-1}), indicating that MoSx was present in all cases. On the contrary, only a very weak G peak around 1580 cm^{-1}, characteristic of ta-C, was found.

As ta-C coatings usually give a strong Raman signal, it is assumed that the amorphous MoSx on top of it became highly ordered due to shearing. Consequently, the reflectance increased to such a high level that the laser beam was barely able to penetrate and probe the underlying ta-C layer [36].

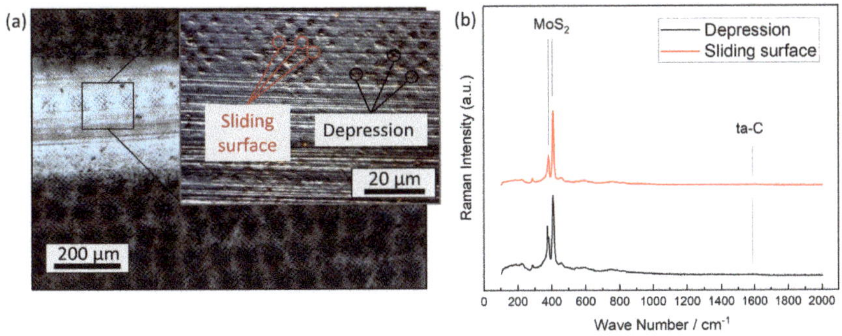

Figure 8. Analysis of the steel/DLIP/ta-C/MoSx surface tested in air showing: (**a**) light microscope image and magnified inset (DIC), exemplarily indicating depression and sliding surface positions for Raman analysis, and (**b**) one representative Raman spectrum for each of the locations.

From an application point of view, the combination of the DLIP structure with the ta-C support layer and MoSx sliding top layer opens up the possibility to create tribological contacts which can endure very high loads in both the atmosphere and vacuums without compromising friction and wear performance. While in this work, the two coatings were deposited in two different batches, it is possible in principle to deposit such bilayer coatings in a single batch using two separate arc sources.

Given the stable wear structure formed during tribological loading, both long-term testing as well as testing in changing atmospheres, for example in air and a vacuum, mimicking multi-use space applications, should be studied in the next step.

5. Conclusions

In this work, the combination of DLIP structured steel substrates and deposition of ta-C and MoSx mono- and bilayers using LaserArc™ technology was investigated, resulting in the following surfaces: steel/DLIP, steel/DLIP/ta-C, steel/DLIP/MoSx, steel/DLIP/ta-C/MoSx, and steel/MoSx. Their tribological properties were studied using a pin-on-disk-tribometer using a steel ball as the counter body and without any lubrication. All the surfaces were studied in a vacuum at a pressure of $p < 5 \times 10^{-7}$ mbar. Furthermore, the steel/DLIP/ta-C/MoSx and steel/MoSx were studied in ambient air with the same setup. The wear surfaces on both the disk and ball were examined to identify the wear mechanisms and topographic changes.

In vacuum conditions, all the coatings with a MoSx top layer had similar good performances with a very low friction ($\mu < 0.02$) and low wear, regardless of the underlying structure or underlying coatings. In contrast, very high friction and wear was observed when steel or ta-C were the topmost materials of the disk, which was attributed to their loss of initial passivation and the resulting strong adhesion with the steel counter body.

In ambient air, the ball and disk wear rates as well as the friction coefficients were similar for both the steel/MoSx and steel/DLIP/ta-C/MoSx when compared quantitatively. However, while on the steel/MoSx surface, the coating started to delaminate along the wear tracks, the steel/DLIP/ta-C/MoSx surface was able to endure a much higher contact pressure without delamination. After the soft MoSx layer was partially worn on the steel/DLIP/ta-C/MoSx, the underlying ta-C coating with the DLIP structure was able to provide mechanical support to the top of the structure while providing a MoSx reservoir in between the structure. The load was confined to a small and well-defined contact area with a stable wear mechanism.

The combination of a structured substrate with a hard sub-layer and soft friction layer enables the design of high-load tribological surfaces, which might result in smaller and more durable dry-running bearings that can be used in both vacuums and in air, for example, in machines that temporarily operate in a technical vacuum or for multi-use space applications.

Author Contributions: Conceptualization, F.H. and A.F.L.; data curation, S.M. and F.H.; funding acquisition, S.M. and A.F.L.; investigation, F.H., M.S., M.O. and L.L.; project administration, S.M. and M.S.; resources, F.H., M.S. and F.K.; software, F.H. and L.L.; supervision, S.M. and A.F.L.; validation, F.H. and M.O.; visualization, S.M., F.H. and M.S.; writing—original draft, S.M.; writing—review and editing, F.H., M.S., F.K. and A.F.L. All authors have read and agreed to the published version of the manuscript.

Funding: This work was supported by the Deutsche Forschungsgemeinschaft (DFG—German Research Foundation) under grant agreement 415726702 and Sächsische Aufbaubank (SAB) within project LUBRICOAT—simulation-aided design of solid lubricant coatings.

Data Availability Statement: The data will be made available upon request.

Acknowledgments: The authors acknowledge Falko Hofmann and Marion Franke for the coating deposition, Tanmaya Milind Angane and Yelvin Ragimov for conducting the tribological experiments and imaging of the wear track, and Susann Kleber and Beate Leupoldt for the Raman spectroscopy and helpful discussions.

Conflicts of Interest: The authors declare no conflicts of interest.

References

1. Serles, P.; Gaber, K.; Pajovic, S.; Colas, G.; Filleter, T. High Temperature Microtribological Studies of MoS_2 Lubrication for Low Earth Orbit. *Lubricants* **2020**, *8*, 49. [CrossRef]
2. Gradt, T.; Schneider, T. Tribological Performance of MoS_2 Coatings in Various Environments. *Lubricants* **2016**, *4*, 32. [CrossRef]
3. Chan, T.Y.; Hu, Y.; Gharbi, M.M.; Politis, D.J.; Wang, L. The friction coefficient evolution of a MoS_2/WC multi-layer coating system during sliding wear. *J. Phys. Conf. Ser.* **2016**, *734*, 32081. [CrossRef]
4. Martin, J.M.; Donnet, C.; Le Mogne, T.; Epicier, T. Superlubricity of molybdenum disulphide. *Phys. Rev. B Condens. Matter* **1993**, *48*, 10583–10586. [CrossRef] [PubMed]
5. Seynstahl, A.; Krauß, S.; Bitzek, E.; Meyer, B.; Merle, B.; Tremmel, S. Microstructure, Mechanical Properties and Tribological Behavior of Magnetron-Sputtered MoS_2 Solid Lubricant Coatings Deposited under Industrial Conditions. *Coatings* **2021**, *11*, 455. [CrossRef]
6. Erdemir, A.; Eryilmaz, O. Achieving superlubricity in DLC films by controlling bulk, surface, and tribochemistry. *Friction* **2014**, *2*, 140–155. [CrossRef]
7. Kato, K.; Umehara, N.; Adachi, K. Friction, wear and N2-lubrication of carbon nitride coatings: A review. *Wear* **2003**, *254*, 1062–1069. [CrossRef]
8. Konicek, A.R.; Grierson, D.S.; Sumant, A.V.; Friedmann, T.A.; Sullivan, J.P.; Gilbert, P.U.P.A.; Sawyer, W.G.; Carpick, R.W. Influence of surface passivation on the friction and wear behavior of ultrananocrystalline diamond and tetrahedral amorphous carbon thin films. *Phys. Rev. B* **2012**, *85*, 155448. [CrossRef]
9. Andersson, J.; Erck, R.; Erdemir, A. Frictional behavior of diamondlike carbon films in vacuum and under varying water vapor pressure. *Surf. Coat. Technol.* **2003**, *163–164*, 535–540. [CrossRef]
10. Meunier, C.; Alers, P.; Marot, L.; Stauffer, J.; Randall, N.; Mikhailov, S. Friction properties of ta-C and a-C:H coatings under high vacuum. *Surf. Coat. Technol.* **2005**, *200*, 1976–1981. [CrossRef]
11. Chen, Q.; Wu, G.; Li, D.; Li, A.; Shang, L.; Lu, Z.; Zhang, G.; Wu, Z.; Tian, G. Understanding the unusual friction behavior of TiN films in vacuum. *Tribol. Int.* **2019**, *137*, 379–386. [CrossRef]
12. Kara, L.; Özkan, D.; Barış Yağcı, M.; Sulukan, E.; Sert, Y.; Sonsuz Sert, T. Friction and Wear Behaviors of TiN Coatings under Dry and Vacuum Conditions. *Tribol. Trans.* **2019**, *62*, 362–373. [CrossRef]
13. Yu, G.; Gong, Z.; Jiang, B.; Wang, D.; Bai, C.; Zhang, J. Superlubricity for hydrogenated diamond like carbon induced by thin MoS_2 and DLC layer in moist air. *Diam. Relat. Mater.* **2020**, *102*, 107668. [CrossRef]
14. Polcar, T.; Cavaleiro, A. Review on self-lubricant transition metal dichalcogenide nanocomposite coatings alloyed with carbon. *Surf. Coat. Technol.* **2011**, *206*, 686–695. [CrossRef]
15. Vuchkov, T.; Evaristo, M.; Yaqub, T.B.; Polcar, T.; Cavaleiro, A. Synthesis, microstructure and mechanical properties of W–S–C self-lubricant thin films deposited by magnetron sputtering. *Tribol. Int.* **2020**, *150*, 106363. [CrossRef]
16. Yaqub, T.B.; Hebbar Kannur, K.; Vuchkov, T.; Pupier, C.; Héau, C.; Cavaleiro, A. Molybdenum diselenide coatings as universal dry lubricants for terrestrial and aerospace applications. *Mater. Lett.* **2020**, *275*, 128035. [CrossRef]

17. Zhang, J.; Jiang, D.; Wang, D.; Yu, Q.; Bai, Y.; Cai, M.; Weng, L.; Zhou, F.; Liu, W. MoS_2 Lubricating Film Meets Supramolecular Gel: A Novel Composite Lubricating System for Space Applications. *ACS Appl. Mater. Interfaces* **2021**, *13*, 58036–58047. [CrossRef] [PubMed]
18. Torres, H.; Rojacz, H.; Čoga, L.; Kalin, M.; Rodríguez Ripoll, M. Local mechanical and frictional properties of Ag/ MoS_2-doped self-lubricating Ni-based laser claddings and resulting high temperature vacuum performance. *Mater. Des.* **2020**, *186*, 108296. [CrossRef]
19. Zhao, D.; Li, S.; Zhao, X.; Hao, E.; An, Y.; Zhou, H.; Chen, J. Preparation and vacuum tribological properties of composite coatings fabricated by effective introduction of soft metal Ag into spray-formed YSZ templates. *Appl. Surf. Sci.* **2020**, *518*, 146176. [CrossRef]
20. Cheng, J.; Zhen, J.; Zhu, S.; Yang, J.; Ma, J.; Li, W.; Liu, W. Friction and wear behavior of Ni-based solid-lubricating composites at high temperature in a vacuum environment. *Mater. Des.* **2017**, *122*, 405–413. [CrossRef]
21. Härtwig, F.; Lorenz, L.; Makowski, S.; Krause, M.; Habenicht, C.; Lasagni, A.F. Low-Friction of ta-C Coatings Paired with Brass and Other Materials under Vacuum and Atmospheric Conditions. *Materials* **2022**, *15*, 2534. [CrossRef] [PubMed]
22. Theiler, G.; Gradt, T.; Österle, W.; Brückner, A.; Weihnacht, V. Friction and endurance of MoS_2/ta-C coatings produced by Laser Arc deposition. *Wear* **2013**, *297*, 791–801. [CrossRef]
23. Rosenkranz, A.; Hans, M.; Gachot, C.; Thome, A.; Bonk, S.; Mücklich, F. Direct Laser Interference Patterning: Tailoring of Contact Area for Frictional and Antibacterial Properties. *Lubricants* **2016**, *4*, 2. [CrossRef]
24. Segu, D.Z.; Kim, S.S. Influence on friction behavior of micro-texturing under lubricated non-conformal contact. *Meccanica* **2014**, *49*, 483–492. [CrossRef]
25. Vilhena, L.M.; Sedlaček, M.; Podgornik, B.; Vižintin, J.; Babnik, A.; Možina, J. Surface texturing by pulsed Nd:YAG laser. *Tribol. Int.* **2009**, *42*, 1496–1504. [CrossRef]
26. Ränke, F.; Baumann, R.; Voisiat, B.; Fabián Lasagni, A. High throughput laser surface micro-structuring of polystyrene by combining direct laser interference patterning with polygon scanner technology. *Mater. Lett. X* **2022**, *14*, 100144. [CrossRef]
27. Müller, D.W.; Fox, T.; Grützmacher, P.G.; Suarez, S.; Mücklich, F. Applying Ultrashort Pulsed Direct Laser Interference Patterning for Functional Surfaces. *Sci. Rep.* **2020**, *10*, 3647. [CrossRef] [PubMed]
28. Gachot, C.; Rosenkranz, A.; Reinert, L.; Ramos-Moore, E.; Souza, N.; Müser, M.H.; Mücklich, F. Dry Friction Between Laser-Patterned Surfaces: Role of Alignment, Structural Wavelength and Surface Chemistry. *Tribol. Lett.* **2013**, *49*, 193–202. [CrossRef]
29. Roch, T.; Weihnacht, V.; Scheibe, H.-J.; Roch, A.; Lasagni, A.F. Direct Laser Interference Patterning of tetrahedral amorphous carbon films for tribological applications. *Diam. Relat. Mater.* **2013**, *33*, 20–26. [CrossRef]
30. Maclucas, T.; Daut, L.; Grützmacher, P.; Guitar, M.A.; Presser, V.; Gachot, C.; Suarez, S.; Mücklich, F. Influence of structural depth of laser-patterned steel surfaces on the solid lubricity of carbon nanoparticle coatings. *Friction* **2023**, *11*, 1276–1291. [CrossRef]
31. Maclucas, T.; Leonhard-Trautmann, P.; Suarez, S.; Mücklich, F. Long-Term Lubricity of Carbon Nanoparticle Coatings on Periodically Laser-Patterned Metallic Surfaces. *Tribol. Lett.* **2022**, *70*, 123. [CrossRef]
32. Fu, Y.; Soldera, M.; Wang, W.; Voisiat, B.; Lasagni, A.F. Picosecond Laser Interference Patterning of Periodical Micro-Architectures on Metallic Molds for Hot Embossing. *Materials* **2019**, *12*, 3409. [CrossRef] [PubMed]
33. Kaulfuss, F.; Weihnacht, V.; Zawischa, M.; Lorenz, L.; Makowski, S.; Hofmann, F.; Leson, A. Effect of Energy and Temperature on Tetrahedral Amorphous Carbon Coatings Deposited by Filtered Laser-Arc. *Materials* **2021**, *14*, 2176. [CrossRef] [PubMed]
34. Krülle, T.; Kaulfuß, F.; Weihnacht, V.; Hofmann, F.; Kirsten, F. Amorphous Carbon Coatings with Different Metal and Nonmetal Dopants: Influence of Cathode Modification on Laser-Arc Evaporation and Film Deposition. *Coatings* **2022**, *12*, 188. [CrossRef]
35. Bichsel, R.; Buffat, P.; Levy, F. Correlation between process conditions, chemical composition and morphology of MoS_2 films prepared by RF planar magnetron sputtering. *J. Phys. D Appl. Phys.* **1986**, *19*, 1575–1585. [CrossRef]
36. Niu, Y.; Gonzalez-Abad, S.; Frisenda, R.; Marauhn, P.; Drüppel, M.; Gant, P.; Schmidt, R.; Taghavi, N.S.; Barcons, D.; Molina-Mendoza, A.J.; et al. Thickness-Dependent Differential Reflectance Spectra of Monolayer and Few-Layer MoS_2, $MoSe_2$, WS_2 and WSe_2. *Nanomaterials* **2018**, *8*, 725. [CrossRef]

lubricants

MDPI

Article

Lubrication Characteristics of a Warhead-Type Irregular Symmetric Texture on the Stator Rubber Surfaces of Screw Pumps

Xinfu Liu [1,*,†], Yi Sun [1,*,†], Chunhua Liu [2], Xiangzhi Shi [1], Xinglong Niu [1], Gang Zheng [3,4], Wei Wei [3,4], Songbo Wei [5] and Shouzhi Huang [5]

1 Key Lab of Industrial Fluid Energy Conservation and Pollution Control (Ministry of Education), Qingdao University of Technology, Qingdao 266520, China; xzshi1999@163.com (X.S.); nl18464268306@163.com (X.N.)
2 College of Mechanical and Electronic Engineering, China University of Petroleum (East China), Qingdao 266580, China; 20090053@upc.edu.cn
3 Oil and Gas Technology Institute, PetroChina Changqing Oilfield Company, Xi'an 710018, China; 15874697883@163.com (G.Z.); liuwei_gmcl@163.com (W.W.)
4 National Engineering Laboratory of Low-Permeability Oil & Gas Exploration and Development, Xi'an 710018, China
5 Research Institute of Exploration & Development, PetroChina, Beijing 100083, China; wsbthu@163.com (S.W.); huangszl@petrochina.com.cn (S.H.)
* Correspondence: upcdoctor@126.com (X.L.); 17866926604@163.com (Y.S.)
† These authors contributed equally to this work.

Abstract: A theoretical model for the micro-texture on the inner wall of the stator rubber in screw pumps was developed. The finite element analysis method was employed. The pressure and streamline distributions for warhead-type, concentric circle-type, and multilayer rectangular-type textured surfaces were calculated. The effects of textured morphology, groove depth, groove width, and other parameters on the lubrication field were systematically investigated and analyzed. A nanosecond laser was employed to process the textured rubber surface of the stator in the screw pump. Subsequently, a micro-texture friction performance test was conducted on the rubber surface of the stator in actual complex well fluids from shale oil wells. Given the results of the simulation analysis and experimental tests, the lubrication characteristics of textured rubber surfaces with varying texture morphologies, rotational speeds, and mating loads were revealed. Furthermore, it indicated that the irregular symmetric warhead-type micro-texture exhibited excellent dynamic pressure lubrication performance compared with concentric circle-type and multilayer rectangular-type textures. The irregular symmetry enhanced the dynamic pressure lubrication effect, enhanced the additional net load-bearing capacity of the oil film surface, and reduced friction. As the groove depth increased, the volume and number of vortices within the groove also increased. The fluid kinetic energy was transformed into vortex energy, leading to a reduction in wall stress on the surface of the oil film, thereby affecting its bearing capacity. Initially, the maximum pressure on the wall surface of the oil film increased and then decreased. The optimal dynamic pressure lubrication effect was achieved with a warhead-type texture size of 3 mm, a groove width of 0.2 mm, and a groove depth of 0.1 mm. Well-designed texture morphology and depth parameters significantly enhanced the oil film-bearing capacity of the stator rubber surface, improving the dynamic pressure lubrication effect, and consequently extending the service life of the stator–rotor interface in the screw pump.

Keywords: screw pump; stator rubber; irregular symmetry; warhead-type texture; lubrication property

1. Introduction

Screw pump oil extraction has emerged as a pivotal method for lifting oil and gas in ultradeep wells, highly deviated wells, and horizontal wells. This is attributable to its high efficiency in medium conveyance, low energy consumption, and strong adaptability [1–5]. The mechanical wear behavior between stator–rotor contact surfaces critically impacts the service life and operational efficiency of screw pumps. Consequently, mitigating the wear between the stator and rotor is essential to prolonging the operational life of screw pumps, enhancing the efficiency of oilfield recovery operations.

Recently, studies have proposed surface texture technology to enhance the friction and wear performance of contact surfaces in relative motion. This involves designing and processing various micro-textures, such as micro-pits, micro-bulges, and micro-grooves, with specific patterns on the material surface. These micro-textures significantly reduce surface friction and enhance the load-bearing capacity of the oil film (Figure 1) [6–13]. Wang and Yu et al. engineered pit and groove micro-textures on bearing surfaces, demonstrating that the average friction coefficient of pit micro-textured surfaces decreased by 41.4% compared with non-textured surfaces. Furthermore, they emphasized that the careful selection of groove morphology is critical for optimizing bearing lubrication performance [14,15]. Additionally, Jia et al. developed a plunger texture lubrication model that accounts for plunger deformation and eccentricity. Their findings indicated that laser-etched pit-type micro-textures on the plunger surface significantly enhance lubrication performance, resulting in a 27.8% reduction in the average friction coefficient [16]. Zhang et al. analyzed the surface texturing of polymer polyethylene materials by establishing a theoretical model aimed at optimizing the load-carrying capacity, friction, and drag reduction of four texture shapes: circular, rectangular, square, and triangular. It was observed that the texture area rate, width, and depth strongly influence the load-carrying capacity of the oil film, with the optimal texture area rate for superior tribological performance ranging between 12% and 36% [17]. Bai et al. conducted hydrodynamic simulations to investigate the lubrication properties of micro-groove textures on surfaces at various flow rates. Their results revealed that vortices in the near-wall flow field on textured surfaces exhibited greater stability compared with those on smooth surfaces [18]. Ma et al. designed a sinusoidal groove texture employing both simulation analysis and friction experiments. It was observed that the optimal groove spacing enhanced the dynamic pressure effect and improved the oil film condition, with a texture area rate of 25%, resulting in optimal tribological performance [19]. Han and Shen employed theoretical analysis and numerical simulation. It was observed that asymmetric rectangular micro-textures exhibited exceptional lubrication performance compared with symmetric rectangular textures [20,21]. Wang et al. investigated the stator–rotor interference fit conditions in screw pumps and introduced micro-textures to the rotor surface to analyze the influence of texture, shape, and angle on the tribological performance between the stator and rotor. The results indicated that the average coefficient of friction was reduced to ~62% compared with non-textured surfaces. Additionally, the tribological performance under various lubrication conditions was examined by applying micro-textures to both stator and rotor surfaces. The study revealed that at a texture angle of 0° under dry friction conditions, the presence of micro-textures significantly enhanced the tribological performance of the stator–rotor interface [22,23]. Zhang Baoling employed fluid dynamics methods to analyze and compare the friction characteristics, pressure distribution, and flow field distribution of surfaces with no texture, as well as circular, spherical, or V-shaped textures. By conducting tests under water lubrication, oil lubrication, and oil–water mixing lubrication conditions, Zhang verified the theory of stator rubber surface texturing lubrication for screw pump stators. This study highlighted that optimizing texturing parameters significantly improved the friction characteristics of stator–rotor pairs, thereby demonstrating a significant enhancement in screw pump performance [24].

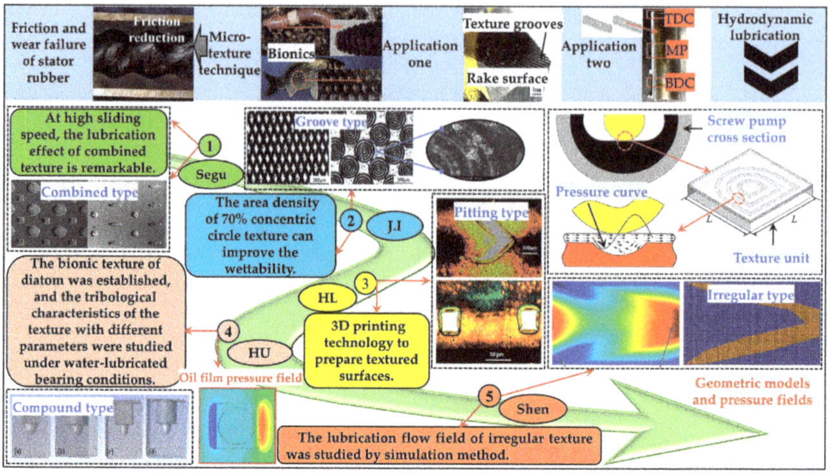

Figure 1. Research status of surface micro-texture technology: (**a**) Rectangular semicircle combination form; (**b**) Cylindrical semicircle combination form; (**c**) Rectangular ellipse combination form; (**d**) Cylindrical ellipse combination form.

Surface texturing technology has shown significant promise in enhancing the tribological properties of materials. However, current studies predominantly focused on hard materials such as metals, commonly employed in bearings, gears, and cutting tools. Soft materials such as the rubber stator found in screw pumps, have received considerably less attention [25–34]. Therefore, this study addressed this gap by proposing the design of an irregular high-lubrication and friction-reducing texture, drawing inspiration from concentric circular patterns and multilayer rectangular textures. Based on fluid dynamic pressure lubrication theory, a mathematical model for stator rubber lubrication was established. Subsequently, finite element analysis was employed to simulate the impact of texture type, groove depth, and groove width on lubrication performance under actual shale oil well conditions. This study offered guidance for the optimal design of the textured surface of screw pump stator rubber, thereby advancing texturing design theory.

2. Theoretical Model of the Textured Surface of the Stator Rubber

2.1. Geometric Modeling of the Stator Rubber Textured Surface

Screw pumps, integral to rodless lift oil recovery systems, often face mechanical wear at the stator–rotor interface due to prolonged exposure to oil pressure, temperature, and viscosity. This wear can lead to pump failure. The application of surface texturing on the stator rubber can significantly enhance lubrication between the stator and rotor, thereby enhancing pump performance and longevity. In this study, a textured unit on the rubber surface of the screw pump stator was selected for analysis.

The warhead-type texture is designed based on a combination of warhead-type, concentric circle-type, and multilayer rectangular-type texture morphology, incorporating irregular symmetry to enhance the dynamic pressure lubrication effect in the grooves. Geometric models of the warhead-type, concentric circle-type, and multilayer rectangular-type textures were established sequentially (Figure 2). Textures were arranged in layers, starting from the center of the $L \times L$ unit area and expanding outward. The grooves of the textures have a width of w and a depth of h, with the overall texture size controlled by adjusting the dimensions of the outermost layer.

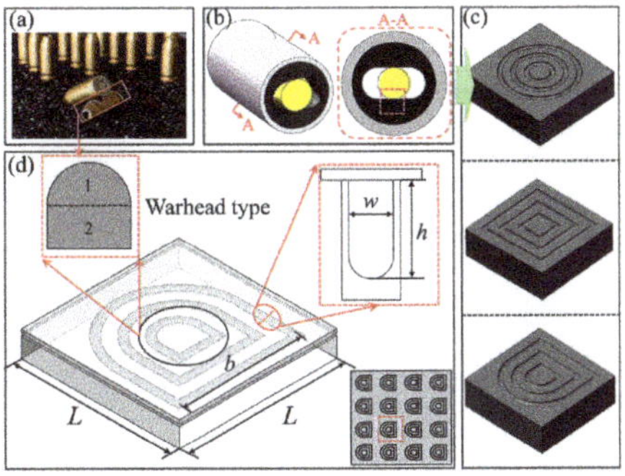

Figure 2. **(a–d)** Geometric model of textured rubber with different morphologies.

2.2. Mathematical Modeling of the Stator Rubber Textured Surface

Assuming a constant fluid flow, the Reynolds number (R_e) is employed to determine the flow regime. The flow is considered laminar when $R_e \leq 2300$ and turbulent when $R_e \geq 4000$ [35,36].

The actual Reynolds number of the fluid within the stator–rotor friction pair of a screw pump is calculated using the following formula:

$$R_e = \frac{\rho U h_0}{\eta} \tag{1}$$

where U is the sliding velocity on the upper surface of the oil film (m/s), h_0 is the thickness of the oil film (m), ρ is the density of the shale oil (kg/m^3), and η is the coefficient of dynamic viscosity of the shale oil (m^2/s).

When the Reynolds number indicates turbulent flow, the k–epsilon model [37,38] was selected, introducing the turbulent kinetic energy k and dissipation rate ε to enhance computational accuracy.

$$\frac{\partial}{\partial t}(\rho k) + \frac{\partial}{\partial x_i}(\rho k u_i) = \frac{\partial}{\partial x_j}\left[\left[\mu + \frac{\mu_t}{\sigma_k}\right]\frac{\partial k}{\partial x_j}\right] + G_k + G_b - \rho\varepsilon - Y_M + S_k \tag{2}$$

$$\frac{\partial}{\partial t}(\rho\varepsilon) + \frac{\partial}{\partial x_i}(\rho\varepsilon u_i) = \frac{\partial}{\partial x_j}\left[\left(\mu + \frac{\mu_t}{\sigma_\varepsilon}\right)\frac{\partial \varepsilon}{\partial x_j}\right] + C_{1\varepsilon}\frac{\varepsilon}{k}(G_k + C_{3\varepsilon}G_b) - C_{2\varepsilon}\rho\frac{\varepsilon^2}{k} + S_\varepsilon \tag{3}$$

where $C_{1\varepsilon}$, $C_{2\varepsilon}$ are empirical constants, and σ_k as well as σ_ε are the Prandtl numbers corresponding to the turbulent kinetic energy and dissipation rate, respectively.

Based on the theory of elastohydrodynamic lubrication, and considering the influence of inertial forces, the Navier–Stokes (N–S) equations describe incompressible Newtonian fluids in a Cartesian coordinate system [39–44]:

$$\frac{\partial(\rho u)}{\partial t} + div(\rho\eta u) = \frac{\partial \tau xx}{\partial x} + \frac{\partial \tau_{yx}}{\partial y} + \frac{\partial \tau_{zx}}{\partial z} - \frac{\partial P}{\partial x} + F_x \tag{4}$$

$$\frac{\partial(\rho v)}{\partial t} + div(\rho\eta v) = \frac{\partial \tau_{xy}}{\partial x} + \frac{\partial \tau_{yy}}{\partial y} + \frac{\partial \tau_{zy}}{\partial z} - \frac{\partial P}{\partial y} + F_y \tag{5}$$

$$\frac{\partial(\rho w)}{\partial t} + div(\rho\eta w) = \frac{\partial \tau_{xz}}{\partial x} + \frac{\partial \tau_{yz}}{\partial y} + \frac{\partial \tau_{zz}}{\partial z} - \frac{\partial P}{\partial z} + F_z \tag{6}$$

where η is the dynamic viscosity of the fluid (Pa·s); ρ is the density of the fluid (kg/m^3); u, v, and w are the components of the velocity vector in the x-, y-, and z-axis directions (m/s), respectively. τ_{xx}, τ_{xy}, and τ_{xz} are the components of viscous action at a point, respectively.

P is the pressure of the fluid at a point (Pa). F_x, F_y, and F_z are the components of the volumetric force of the fluid in the x-, y-, and z-axis directions (N), respectively.

The oil film-bearing capacity was obtained by integrating the pressure distribution region of the wall shear stress over the region [45–47]:

$$F = \int_0^L \int_0^L P dx dy \tag{7}$$

where P is the pressure of the shale oil film in Pascals (Pa), L is the boundary for calculations in the x and y-axis directions (m).

The oil film friction was obtained by integrating the region of wall shear stress on the oil film [48–50]:

$$f = \int_0^L \int_0^L \tau dx dy \tag{8}$$

where τ is the shear stress of the oil film (Pa).

The coefficient of friction on the rubber surface of the textured stator in the screw pump is calculated as the ratio of the friction force to the load-carrying capacity:

$$\mu = \frac{f}{F} \tag{9}$$

3. Simulation Analysis of Stator Micro-Texture Surface Lubrication Flow Field

3.1. Simulation Model of the Rubber Texture Surface Flow Field of a Screw Pump Stator

The number of mesh elements directly affects the accuracy of simulation analysis, which is crucial for determining the distribution of the wall pressure field and lubrication flow field on the oil film. Therefore, maintaining the number of mesh elements within a reasonable range is essential [51,52]. Computational fluid dynamics was employed to delineate the stator rubber surface warhead-type micro-texture geometric model mesh, solving for the oil film wall pressure field and lubrication flow field. Furthermore, for the warhead-type texture model, mesh independence validation was conducted using the radial pressure on the oil film wall as the target parameter. The validation of mesh independence for the stator rubber warhead-type micro-texture is illustrated in Figure 3a. Subsequently, with varying numbers of meshes, the radial pressure along the oil film wall was relatively constant. Given both the calculation accuracy and computation time, the mesh was determined at 470,000 (Figure 3b). Under varying tolerance values, the maximum pressure on the upper surface of the stator rubber microstructure oil film decreases as the tolerance value is reduced, initially declining rapidly before reaching a more gradual trend. When the tolerance values are set to 1×10^{-5} and 1×10^{-8}, the calculation error for the maximum pressure on the microtextured oil film's upper surface remains below 5%. Considering both computational cost and convergence rate, a final tolerance value of 1×10^{-5} is selected.

In this simulation, the upper wall of the micro-texture simulation model was set with slip conditions; the lower wall was set with stationary boundary conditions; the front and rear walls were set with symmetric boundary conditions; and the pressure inlet and outlet boundary conditions are illustrated in Figure 3c. The lubricating medium utilized was shale oil with a density of 870 kg/m^3 and a viscosity of 0.08 Pa·s, which flows along the negative x-axis direction at 1 m/s. The outlet and inlet pressures were maintained at 101.325 kPa. The three-dimensional steady-state challenge within the computational domain was analyzed using a pressure solver combined with PISO coupling to solve the pressure and velocity fields.

The texture simulation model includes warhead-type, concentric circle-type, and multilayer rectangular-type textures. The texture dimensions are as follows: sizes of 2.5 mm, 3 mm, and 3.5 mm; groove widths of 0.1 mm and 0.2 mm; and depths ranging from 0.05 mm to 0.5 mm. Ultimately, the maximum pressure on the oil film surface served as the final evaluation criterion for texture performance, guiding the selection of surface texture for the rubber stator.

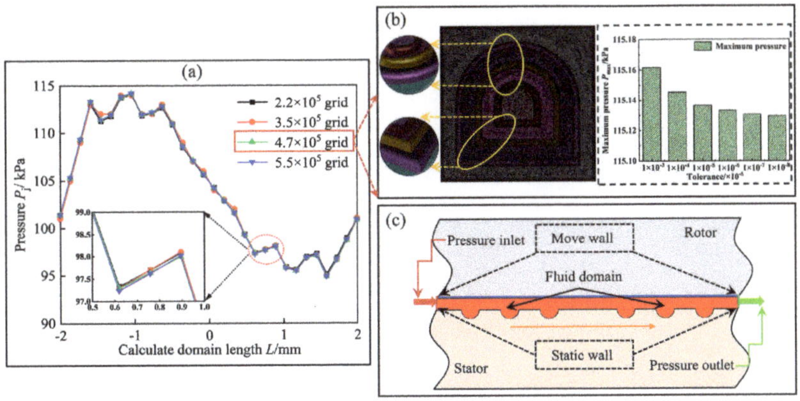

Figure 3. Stator rubber warhead-type micro-texture mesh division and independence validation: (**a**) Mesh independence validation; (**b**) Effect of different tolerance values; (**c**) Boundary condition.

3.2. *Influence of Texture Morphology on the Lubricating Properties of Stator Rubber Surfaces*

Figure 4a–c depict the pressure distribution of the oil film on the stator rubber surface with different texture morphologies. Shale oil transits from a small convergence wedge to a large convergence wedge. With the increase in wedge angle, the volume of the fluid channel increases, and the volume of shale oil decreases instantaneously in the transition zone. In this process, the relative volume of shale oil changes, resulting in the emergence of negative pressure. The alternating converging and diverging wedges created an irregularly symmetric pressure distribution on the oil film, thereby enhancing the net load-bearing capacity. Thus, with varying texture morphology, the positive and negative pressure areas on the oil film wall exhibited changes. Upon comparing Figure 4a–c, it was observed that the maximum pressure and positive pressure areas of the asymmetric warhead-type micro-texture exceeded those of the symmetric concentric circle-type and multilayer rectangular-type textures. In Figure 4d, the grooves on the stator rubber surface formed a wedge gap, exhibiting a dynamic pressure effect, leading to fluctuations in the pressure field. Figure 4e illustrates the net pressure distribution curve of the warhead-type texture oil film. The grooves exhibited distinct pressure peaks, with the positive pressure region exceeding those of the negative pressure region. At this point, the dynamic pressure effect dominated the flow field. The asymmetry effectively prevented the positive and negative pressures from canceling each other out, enhancing the texture fluid dynamic effect, improving the load-bearing capacity of the film, and augmenting the friction and drag reduction effects.

Figure 4f illustrates the maximum pressure on the oil film surface of the textured surface under working conditions with groove widths of 0.1 mm and 0.2 mm for three different texture shapes. The oil film of the warhead-type texture exhibited a higher maximum pressure under the same groove width, indicating that irregularly symmetric textures exhibited superior dynamic pressure lubrication compared with concentric circle-type and multilayer rectangular-type textures.

The upper wall of the oil film exhibited a higher maximum pressure for a groove width of 0.2 mm compared with a groove width of 0.1 mm. At a groove width of 0.2 mm, the texture convergence interval increased, and the stator–rotor mutual sliding process enhanced the hydrodynamic pressure, indicating dynamic pressure lubrication performance. Initially, as the texture size increased, the maximum pressure tended to increase and then decrease, peaking at an outer texture size of 3 mm. Subsequently, as the texture size increased, the maximum pressure on the oil film wall decreased. This resulted in a reduced net bearing capacity, diminished friction, and drag reduction effects.

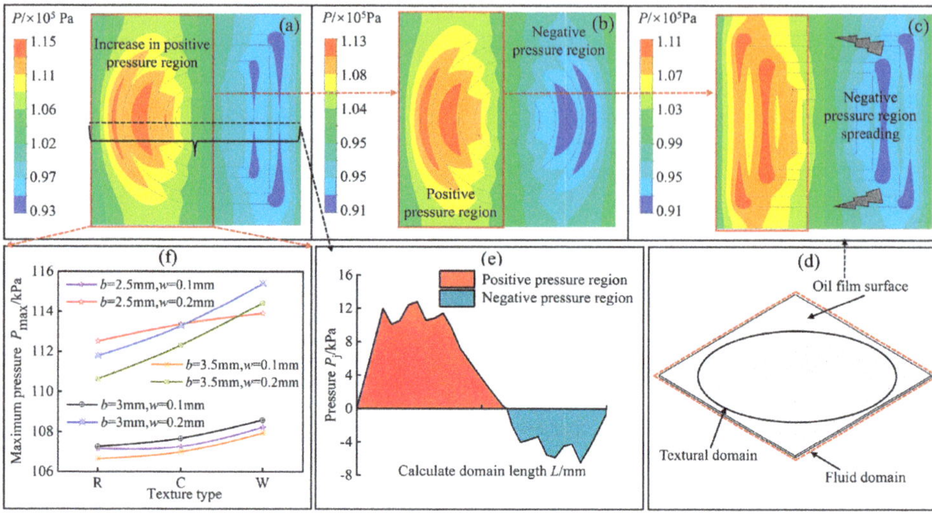

Figure 4. The influence of different texture morphologies on the pressure field and the maximum pressure distribution on the upper wall of the oil film: (**a**–**c**) Pressure field on the upper surface of the oil film under different weave morphology conditions; (**d**) Simulation model fluid domain; (**e**) Cross-sectional pressure curve; (**f**) The maximum pressure change under different texture size conditions.

3.3. Influence of Texture Depth on the Lubricating Properties of the Stator Rubber Surface

To analyze the influence of the texture groove depth on the lubrication performance of the rubber surface, three types of textures with varying morphologies were selected and simulated across groove depths from 50 μm to 500 μm. The fluid domains for groove depths of 100 μm, 300 μm, and 500 μm are shown in Figure 5a. The pressure distribution across the cross-section of the microtextured grooves was analyzed, showing distinct variations as shale oil flows through the grooves. The gradual lightening of pressure color from the outer to the inner grooves indicates an enhancement in hydrodynamic lubrication. In regions with darker pressure color, an increase in additional load-bearing capacity is observed, leading to improved lubrication performance.

The maximum pressure on the upper wall of the textured oil film versus the groove depth for different textured morphologies and groove width is shown in Figure 5b. The curve code initials represent different textured morphologies: W for warhead-type textures, C for concentric circle-type textures, and R for multilayer rectangular textures. As the groove depth increased, the maximum pressure on the oil film surface of the three textures initially increased and then decreased. The pressure on the upper surface of the oil film within the unit area was integrated across various depth conditions to assess the variation in average pressure. As illustrated in Figure 5c, the trend of average pressure aligns closely with that of the maximum pressure. The threshold groove depth was 100 μm. Beyond this threshold, the vortex volume inside the texture expanded. Thus, when the groove depth reached 500 μm, a new vortex formed inside the texture, consuming part of the fluid dynamic pressure energy. Thus, energy loss increased. Under different depth conditions, when the micro-texture groove width was 0.2 mm for the three different texture morphologies, the maximum positive pressure on the upper wall of the oil film was greater, resulting in a higher net bearing capacity and enhancing friction and drag reduction effects.

Figure 5d shows the dynamic pressure lubrication field distribution at the stator rubber textured groove depths of 100 μm, 300 μm, and 500 μm for the warhead-type textured groove with a width of 0.2 mm condition. Furthermore, when the groove depth was 100 μm, the vortex did not fully form, the distribution of flow traces was dense, and the dynamic pressure lubrication effect inside the groove was dominant. As the groove depth increased, the flow state of the internal flow field inside the groove changed dramatically, accompanied by varying degrees of reflux and vortex phenomena as well as energy changes. Groove depth increased vortex volume and intensity, increasing the maximum pressure on the oil film wall. However, the vortex also led to energy loss, inhibiting the increase in the oil film-carrying capacity. As vortex energy consumption increased, the flow trace line distribution became sparse, reducing the impact of the fluid dynamic pressure lubrication effect. Therefore, the groove depth was selected such that the vortex was not fully formed, preventing excessive vortex intensity.

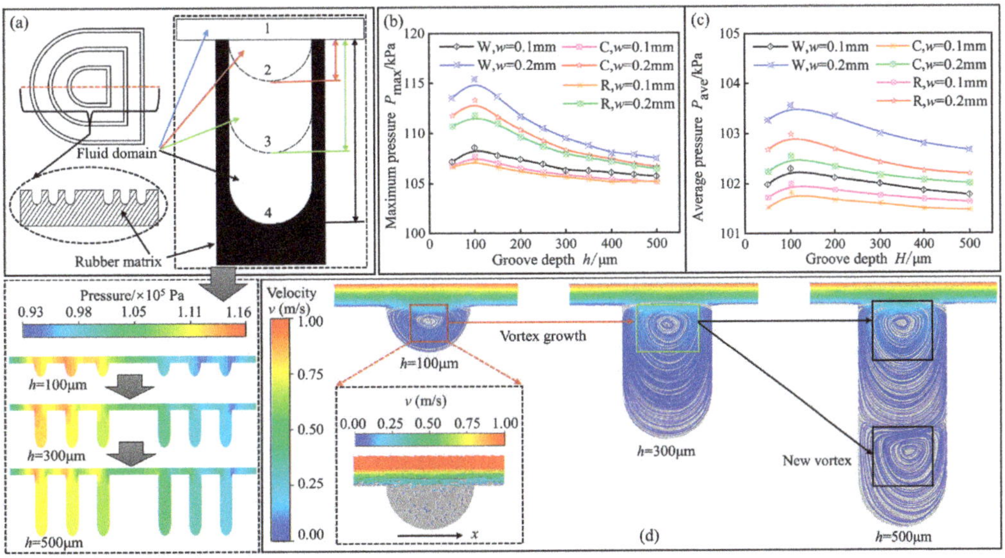

Figure 5. The influence of groove depth on the maximum pressure on the upper wall of the oil film and the lubrication flow field: (**a**) Fluid domain; (**b**,**c**) The maximum pressure and average change under different texture depth conditions; (**d**) Velocity field.

3.4. Influence of Irregular Symmetry on the Lubrication Performance of Stator Rubber Surface

The irregular symmetrical properties of the warhead-type microstructure led to differences in the direction of lubricant flow direction and different dynamic pressure lubrication effects on the stator rubber surface. To investigate this influence, the positive direction along the positive x-axis and the negative direction along the negative x-axis were defined. Simulations were conducted for both positive and negative directional textures.

Furthermore, when shale oil flows from the positive and negative directions toward the warhead-type texture, the pressure distribution on the upper wall of the oil film is depicted in Figure 6a,b. When shale oil flows in the positive direction of the x-axis, the texture head exhibits a negative pressure area, and the tail exhibits a positive pressure area. Conversely, when shale oil flows in the negative direction of the x-axis, the texture head exhibited positive pressure, and the tail exhibited negative pressure. Upon comparing Figure 6a,b, it was observed that when the shale oil flows along the negative direction of the x-axis, the maximum pressure on the upper wall of the stator rubber surface texture oil film was higher, and the positive pressure region was larger. This was attributable to the front of the warhead-type texture, which exhibited exceptional convergence compared to the tail, enhancing the dynamic pressure effect. The irregular symmetry of the warhead-type texture inhibited positive pressure loss and enhanced the additional net load capacity obtained at the wall on the oil film compared with the regular symmetrical texture.

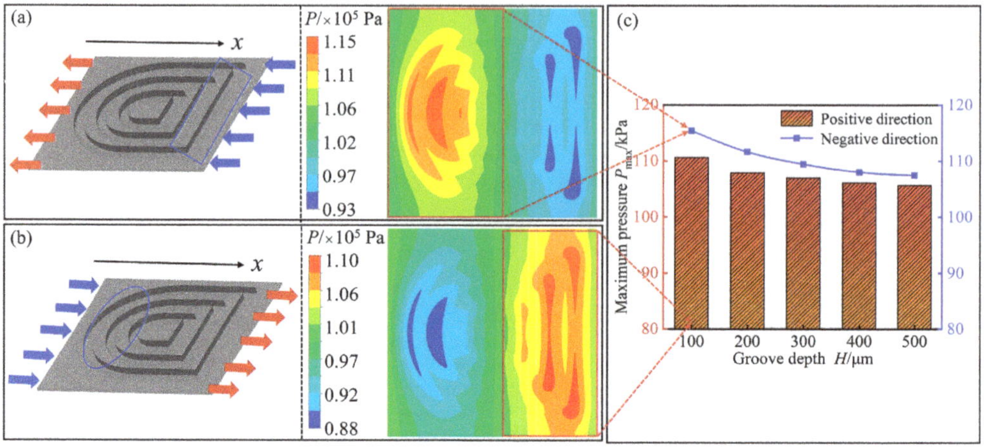

Figure 6. Pressure field and maximum pressure distribution on the upper wall of the warhead-type texture in both positive and negative directions: (**a**,**b**) The pressure field changes of oil along the x-axis in positive and negative directions; (**c**) The change of pressure field on the upper surface of oil film.

Figure 6c illustrates the influence of shale oil flow along the x-axis in both the positive and negative directions on the maximum pressure on the upper wall of the oil film at the different groove depths. As the groove depth increased, the maximum pressure on the upper wall of the oil film decreased in both directions. When the groove depth was 100 μm, the maximum pressure on the upper wall of the oil film increased by 4.10% compared with the positive direction. As the groove depth increased, the rate of the maximum pressure increases on the upper wall of the oil film decreased. At a groove depth of 500 μm, the maximum pressure increases by 1.64%. At each depth, the maximum pressure on the upper wall of the oil film increased by an average of 2.64%, indicating that under reasonable depth conditions, the dynamic pressure effect dominated. Thus, the irregular symmetry significantly affected the texture lubrication performance. As the depth increased, the volume and number of vortices increased, the vortex effect dominated, and the influence of the irregular symmetry texture on the lubrication performance was weakened. Therefore, given the irregular symmetry characteristics, designing the texture with a reasonable depth effectively enhanced the hydrodynamic lubrication effect.

4. Screw Pump Textured Stator Rubber Friction Test

4.1. Preparation of the Textured Rubber Surface

High acrylonitrile rubber was selected as the test material, and a cuboid with dimensions of 60 mm × 20 mm × 3 mm was prepared. The stator rubber was textured using IPG nanosecond laser equipment. As shown in Figure 7, the texture laser processing parameters were as follows: speed of 800 mm/s, power of 10%, frequency of 200 kHz, and five engraving passes. After processing, the sample was cleaned using an ultrasonic cleaner combined with anhydrous ethanol. After drying, the surface texture morphology of the rubber sample was observed using three-dimensional confocal microscope (VK-X1050), KEYENCE, Osaka, Japan.

Figure 8a shows the surface morphology of three textures after preparation under Olympus 3D confocal microscope: the warhead type, concentric circle type, and multilayer rectangle type. A slight laser ablation phenomenon was observed at the starting point of the adjacent groove. This was attributable to the energy retention of the initial laser pulse during processing, prolonging the laser retention time within the unit pulse time, and resulting in unavoidable slight thermal damage at the starting point. The overall morphology of the texture was a layered groove array. The texture was uniform, and there were no micro-convex peaks. Figure 8b shows the change in hardness value of screw pump stator rubber under the condition of 70 °C shale oil immersion for 7 days. Under this working condition, the hardness of the rubber sample was tested every 24 h, and the average hardness of the rubber sample was recorded. As the number of soaking days increased, the hardness of the rubber sample decreased first and then eased, and the original hardness decreased from 75.8 HA to 72.4 HA. Figure 8c shows how the tensile sample of the soaked rubber sample was subjected to tensile test. As the tensile stress increased, the rubber sample broke at 85.7 mm, and the elongation at break reached 285.7%.

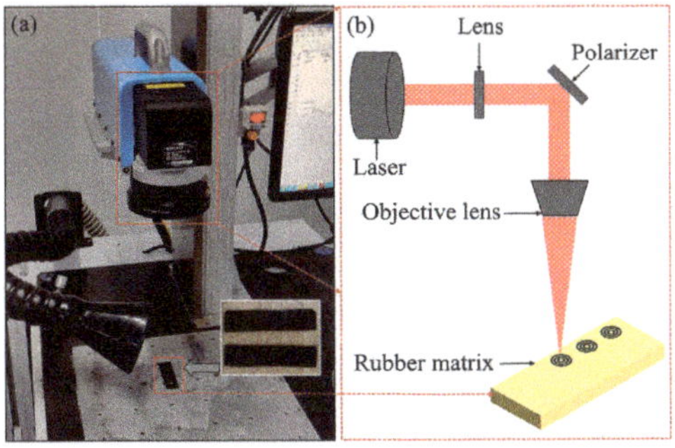

Figure 7. Schematic diagram of nanosecond laser processing: (**a**) Nanosecond laser; (**b**) Schematic of nanosecond laser.

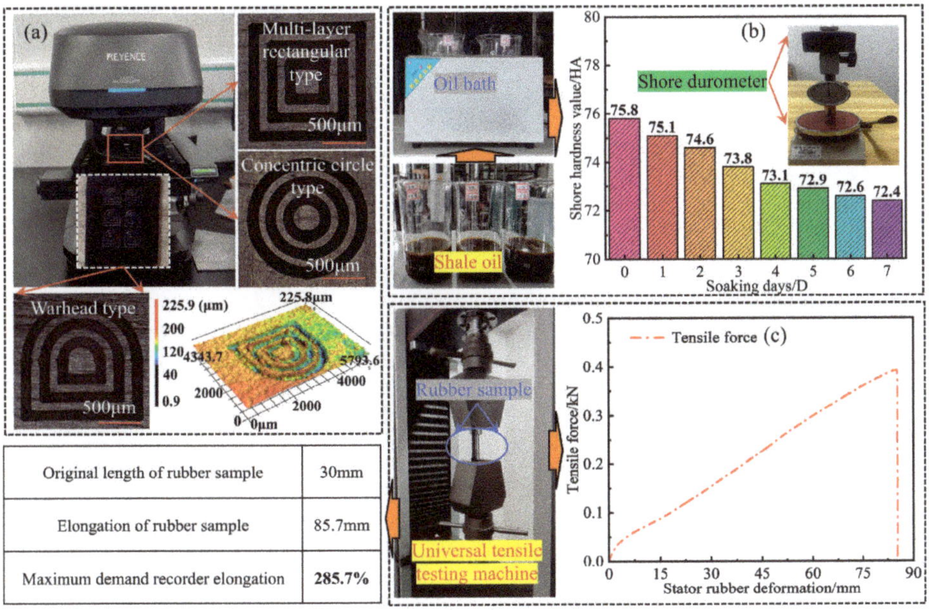

Figure 8. Prepared rubber samples with different textures and morphologies, and the hardness and elongation at break were tested: (**a**) Different texture morphologies; (**b**) Hardness test; (**c**) Tensile test.

4.2. Actual Complex Well Fluid Textured Rubber Friction Test

The friction and wear tests on the surface of texturized rubber were carried out on the UMT-3 reciprocating friction and wear tester based on the actual operating conditions of screw pumps in shale oil wells. Using the UMT-3 reciprocating friction and wear testing machine, the experimental study on the surface of texturized rubber is carried out, and the schematic diagram of the friction and wear testing machine is given, as shown in Figure 9a,b. A GCr15-bearing steel ball served as the upper sample, sliding relative to the lower sample at a constant speed. The test was conducted in an oil-rich state to simulate the real-time operating conditions of the screw pump, ensuring the accuracy and effectiveness of the test results. The actual complex well-conditioned fluid of a shale oil well served as the lubricating medium upon fluid addition through dripping. The test environment temperature was 18 °C, the friction frequency was 3 Hz, the reciprocating distance was 6 mm, the loads were 5 N, 10 N, and 15 N, the test duration was 1200 s.

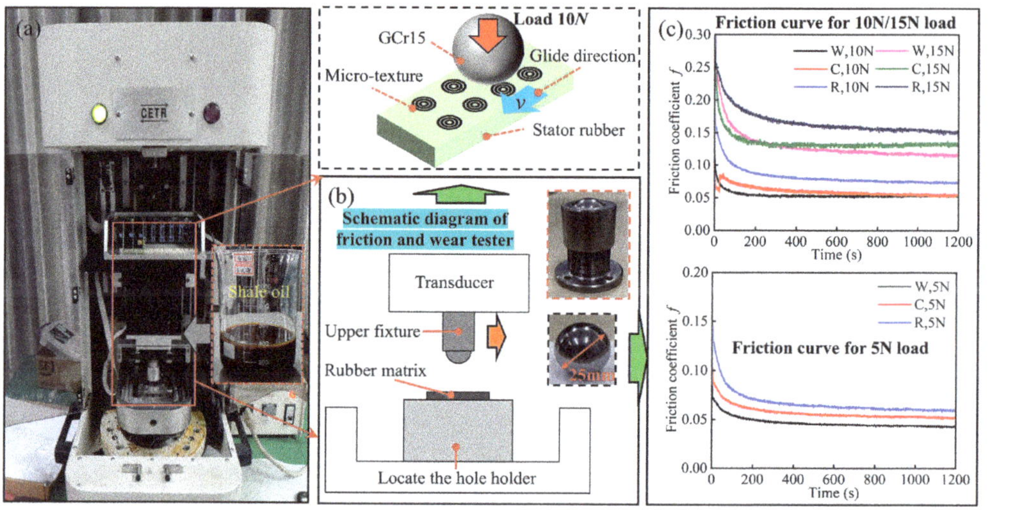

Figure 9. Schematic diagram of the UMT-3 friction and wear testing machine and the variation curve of the friction coefficient of textured rubber: (**a**) UMT-3 friction and wear testing machine; (**b**) UMT-3 friction testing machine schematic diagram; (**c**) The change of friction coefficient under different load conditions.

Figure 9c illustrates the tribological characteristics of textured stator rubbers in screw pumps under actual well conditions with liquid lubrication at applied loads of 5 N, 10 N, and 15 N. As the load increases, the friction coefficients of warhead-type, concentric-circle type, and multilayer-rectangular type textured stator rubbers exhibit an upward trend. Among these, the warhead-type shows enhanced adaptability across varying load conditions and demonstrates superior lubrication performance. Under a 10 N applied load, the friction coefficient of the warhead-type and multilayer rectangular-type textured stator rubber decreased significantly within 200 s and gradually stabilized thereafter. Furthermore, during the test, the upper sample continued to reciprocate, leading to the actual well-conditioned fluid in the reciprocating interval, exhibiting a certain degree of temperature increase. This temperature increase led to a decrease in the fluid viscosity, thereby promoting a reduction in the friction coefficient. In the initial stage of the running-in period, the concentric circular micro-texture does not immediately form a complete lubricant film, leaving numerous contact points on the friction surface and resulting in a higher friction coefficient. Over time, as the lubricant fluid gradually accumulates and diffuses within the texture, the lubricating effect improves, causing the friction coefficient to initially increase, then decrease, and eventually stabilize. After 800 s, the friction coefficients of the warhead-type and concentric circular textures became similar within a small range. Additionally, due to the extrusion effect, the lubricating oil at the groove of the warhead-type texture exhibited side leakage, resulting in a slight overlap of the friction coefficients of the two textures. The geometric properties possessed by the rectangular micro-texture make the micro-texture edges and sharp corners produce stress concentrations during friction, making it difficult for the lubricant to maintain a uniform distribution in these areas. At the same time, the edges of the multilayer rectangular-type micro-texture restrict the recirculation and distribution of the lubricant, making it more difficult for the lubricant to return to the friction surface. This restriction leads to a higher coefficient of friction in the friction test results for the multilayer rectangular-type micro-texture. The results indicated that under loads of 5 N, 10 N and 15 N, the rubber surface with a warhead-type texture exhibited the smallest friction coefficient. Consequently, the warhead-type texture exhibited enhanced lubrication and friction reduction effect, which was followed by the concentric circular-type and the multilayer rectangular-type texture.

5. Conclusions

1. The friction performance of the warhead-type micro-texture was superior compared with the concentric circle and multilayer rectangular micro-textures. The head of the warhead-type texture featured a sharp converging wedge along the flow direction, with its irregular symmetry enhancing the hydrodynamic lubrication effect within the texture groove. This improvement significantly enhanced the lubrication characteristics between the stator rubber and the rotor of the screw pumps.
2. The hydrodynamic lubrication performance of the textured stator rubber surface was highly influenced by the groove width. As the groove width increased, the pressure distribution on the upper wall of the oil film changed significantly, leading to an increase in additional net bearing capacity. This effectively enhanced the film pressure between the stator and rotor friction pairs.
3. Under different textured morphology conditions, the maximum pressure on the wall surface of the textured stator rubber oil film initially increased with depth, followed by a decrease. As depth increased, the vortex volume within the grooves increased, which was accompanied by energy changes that further diminished the effect of dynamic pressure lubrication.
4. Friction tests on the textured stator rubber of a progressive cavity pump show that under a load of 5 N, 10 N and 15 N, a friction frequency of 3 Hz, and an oil-rich condition, the average friction coefficient of the three types of textured stator rubber initially decreased and then gradually stabilized over time. The warhead texture exhibited an enhanced lubrication effect. Given all texture parameters, the optimal combination included a warhead-type texture size of 3 mm, a groove width of 0.2 mm, and a depth of 0.1 mm, resulting in an excellent dynamic pressure lubrication effect.

Author Contributions: Conceptualization, X.L. and Y.S.; methodology, X.L. and Y.S.; software, Y.S., X.S. and X.N.; validation, Y.S., C.L., X.S. and X.N.; formal analysis, Y.S., C.L., X.S., G.Z., S.W. and S.H.; investigation, X.L., Y.S., X.S., X.N., G.Z., W.W. and S.W.; resources, X.L. and C.L.; data curation, X.L., C.L. and X.N.; writing—original draft preparation, X.L. and Y.S.; writing—review and editing, X.L., C.L. and Y.S.; visualization, Y.S.; supervision, X.L., X.S. and S.H.; project administration, X.L., C.L., G.Z., W.W., S.W. and S.H.; funding acquisition, X.L. and C.L. All authors have read and agreed to the published version of the manuscript.

Funding: Supported by National Natural Science Foundation of China (Grant Nos. 52074161 and 52005281), Taishan Scholar Project of Shandong Province (Grant No. tsqn202211177), Shandong Provincial Plan for Introduction and Cultivation of Young Pioneers in Colleges and Universities (Grant No. 2021-Qing Chuang-30613019) and Natural Science Foundation of Shandong Province (Grant Nos. ZR2022ME173 and ZR2023QE011).

Data Availability Statement: The raw data supporting the conclusions of this article will be made available by the authors on request.

Acknowledgments: The authors sincerely thank the anonymous reviewers for their insightful comments and suggestions.

Conflicts of Interest: Gang Zheng and Wei Wei are employed by PetroChina Changqing Oilfield Company and National Engineering Laboratory of Low-Permeability Oil & Gas Exploration and Development. Songbo Wei and Shouzhi Huang are employed by Research Institute of Exploration & Development, PetroChina. The remaining authors declare that the research was conducted in the absence of any commercial or financial relationships that could be construed as a potential conflict of interest.

References

1. Zhu, J.; Zhang, H. A review of experiments and modeling of gas-liquid flow in electrical submersible pumps. *Energies* **2018**, *11*, 180. [CrossRef]
2. Hao, Z.; Zhu, S.; Pei, X.; Huang, P.; Tong, Z.; Wang, B.; Li, D. Submersible direct-drive progressing cavity pump rodless lifting technology. *Pet. Explor. Dev.* **2019**, *46*, 621–628. [CrossRef]
3. Rong, X.; Zhu, H.; Hu, B. Performance Research and Structure Optimization of Labyrinth Screw Pump. *Micromachines* **2021**, *12*, 790. [CrossRef]
4. Shishlyannikov, D.; Zverev, V.; Ivanchenko, A.; Zvonarev, I. Increasing the Time between Failures of Electric Submersible Pumps for Oil Production with High Content of Mechanical Impurities. *Appl. Sci.* **2022**, *12*, 64. [CrossRef]
5. Li, Y.; Nie, D.; Cai, Z. The performance analysis of screw pump stator elastomers: Polyamide 6/hydrogenated nitrile blends: Mechanical, oil resistance and tribological properties. *Polym. Test.* **2023**, *128*, 108–226. [CrossRef]
6. Segu, D.Z.; Hwang, P. Friction control by multi-shape textured surface under pin-on-disc test. *Tribol. Int.* **2015**, *91*, 111–117. [CrossRef]
7. Costa, H.L.; Hutchings, I.M. Some innovative surface texturing techniques for tribological purposes. *Proc. Inst. Mech. Eng. Part J J. Eng. Tribol.* **2014**, *229*, 429–448. [CrossRef]
8. Arslan, A.; Masjuki, H.H.; Kalam, M.; Varman, R.A.; Mufti, M.H.; Mosarof, L.S. Surface texture manufacturing techniques and tribological effect of surface texturing on cutting tool performance: A review. *Crit. Rev. Solid State Mater. Sci.* **2016**, *41*, 447–481. [CrossRef]
9. Ahuir-Torres, J.I.; Arenas, M.A.; Perrie, W.; Dearden, G.; de Damborenea, J. Surface texturing of aluminium alloy AA2024-T3 by picosecond laser: Effect on wettability and corrosion properties. *Surf. Coat. Technol.* **2017**, *321*, 279–291. [CrossRef]
10. Lu, P.; Wood, R.J.K. Tribological performance of surface texturing in mechanical applications—A review. *Surf. Topogr. Metrol. Prop.* **2020**, *8*, 043001. [CrossRef]
11. Lei, P.X.; Zhang, P.L.; Song, S.J.; Liu, Z.Y.; Yan, H.; Sun, T.Z.; Lu, Q.H.; Chen, Y. Research Status of Laser Surface Texturing on Tribological and Wetting Properties of Materials: A Review. *Optik* **2024**, *298*, 171581. [CrossRef]
12. Zhang, J.; Wang, H.; Chen, W.; Yin, H.; Jing, P.R. Research status of biomimetic surface texture in agriculture and forestry machinery. *Agric. Mech. Res.* **2024**, *46*, 1–7.
13. Zu, H.; Sun, J.; Ye, W.; Li, D. Study on dynamic mechanical characteristics and Fatigue life prediction of single progressing cavity pump. *China Mech. Eng.* **2024**, *35*, 1358–1365.
14. Yu, M.; Yu, X.; Zheng, X.; Hang, Q.; Yuan, T.; Li, D. Influence of recess shape on comprehensive lubrication performance of high speed and heavy load hydrostatic thrust bearing. *Ind. Lubr. Tribol.* **2019**, *71*, 301–308. [CrossRef]
15. Wang, Y.; Zhang, Y.; Long, R. Effect of compound pit textures on the friction and wear of thrust cylindrical roller bearings under starved lubrication. *Ind. Lubr. Tribol.* **2023**, *75*, 343–351. [CrossRef]
16. Jia, H.; Zhou, Z.; Yin, B.; Zhou, H.; Xu, B. Influence of microdimple on lubrication performance of textured plunger pump. *Ind. Lubr. Tribol.* **2021**, *73*, 563–571. [CrossRef]
17. Zhang, Y.; Zhang, X.; Matsoukas, G. Numerical study of surface texturing for improving tribological properties of ultra-high molecular weight polyethylene. *Biosurface Biotribol.* **2015**, *1*, 270–277. [CrossRef]
18. Bai, Q.; Bai, J.; Meng, X.; Ji, C.; Liang, Y. Drag reduction characteristics and flow field analysis of textured surface. *Friction* **2016**, *4*, 165–175. [CrossRef]
19. Ma, J.; Liu, Y.; Yi, P.; Jia, H.; Zhang, N.; Sun, J. Anti-friction mechanism of sinusoidal texture with various intervals: The synergistic effect of dynamic pressure and tribofilm. *Tribol. Int.* **2022**, *173*, 107635. [CrossRef]
20. Han, J.; Fang, L.; Sun, J.; Wang, Y.; Ge, H.; Zhu, H. Hydrodynamic lubrication of surfaces with asymmetric microdimple. *Tribol. Trans.* **2011**, *54*, 607–615. [CrossRef]
21. Shen, C.; Khonsari, M. Numerical optimization of texture shape for parallel surfaces under unidirectional and bidirectional sliding. *Tribol. Int.* **2015**, *82*, 1–11. [CrossRef]
22. Wang, M.; Li, Q.; Huang, Z.; Qian, W.; Chen, X.; Li, Q.; Lai, T. Influence of surface texture parameters of screw pump rotor on tribological properties of its friction pairs. *Ind. Lubr. Tribol.* **2021**, *74*, 964–974. [CrossRef]
23. Wang, M.; Huang, Z.; Qian, W.; Li, G. Study of the effect of laser textured rotors on the starting performance of metal–rubber mating pairs under different lubricating media environments. *J. Eng. Tribol.* **2024**, *238*, 450–463. [CrossRef]
24. Zhang, B. Study on Surface Texturing Flow Field Characteristics and Anti-Friction Performance of Screw Pumped. Master's Thesis, Northeast Petroleum University, Daqing, China, 2021. (In Chinese)
25. Brizmer, V.; Kligerman, Y. A laser surface textured journal bearing. *J. Tribol.* **2012**, *134*, 031702-1–031702-9. [CrossRef]

26. Zhang, H.; Hua, M.; Dong, G.; Zhang, D.; Chin, K. A mixed lubrication model for studying tribological behaviors of surface texturing. *Tribol. Int.* **2016**, *93*, 583–592. [CrossRef]
27. Lu, P.; Wood, R.J.K.; Gee, M.G.; Wang, L.; Pfleging, W. A novel surface texture shape for directional friction control. *Tribol. Lett.* **2018**, *66*, 51. [CrossRef]
28. Vencl, A.; Ivanović, L.; Stojanović, B.; Zadorozhnaya, E.; Miladinović, S.; Svoboda, P. Surface texturing for tribological applications: A review. *Proc. Eng. Sci.* **2019**, *1*, 227–239. [CrossRef]
29. Wei, S.; Shang, H.; Liao, C.; Huang, J.; Shi, B. Tribology performance of surface texturing plunger. *Biomimetics* **2019**, *4*, 54. [CrossRef]
30. Chen, K.; Yang, X.; Zhang, Y.; Yang, H.; Lv, G.; Gao, Y. Research progress of improving surface friction properties by surface texture technology. *Int. J. Adv. Manuf. Technol.* **2021**, *116*, 2797–2821. [CrossRef]
31. Vishnoi, M.; Kumar, P.; Murtaza, Q. Surface texturing techniques to enhance tribological performance: A review. *Surf. Interfaces* **2021**, *27*, 101463. [CrossRef]
32. Wang, L.; Duan, J.; He, M.; Liu, Y.; Bao, Y. Study on antifriction mechanism of surface textured elliptical bearings. *J. Tribol.* **2023**, *145*, 011702. [CrossRef]
33. Shi, J.; Zhao, B.; He, T.; Tu, L.; Lu, X.; Xu, H. Tribology and dynamic characteristics of textured journal-thrust coupled bearing considering thermal and pressure coupled effects. *Tribol. Int.* **2023**, *180*, 108292. [CrossRef]
34. Guo, P.; Wei, Z.; Zhang, S.; Xiong, Z.; Liu, M. Feature-adaptive toolpath planning with enhanced surface texture uniformity for ultra-precision diamond milling of freeform optics. *J. Mater. Process. Technol.* **2024**, *323*, 118220. [CrossRef]
35. Morris, N.J.; Shahmohamadi, H.; Rahmani, R.; Rahnejat, H.; Garner, C.P. Combined experimental and multiphase computational fluid dynamics analysis of surface textured journal bearings in mixed regime of lubrication. *Lubr. Sci.* **2018**, *30*, 161–173. [CrossRef]
36. Liu, X.; Liu, C.; Hao, Z.; Zhang, Y.; Zhang, K.; Wang, J.; Wei, S.; Hao, A.; Tao, J.; Chen, H. Non-symmetric distributions of solids deposition for solid-water stratified flow in deviated tubing strings. *Pet. Sci.* **2023**, *20*, 3048–3061. [CrossRef]
37. Kirubakaran, V.; Bhatt, D. Effect of different turbulence model on prediction of lean blowout limit for micro gas turbine combustor. *Adv. Mater. Process. Technol.* **2020**, *8*, 1324–1335.
38. Shaheed, R.; Mohammadian, A.; Kheirkhah Gildeh, H. A comparison of standard k–ε and realizable k–ε turbulence models in curved and confluent channels. *Env. Fluid Mech* **2019**, *19*, 543–568. [CrossRef]
39. Liang, X.; Liu, Z.; Wang, H.; Zhou, X.; Zhou, X. Hydrodynamic lubrication of partial textured sliding journal bearing based on three-dimensional CFD. *Ind. Lubr. Tribol.* **2016**, *68*, 106–115. [CrossRef]
40. Hua, X.; Puoza, J.; Zhang, P.; Yin, B.; Xie, X.; Din, J. Numerical simulation and experimental analysis of grease friction properties on textured surface. *Iran. J. Sci. Technol. Trans. Mech. Eng.* **2019**, *43*, 357–369. [CrossRef]
41. Liu, W.; Ni, H.; Chen, H.; Wang, P. Numerical simulation and experimental investigation on tribological performance of micro-dimples textured surface under hydrodynamic lubrication. *Int. J. Mech. Sci.* **2019**, *163*, 105095. [CrossRef]
42. Xiao, C.; Tang, H. Elastohydrodynamic lubrication characteristics of surface-textured slipper bearing in an axial piston pump. *J. Braz. Soc. Mech. Sci. Eng.* **2020**, *42*, 199. [CrossRef]
43. HU, Y.; Wang, Y.; Jian, G.; Zuo, M.; Fang, Y.; Mo, J. Study on lubrication properties of parallel slider surface with different shape compound texture. *Surf. Technol.* **2022**, *51*, 43–51.
44. Jiang, S.; Liu, Q.; Wang, J.; Hua, J.; Dong, G. Numerical optimization of asymmetric surface texturing under reciprocating sliding conditions. *Tribol. Int.* **2023**, *180*, 108310. [CrossRef]
45. Wang, L.; Zhao, X.; Guo, S.; Wang, M. Tribological properties of surface microtexture friction pairs under different lubrication conditions. *Adv. Mech. Eng.* **2019**, *11*, 1687814019881569. [CrossRef]
46. Zhang, Y.; Yang, X.; Cong, J.; Sun, J.; Shao, S.; Gao, Y.; Chen, K.; Lv, G.; Yang, H. In Situ Microtexture Hydrodynamic Lubrication Characteristics of Ductile Cast Iron Crankshaft Surface. *J. Mater. Eng. Perform.* **2022**, *32*, 7390–7405. [CrossRef]
47. Yin, H.; Yang, J.; Gu, Q. Numerical study on the hydrodynamic lubrication performance improvement of bio-inspired peregrine falcon wing-shaped microtexture. *Tribol. Int.* **2024**, *191*, 109049. [CrossRef]
48. Kang, Z.; Fu, Y.; Ji, J.; Tian, L. Numerical investigation of microtexture cutting tool on hydrodynamic lubrication. *J. Tribol.* **2017**, *139*, 054502. [CrossRef]
49. Chen, X.; Zeng, L.; Chen, X. Eccentric correction of piston based on bionic micro-texture technology for the gap seal hydraulic cylinder. *Micro Nano Lett.* **2019**, *14*, 33–37. [CrossRef]
50. Wang, L.; Liu, Y.; Duan, J.; Bao, Y. Performance analysis of columnar convex-concave compound microtexture thrust bearing. *Ind. Lubr. Tribol.* **2024**, *76*, 374–381. [CrossRef]
51. Liang, Y.; Wang, C.; Wang, W.; Xing, H.; Zhang, Z.; Gao, D. Effect of composite bionic micro-texture on bearing lubrication and cavitation characteristics of slipper pair. *J. Mar. Sci. Eng.* **2023**, *11*, 582. [CrossRef]
52. Liu, X.; Niu, X.; Liu, C.; Shi, X.; Sun, Y.; Hao, Z.; Huang, S.; Wang, Y.; Tao, H. Study on the Tribological Properties of Multilayer Concentric Hexagonal Laser Texturing on Rubber Surfaces of Screw Pumps. *Materials* **2024**, *17*, 3708. [CrossRef] [PubMed]

lubricants

Article

Laser Surface Texturing for Ground Surface: Frictional Effect of Plateau Roughness and Surface Textures under Oil Lubrication

Hongzhi Yue [1,2,*], Johannes Schneider [1] and Jianxin Deng [2,*]

[1] Institute for Applied Materials—Reliability and Microstructure (IAM-ZM), Karlsruhe Institute of Technology (KIT), 76131 Karlsruhe, Germany; johannes.schneider@kit.edu
[2] Key Laboratory of High Efficiency and Clean Mechanical Manufacture of MOE, School of Mechanical Engineering, Shandong University, Jinan 250061, China
* Correspondence: hongzhi.yue@kit.edu (H.Y.); jxdeng@sdu.edu.cn (J.D.)

Abstract: Laser surface texturing has proven beneficial in improving tribological performance in different lubrication regimes. However, the interaction between plateau roughness and surface texture remains to be further investigated, even though rough surfaces are common in engineering applications. In the present study, we investigated the frictional influence of surface texturing of ground surfaces under different lubrication conditions. Channel textures with different depths and area ratios were fabricated on ground surfaces, and their friction was tested in reciprocating tests. The experimental findings indicate that the textures caused increased friction for ground surfaces under boundary or mixed lubrication when the interface is well lubricated. Nevertheless, when the oil supply was limited, an up to 40% friction reduction was observed under test conditions.

Keywords: laser surface texturing; oil lubrication; ground surface; surface roughness; boundary or mixed lubrication

1. Introduction

Energy saving has become an increasingly important issue due to the high demand for energy and environmental concerns. In many industrial applications, friction is a major contributor to energy waste, leading to high levels of energy consumption and carbon emissions [1]. Therefore, reducing friction can considerably contribute to energy conservation efforts.

Laser surface texturing has been proven effective in reducing friction in different applications [2–4]. Theoretical and experimental approaches have been applied to investigate the influences of texture [5–7]. It is believed that the main effect of textures is the generation of hydrodynamic lift under full-film or mixed lubrication, while, under boundary or near-boundary mixed lubrication, edge stresses may cause the main negative effects, and the balance between edge stress and hydrodynamic/static pressure should be considered [4]. Studies have been conducted to optimize parameters including texture shape, depth, and area density; however, the interaction between texture and plateau roughness has been investigated to a lesser extent. In most studies, polished surfaces are used, which are mostly assumed to be perfectly smooth in simulation, although rough surfaces are more commonly used in engineering applications.

The surface roughness can dramatically influence the frictional behavior of a surface. Flow factors [8,9] and homogenization [10–12] are widely used to analyze the influence of roughness on frictional behavior. When textures are introduced to the rough surface, the interaction between surface roughness and texture may lead to different behavior as the effect of the roughness pattern may be changed. Surface texture and plateau roughness may act as multi-scale textures and have shown potential in improving tribological behavior [13,14].

However, some contradictory results have also been reported in the literature. Brunetiere et al. [15] presented a numerical study on rough-textured surfaces using Reynold's equation coupled with a mass-conservative cavitation algorithm. Interestingly, they found the interaction of roughness and texture to be more crucial than the texture alone in generating hydrodynamic lift. The effectiveness of texture is highly dependent on the roughness, and, if the initial friction performance of a rough surface is sufficient, the texture may offer little improvement. If the surface is smooth, texture alone cannot produce significant hydrodynamic pressure. Gu et al. [16] conducted a numerical study on a rough-textured ring–liner conjunction and found that the tribological performance improvement realized by surface texturing can also be achieved or hindered by the effects of skewness and kurtosis. Studies by Podgornik and Sedlacek et al. [17–19] on rough surfaces showed that the beneficial effects of textures can be affected by skewness and kurtosis parameters. Meylan et al. [20] tested different textures on rough surfaces and found that most of the textures had similar or worse tribological properties than those of the untextured references. This suggests that the frictional behavior of textured rough surfaces remains to be investigated and that their friction mechanisms are poorly understood. This is where our research comes in.

This study focuses on the frictional effects of textures on rough ground surfaces. To this end, grinding traces were fabricated perpendicular to the sliding orientation to maximize their hydroeffect, and channel textures parallel to the grinding traces were chosen to ensure that the texture and grinding traces had a similar shape, minimizing the shape effect. The influences of texture depth and area density were investigated under immerged and limited-lubrication conditions.

2. Materials and Methods

2.1. Sample Preparation

Tribological tests were performed in reciprocating motion with carbon-steel (AISI 1045, Eisen Schmidt, Karlsruhe, Germany) pins on a chromium-steel (AISI 5140, ArcelorMittal Warszawa, Poland) disk. The 8 mm diameter pins were hardened and tempered to a hardness of about HRC 40; then, they were ground to a final roughness (R_q) of around 0.3 μm using a cup grinding machine (C&N MPS 2 R300, Erlangen, Germany). The disks were hardened and tempered to a hardness of about HRC 45. The disk surfaces were ground to a base roughness of $R_q = 0.3$ μm with grinding traces quasiperpendicular to the sliding direction. The surface roughness was checked by a tactile roughness measurement device (Hommel T8000, Jenoptic, Jena, Germany). After grinding, laser surface texturing (LST) was conducted to fabricate microchannel patterns on the surfaces using a Piranha II system from Acsys (Kornwestheim, Germany) with a 1064 nm wavelength. The plateau roughness, texture depth, and texture area density were chosen as variables in the test. By adjusting the laser power and channel spacing, the surface textures were designed to vary in depth (1–7 μm) and area density (7.5–30%). The surfaces were named according to the surface textures as US (untextured surface), TS (textured surface), TS-MD (textured surface with medium depth), TS-HD (textured surface with high depth), TS-LA (textured surface with low area ratio) and TS-HA (textured surface with high area ratio), as shown in Table 1. The lasering parameters were as follows: laser pulse length of 10 ns, average power of 2.8–3.6 W, frequency of 290 kHz, and scanning speed of 2 mm/s. Four scans were conducted with 5 μm spacing to form a single texture channel. The texture channel width was set to a constant value of ~30 μm. As the base grinding traces needed to be kept, a short laser pulse of 10 ns was chosen to minimize bulges and burrs. After texturing, the surfaces were cleaned using rubber to remove the debris, then ultrasonically cleaned in isopropanol for 5 min each. The samples were examined by optical, digital, and 3D microscopy using a light microscope (Keyence VHX 600D, Osaka, Japan) and a white-light profilometer (Sensofar Plu neox, Sensofar, Barcelona, Spain), with images recorded before and after processing and tribological testing. The surface roughness parameters of arithmetic average height (S_a) and root-mean-square height (S_q), as well as a parameter set developed in our previous report [21] (including the plateau root-mean-square height (S_{pq}), the reformed surface-bearing index (S^*_{bi}), the reformed core fluid retention index (S^*_{ci}) and the reformed valley fluid retention index S^*_{vi}), were used to characterize the roughness of the surfaces. The algorithms of the parameter set are presented in Appendix A.

Table 1. Texture design parameters.

Surface	Texture	Texture Depth (μm)	Texture Area Density (%)
US	No	--	--
TS	Yes	1	15
TS-MD	Yes	4	15
TS-HD	Yes	7	15
TS-LA	Yes	1	7.5
TS-HA	Yes	1	30

2.2. Tribological Test

The tribological tests were conducted using a UMT-3 tribometer (UMT-3, CETR, Campbell, CA, USA) with a pin-on-disk setup, as shown in Figure 1. The pin was mounted in a self-aligning pin holder to ensure full surface contact. For all tests, the normal force was set as 50 N, corresponding to a nominal contact pressure of 1 MPa. Both the load and friction force sensors had a resolution of 0.01 N. The sliding stroke was 12.5 mm, and the reciprocating frequency ranged from 0.05 to 0.8 Hz, corresponding to the average sliding speed from 1.25 to 20 mm/s. The tests were carried out in a controlled laboratory environment of 25 ± 3 °C and 50 ± 5% RH.

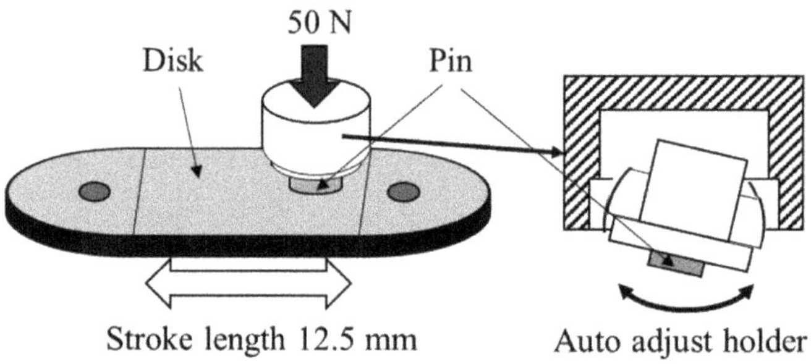

Figure 1. Experimental setup.

The sliding was firstly run at a low load (10 N) and frequency (1 Hz) for 10 s to ensure proper alignment of the pin and disk. Following this, the sliding was momentarily halted. Subsequently, a 50 N load was applied to the surfaces. When the load reached a steady state, the oscillation frequency was increased stepwise from 0.05 Hz to 0.8 Hz, with each frequency step lasting for 300 s. Each full-frequency ramp was repeated 5 times, and only the last 3 ramps were calculated to relieve the running-in effect. As the sliding speed is not a constant, friction force at the turning-back point would reach a peak and then reverse to the opposite peak. In this study, the average coefficient of friction (COF) was calculated ignoring the highest 10% and lowest 10% of the absolute value to minimize the random error at the turning-back point. Each test was repeated 3 times. The tests utilized commercially available L-G68 oil (SW D68, Castrol, UK) with a kinematic viscosity of approximately 68 mm^2/s at 40 °C and 160 mm^2/s at 24 °C. Two oil supply conditions (Figure 2) were used: immerged lubrication and limited lubrication. For immerged lubrication, the contact surfaces were immerged in the oil pool, while limited lubrication means the oil on the lower surface was scratched away using a knife edge, representing the engineering situation where the oil is brushed away by the contact edges.

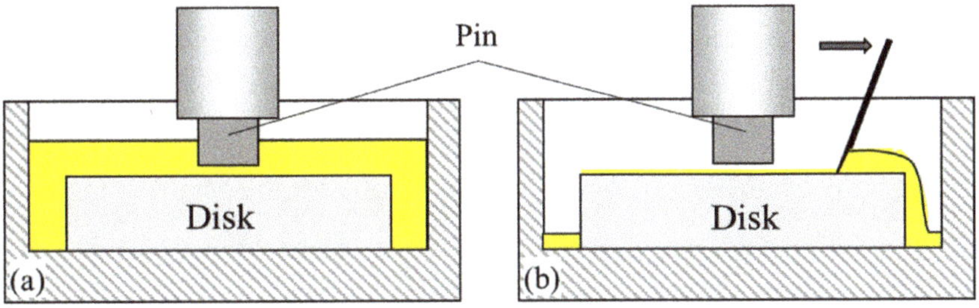

Figure 2. Lubricating conditions: (**a**) immerged lubrication, (**b**) limited lubrication.

3. Results

3.1. Surface Characterization

The surfaces were successfully prepared according to the design parameters listed in Table 1. Figure 3 presents the optical micrographs of the pin and disks with untextured (US) and textured (TS) surfaces. The 3D profiles for different disk surfaces are shown in Figure 4. The grinding traces were well maintained, and the bulges or burrs at the channel edge were insignificant compared to the roughness scale. Figure 5 shows the material ratio curve (MRC), i.e., the Abbott–Firestone curve, of different surfaces. Obviously, the upper parts of the curves show high consistency, while the lower parts vary significantly depending on the surface texture parameters.

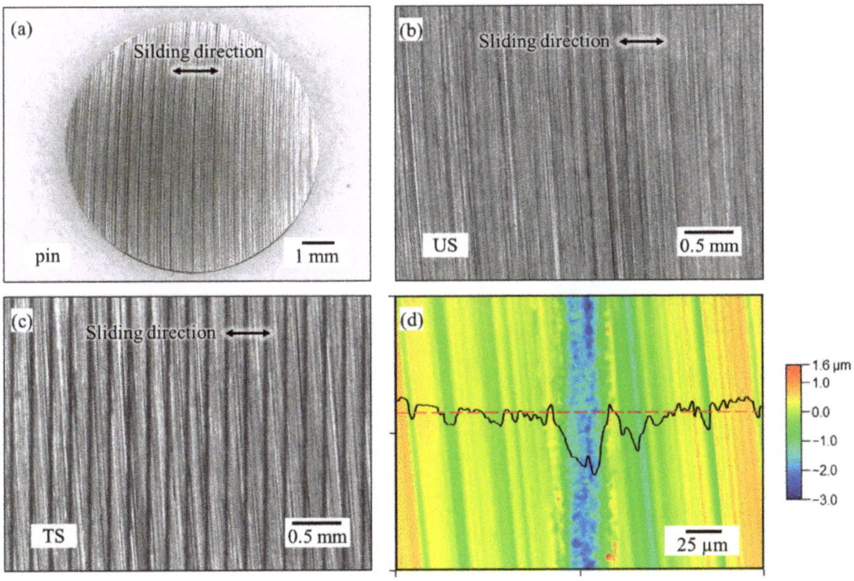

Figure 3. Optical images of (**a**) pin, (**b**) US disk surface and (**c**) TS disk surface. (**d**) White-light profile of a single channel.

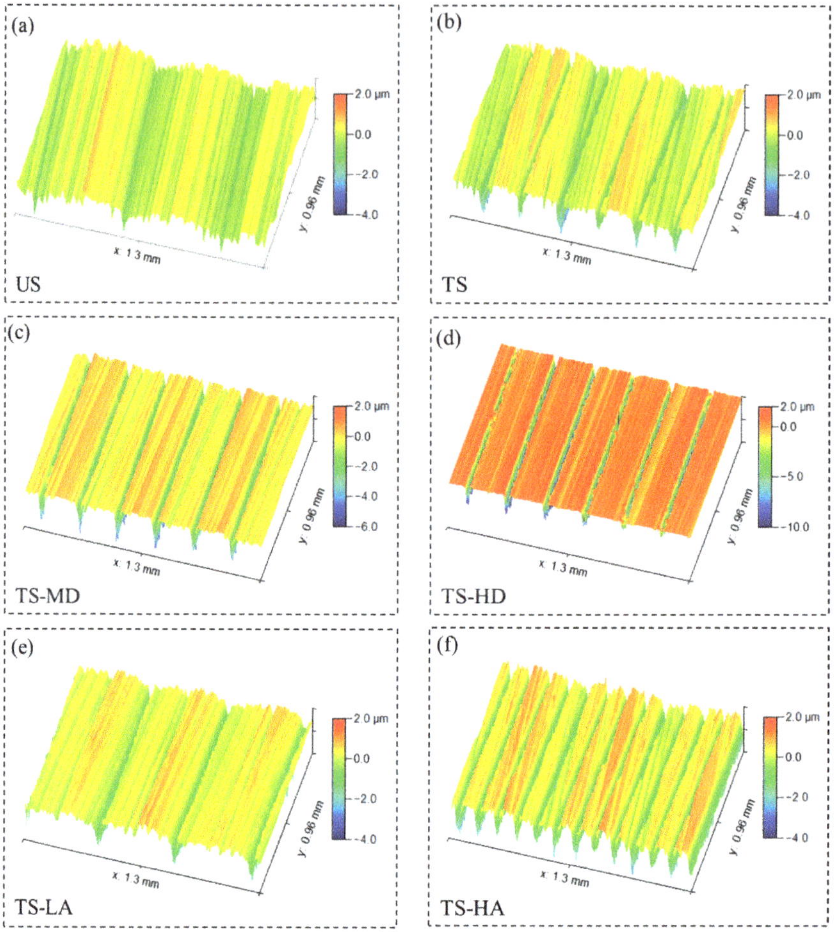

Figure 4. White-light profilometry images of (**a**) US, (**b**) TS, (**c**) TS-MD, (**d**) TS-HD, (**e**) TS-LA, (**f**) TS-HA.

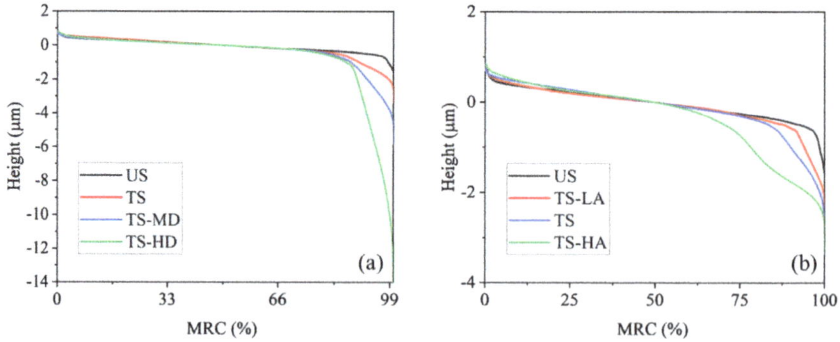

Figure 5. Material ratio curve of surfaces with different (**a**) texture depth and (**b**) texture area density.

The surface roughness parameters measured for the surfaces were as shown in Table 2. Compared to the commonly used parameter S_a, the parameter set gives information about the influence of textures on plateau roughness and the void/material volume at surface peak, core, and valley regions, as shown in Figure 6. The surface roughness parameter arithmetic average height S_a and root-mean-square height S_q increased dramatically with the increase in texture depth and area density, while only a minor increase was observed for the plateau root-mean-square height S_{pq}, which gives a quantitative description to the height distribution of an untextured area. No obvious trend was observed for the reformed surface-bearing index S^*_{bi}. The reformed core fluid retention index S^*_{ci} increased with texture area density but showed low relevance to the texture depth. The reformed valley fluid retention index S^*_{vi} increased both with texture area density and texture depth, except for in the case of TS-HA.

Table 2. Surface roughness parameters.

Surface	S_a (μm)	S_q (μm)	S_{pq} (μm)	S^*_{bi}	S^*_{ci}	S^*_{vi}
US	0.27	0.34	0.24	0.58	1.71	0.18
TS	0.42	0.60	0.28	0.55	1.81	0.49
TS-MD	0.53	0.95	0.26	0.58	1.85	1.02
TS-HD	0.96	2.31	0.29	0.61	1.83	2.31
TS-LA	0.26	0.41	0.25	0.58	1.74	0.32
TS-HA	0.62	0.84	0.32	0.55	2.46	0.38

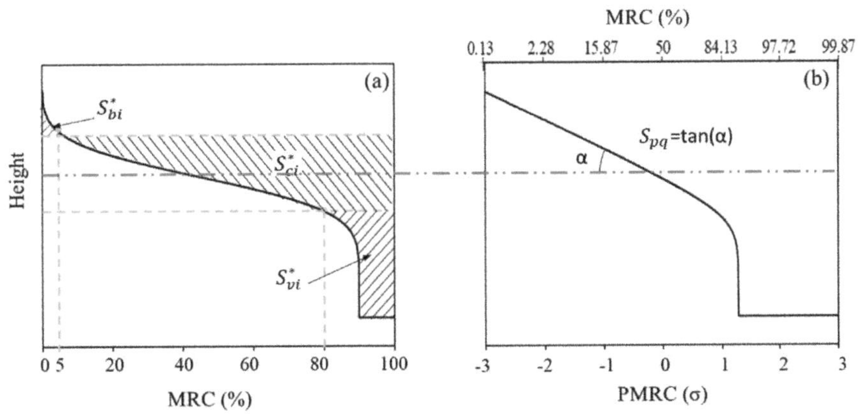

Figure 6. Schematic representation of roughness parameter set in (**a**) material ratio curve, (**b**) the probability material ratio curve for textured surface with rectangular cross section.

3.2. *Influence of the Texture Depth*

The following experiments were conducted under a load of 50 N (correspondingly, 1 MPa) and at room temperature. The average texture depth ranged from 1 to 7 um for TS, TS-MD, and TS-HD. Under immerged lubrication (Figure 7a), all textured surfaces, regardless of depth, showed higher friction compared to the untextured reference. Among all the textured surfaces, the TS surface had the lowest friction, with a friction increase from 1.5% to 72%. The detrimental effect intensified with increasing texture depth, reaching an up to 4.9 times friction increase for the TS-HD sample at an average sliding speed of 20 mm/s.

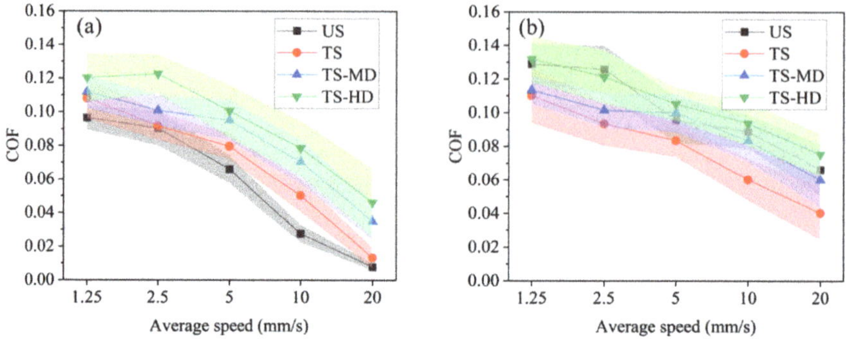

Figure 7. Friction coefficient plotted against the sliding speed for the surfaces with different texture depth under (**a**) immerged lubrication, (**b**) limited lubrication.

Due to the absence of lubricant oil, friction forces under limited lubrication (Figure 7b) increased for all surfaces compared to under immerged lubrication. However, the magnitude of the increment varied across different surfaces. A lower increment was observed for the TS surface, resulting in a friction reduction of up to 39% at 20 mm/s compared to the US surface. The TS-HD surface still had the highest friction, but its friction coefficient was very close to that of the US surface.

3.3. Influence of the Texture Area Density

The textured area density ranged from 7.5% to 30% for TS-LA, TS and TS-HA. All the textured surfaces with varying texture area density exhibited higher friction compared to the untextured US surface under immerged lubrication (Figure 8a). The TS-HA surface had the highest friction of an up to 169% increment compared to the US surface at 20 mm/s. Under limited lubrication (Figure 8b), both TS-LA and TS surfaces showed, at all speeds, friction reduction compared to the US surface, while the TS surface exhibited better behavior in the low-speed region. The maximum friction reduction was still 39% at 20 mm/s for the TS surface.

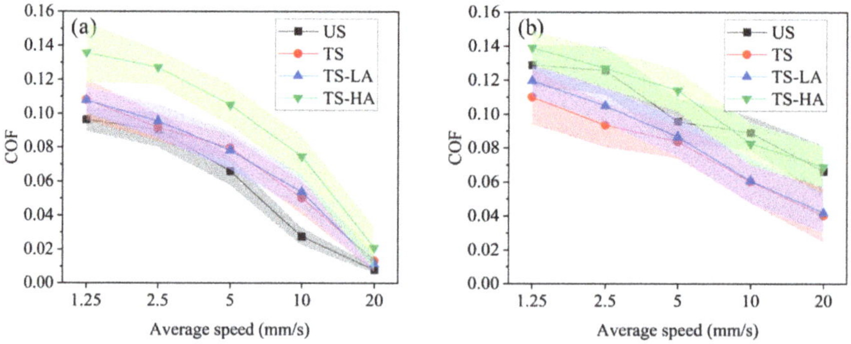

Figure 8. Friction coefficient plotted against the sliding speed for the surfaces with different texture area density under (**a**) immerged lubrication, (**b**) limited lubrication.

4. Discussion

The frictional properties of laser-textured ground surfaces were tested in order to investigate the effect of surface texturing on ground surfaces. The experiments were carried out for different texture depths, area densities and lubricating conditions. All tests were conducted under a 1 MPa load, and commercially available L-G68 oil was used.

In this study, channel textures were used to better compare the size effect between textures and grinding traces. Some researchers have suggested that communicating textures like channels or crossed channels may increase the friction due to a leak of hydrodynamic pressure along the groove [22,23], while some other researches have shown, on the contrary, that channel textures are effective [24–26]. Cross hatching, as a special communicating texture, has also been found beneficial in improving hydrodynamic pressure in cylinder–liner/piston–ring contact [13,27]. These examples indicate the effect of channel surface textures is highly dependent on the texture design parameters and test conditions, including surface roughness, lubricating condition, texture depth, texture area density and so on.

4.1. Surface Roughness

The surface topography plays an important role in tribology. Roughness parameters, as quantitate characterizations of surface topography, are expected to reflect the tribological performance and help understanding of the mechanism of contact, friction and wear, especially under boundary lubrication where the hydrodynamic effect can be ignored. Knowledge about the correlation between roughness parameters and friction is important for surface design and manufacturing control. Figure 9 shows the relation between the friction coefficient and surface roughness parameters S_a and S_{pq} under immerged lubrication at an average sliding speed of 1.25 mm/s. Compared to the widely used parameter S_a, the newly induced parameter S_{pq} shows better potential in reflecting friction behavior. A quasi-linear relation between the friction coefficient and S_{pq} was observed under such test conditions. As shown in Figures 7 and 8, the friction at this speed falls into a boundary or near-boundary mixed lubrication regime. The hydrodynamic effect is comparatively low, so the friction is dominated by the shear of asperity contact. As analyzed in our previous study [21], the peak root-mean-square roughness S_{pq} characterizes the height distribution of the upper part of a textured surface, which is directly related to the deformation in asperity contact, so it is more correlated to friction. However, due to the limitations of tests, this still needs to be confirmed by a future study.

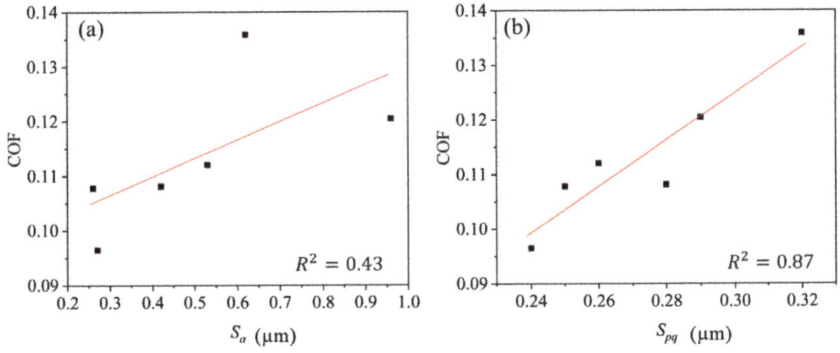

Figure 9. Friction coefficient plotted against the roughness parameters (**a**) S_a and (**b**) S_{pq} (average sliding speed = 1.25 mm/s, immerged lubrication).

Some researchers suggest hydro-static pressure and edge effect around the textures might be crucial to tribology in a boundary regime [4]. However, this may not be applicable under the experimental conditions in this work. First, the channel textures will cause a pressure leak along the groove. Second, the edge of the textures varies in height due to roughness, so it is almost impossible to build a closed oil reservoir when contacting a rough counter surface in face-to-face contact even if discrete textures are adopted. Third, considering the roughness, contact only occurs on the top of asperities, the stress concentration around the edge may be relieved compared to the smooth contact. Thus, the roughness increase caused by plateau area reduction or bulges around textures may have greater influence on the friction under such conditions.

It should be emphasized that all textured surfaces showed higher friction than the untextured reference. An explanation, as suggested by Pettersson et al. [28], is that the textures may have little or even a negative effect under boundary lubrication when the supply of lubricant is already sufficient in the untextured case, and the load-bearing area is decreased due to texturing. In this experimental setup, the contacting surfaces were rough enough for the oil supply along the grinding traces under immerged lubrication. When textures are introduced to the surface, the load-bearing capacity will decrease at the same nominal deformation due to contact area reduction. Thus, the asperity deformation will increase under the same load, resulting in higher stress and friction at the contacting area.

4.2. Immerged Lubrication

In contrast to most of the research [7,29–31], our results show that surface texturing always leads to higher friction under immerged lubrication. Similar behavior was observed in a few reports [4,16,21]. As shown in Figures 7a and 8a, the friction coefficient increased with texture depth and texture area density. Under mixed lubrication, the friction can be divided into two major components that both may cause friction increase. One of these is the increase in the boundary component. As the surface texturing makes the plateau area rougher, the boundary friction coefficient may increase with it. As discussed in Section 4.1, a quasi-linear relation was observed between the friction coefficient and the plateau root-mean-square roughness S_{pq} at an average sliding speed of 1.25 mm/s. However, with increasing speed, the effect of boundary friction will reduce, and the relation between friction and roughness would be more complex.

The other component is the decrease in hydrodynamic lift. The introduction of textures may undermine the hydrodynamic effect of the previous surface, leading to a friction increase. The main factor responsible for this should be the existence of plateau grinding traces. As shown in Figure 4, the plateau grinding traces were well kept and were arranged perpendicular to the sliding direction. The grinding traces, if regarded as a certain kind of texture, will cause a similar beneficial effect as texture grooves, and the laser textures may undermine the effect. The assumption is reasonable because numerical investigations [5,32] show that the highest hydrodynamic effect can be achieved when the texture depth is close to the average film thickness. The contact between two rough surfaces can be transformed into contact between an equivalent rough surface and a smooth, rigid surface [33]. As shown in Figure 10, under boundary or mixed lubrication, where two surfaces come into contact, the average film thickness h_0 is determined by the peak height of the grinding traces, while the grinding trace depth h_r is always of the same size. The texture depth h_t, however, can be much larger than h_0, which will generate lower hydrodynamic lift compared to the texture traces. Thus, the higher the texture depth or area density, the lower the hydrodynamic lift, resulting in higher friction force.

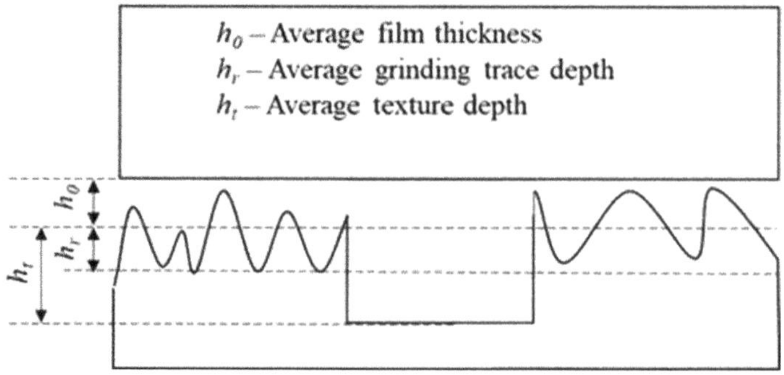

Figure 10. Schematic model for contact under boundary or mixed lubrication.

A potentially misleading conclusion would be that the textures are always detrimental for rough surfaces under boundary or mixed lubrication, which is evidently incorrect as many experiments have shown benefits under such conditions. However, textures can still be beneficial since the practical application is much more complex. Firstly, the real film thickness is always larger than the roughness height due to the flatness deviation, waviness, pitch angle or misalignment. As surface roughness decreases, this effect may become more pronounced, potentially making the influence of roughness lower and textures more effective. Secondly, the textures can be improved by optimizing the depth, area density, shape or distribution, while the roughness trace is always uncontrollable, and sometimes it is not even easy to control the trace orientation. In addition, texturing may bring side effects, including changing physical/chemical properties or introducing bulges around the textures, which might be more helpful than the textures themselves. Meanwhile, the textures have other functions, such as trapping debris or reducing wear, which may also be helpful for reducing friction.

4.3. Limited Lubrication

Another interesting result is that, under limited lubrication, the textured surfaces exhibited improved performance compared to their untextured reference. A simple explanation is that the textured surfaces can store more oil, thus maintaining better friction behavior when the oil is scratched away. Figure 11 gives the comparison ratio of friction coefficient $R_{cof} = f_l / f_m$. The f_l and f_m are the friction coefficient under limited lubrication and immerged lubrication, respectively. Clearly, the textured surfaces are less influenced when the oil level is becoming low. At 1.25 mm/s, the friction coefficient of untextured surface US increased by 34% from immerged lubrication to limited lubrication, while, for all the textured surfaces, the increase was relatively low, with the highest increase of 11% for the low-depth, low-area-ratio surface TS-LA. With the increasing speed, the R_{cof} ratio increased for all surfaces. At 20 mm/s, the friction coefficient of the US surface under limited lubrication became 8.5 times that under immerged lubrication, while the ratio for the TS surface was only 3.0. Higher texture depth seems beneficial to reduce the ratio. With increasing texture depth, the ratio decreased to 1.7 and 1.6 for TS-MD and TS-HD, respectively. Higher texture depth and area density could be more effective in this; however, due to the poor friction behavior of high-depth and high-area-density textures under immerged lubrication, the TS surface showed the best friction performance under limited lubrication (Figures 8b and 9b).

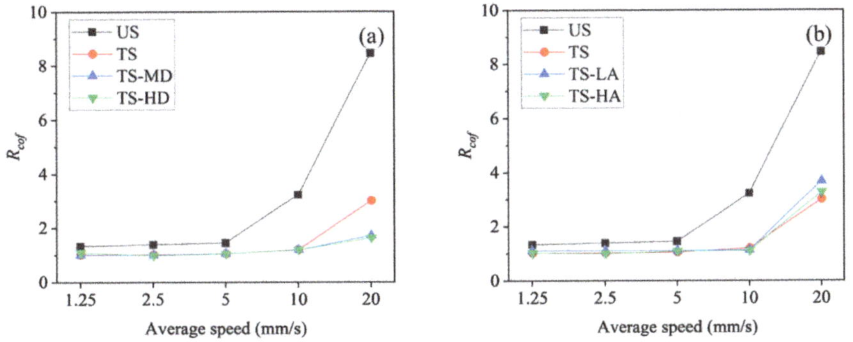

Figure 11. The comparison ratio R_{cof} of friction coefficient plotted against the sliding speed for surfaces with different (**a**) texture depth and (**b**) texture area density.

Although this is a commonly used condition for engineering applications such as sliding guideways, where lubricants are added to the contact surface in a timely fashion and scratched away by the contacting edge, only a few research studies have been reported. A starved lubricating condition is always realized by supplying only a little amount of oil.

Zhang et al. [34] conducted tests where the oil was supplied to just wet the Babbitt alloy disk and found the textures could reduce the friction force. Saeidi et al. [35] carried out tests under starved lubrication and observed a lower COF for textured surfaces. Similar to this study, they observed little or a negative effect on friction for textures on a rough surface under boundary or mixed lubrication in another paper [16]. The explanation could be that the textures could store more oil compared to the untextured references.

The oil retention capacity can be described by the reformed core fluid retention index S^*_{ci} and the reformed valley fluid retention index S^*_{vi}, as shown in Figure 6. As the valley area is defined as the lowest 20% of the surface, it was not feasible for the 30% textured TS-HA surface. Figure 12 shows the comparison ratio of friction coefficient R_{cof} for different surfaces plotted against the fluid retention indices S^*_{ci} and S^*_{vi}, respectively. The comparison ratio decreased with the increase in oil retention capacity at the core and valley area, which means greater oil storage will be beneficial for maintaining the frictional behavior when oil is drying out. However, as the oil retention capacity increased, it became less effective. As shown in Figure 12b, the R_{cof} almost remained the same when S^*_{vi} increased from 1.02 for TS-MD to 2.31 for TS-HD.

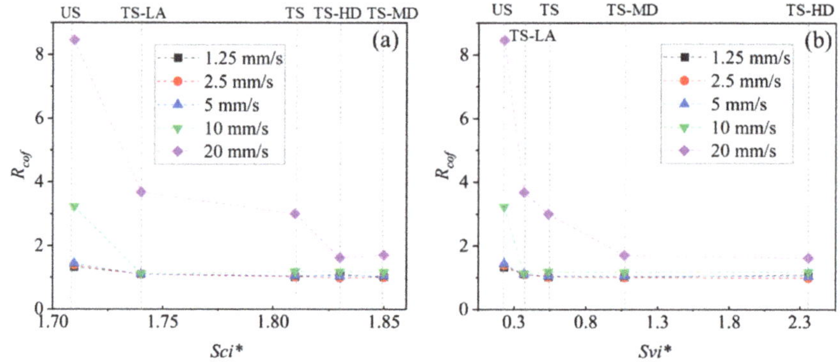

Figure 12. The comparison ratio R_{cof} plotted against (**a**) the reformed core fluid retention index S^*_{ci} and (**b**) the reformed valley fluid retention index S^*_{vi}.

When the sliding speed was low, the hydrodynamic effect could be ignored. Increasing the oil storage will be beneficial to build more robust boundary films between contacting asperities, but the effect is limited as it always stays in the boundary lubrication regime. The oil retention capacity became more effective at the high speed of 20 mm/s. For the untextured reference with low oil retention capacity, the friction under limited lubrication increased dramatically, up to 8.5 times that under immerged lubrication. However, the friction coefficients under the two lubrication conditions became closer with the increase in oil retention capacity. This suggests the oil retention capacity could contribute to the hydrodynamic effect under limited lubrication.

The effect of surface textures under limited lubrication is summarized in Figure 13. Under boundary lubrication, more oil can be supplied to the contact area to reduce shear, while, under mixed lubrication, oil may also generate local hydrodynamic lift at the same time. Increasing the oil retention capacity will be beneficial to these effects. However, the textures may also cause a roughness increase in the plateau area and reduce the hydrodynamic effect of roughness patterns, leading to a friction increase. The effects of textures are similar to the rough valleys in the honed cylinder liners, as reviewed by Pawlus and Reizer [36]. In fact, surface texturing is a method used to create more controllable valleys on the surface. Hence, they share similar design principles. We should balance the benefits and defects in texture design. In this study, a shallow texture in a moderate area density (1 μm and 15% area density) was found to have the best performance.

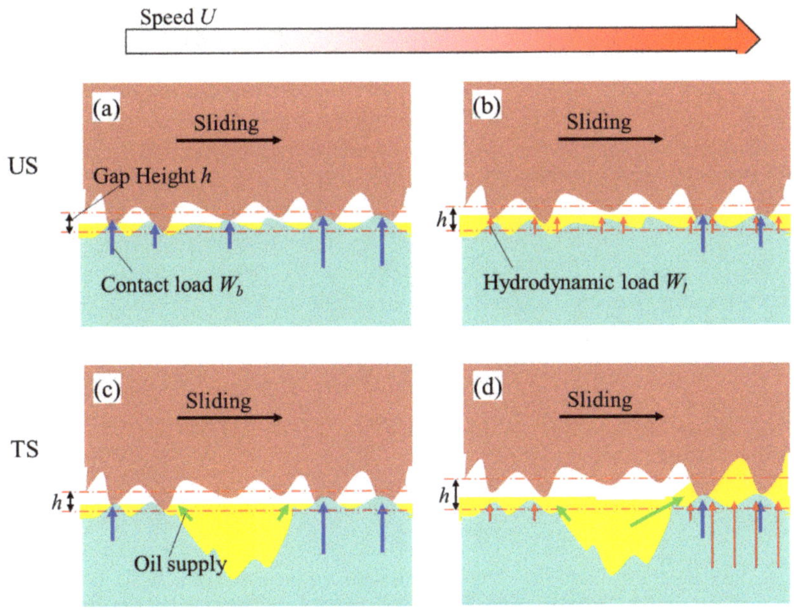

Figure 13. Schematic model for frictional effect of textures under limited lubrication. (**a**) US at low speed, (**b**) US at high speed, (**c**) TS at low speed, (**d**) TS at high speed.

5. Conclusions

In this work, channel textures were fabricated on the ground surface, varying in texture depth and area density. The surface roughness was evaluated using different roughness parameters, including the parameter arithmetic average height S_a, root-mean-square height S_q, the plateau root-mean-square height S_{pq}, the reformed surface-bearing index S^*_{bi}, the reformed core fluid retention index S^*_{ci} and the reformed valley fluid retention index S^*_{vi}. The frictional performance of textured surfaces was investigated under immerged and limited lubricating condition using the reciprocating sliding test.

The experimental results show that surface texturing always leads to higher friction under immerged lubrication. The analysis suggests the textures may cause more severe contact deformation due to lack of contact support at the valley region, presenting a "high roughness" behavior. Compared to the widely used parameter S_a or S_q, the newly introduced S_{pq} could characterize the roughness of the plateau contact area and show better potential in reflecting friction behavior under boundary or near-boundary mixed lubrication.

The friction increased for all surfaces when it was changed from immerged lubrication to limited lubrication due to lack of lubricants. As the textured surfaces could better maintain their performance under such circumstances, they showed improved performance compared to their untextured reference. Shallow textures in moderate area density (1 µm and 15% area density) were found to have the best performance, causing an up to 40% friction reduction.

Author Contributions: Conceptualization, methodology and investigation, H.Y.; writing—original draft preparation, H.Y.; writing—review and editing, J.S. and J.D.; supervision, J.S. and J.D.; project administration, J.D. All authors have read and agreed to the published version of the manuscript.

Funding: This work is supported by the National Natural Science Foundation of China (52275443). Hongzhi Yue acknowledges the China Scholarship Council for a scholarship (202006220107). We acknowledge support by the KIT-Publication Fund of the Karlsruhe Institute of Technology.

Data Availability Statement: Data are available upon reasonable request from the authors.

Conflicts of Interest: The authors declare no conflicts of interest.

Appendix A

Here the algorithms of reformed parameters are given in reference [21].

Reformed surface bearing index:

$$S_{bi}^* = \frac{(1-\delta)S_q^*}{\eta_{0.05}} \tag{A1}$$

where δ is the texture density, $\eta_{0.05}$ is the surface heights at 5% bearing ratio, S_q^* is the reformed RMS deviation of textured surfaces.

$$S_q^* = untextured\ S_q \approx S_{pq} \tag{A2}$$

For a Gaussian surface, the surface-bearing index S_{bi}^* is about 0.608.

Reformed core fluid retention index:

$$S_{ci}^* = \frac{V_v(h_{0.05}) - V_v(h_{0.8})}{S_q^* \Delta A} \tag{A3}$$

where $V_v(h_t)$ means the void volume below the height at t bearing ratio; ΔA is the area of sampling surfaces. For a Gaussian surface, the surface-bearing index S_{ci}^* is about 1.56.

Reformed valley fluid retention index:

$$S_{vi}^* = \frac{V_v(h_{0.8})}{S_q^* \Delta A} \tag{A4}$$

For a Gaussian surface, the valley fluid retention index S_{vi}^* is about 0.11.

References

1. Holmberg, K.; Erdemir, A. Influence of tribology on global energy consumption, costs and emissions. *Friction* **2017**, *5*, 263–284. [CrossRef]
2. Etsion, I.; Kligerman, Y.; Halperin, G. Analytical and experimental investigation of laser-textured mechanical seal faces. *Tribol. Trans.* **1999**, *42*, 511–516. [CrossRef]
3. Braun, D.; Greiner, C.; Schneider, J.; Gumbsch, P. Efficiency of laser surface texturing in the reduction of friction under mixed lubrication. *Tribol. Int.* **2014**, *77*, 142–147. [CrossRef]
4. Gachot, C.; Rosenkranz, A.; Hsu, S.M.; Costa, H.L. A critical assessment of surface texturing for friction and wear improvement. *Wear* **2017**, *372*, 21–41. [CrossRef]
5. Gropper, D.; Wang, L.; Harvey, T.J. Hydrodynamic lubrication of textured surfaces: A review of modeling techniques and key findings. *Tribol. Int.* **2016**, *94*, 509–529. [CrossRef]
6. Rosenkranz, A.; Grützmacher, P.G.; Gachot, C.; Costa, H.L. Surface texturing in machine elements- a critical discussion for rolling and sliding contacts. *Adv. Eng. Mater.* **2019**, *21*, 1900194. [CrossRef]
7. Marian, M.; Almqvist, A.; Rosenkranz, A.; Fillon, M. Numerical micro-texture optimization for lubricated contacts-a critical discussion. *Friction* **2022**, *10*, 1772–1809. [CrossRef]
8. Patir, N.; Cheng, H. An average flow model for determining effects of three- dimensional roughness on partial hydrodynamic lubrication. *ASME J. Lubr. Technol.* **1978**, *100*, 12–17. [CrossRef]
9. Patir, N.; Cheng, H. Application of average flow model to lubrication between rough sliding surfaces. *ASME J. Lubr. Technol.* **1979**, *101*, 220–229. [CrossRef]
10. Phan-Thien, N. Hydrodynamic lubrication of rough surfaces. *Proc. Math. Phys. Eng. Sci.* **1982**, *383*, 439–446. [CrossRef]
11. Allaire, G. Homogenization and two-scale convergence. *SIAM J. Math. Anal.* **1992**, *23*, 1482–1518. [CrossRef]
12. Rom, M.; König, F.; Müller, S.; Jacobs, G. Why homogenization should be the averaging method of choice in hydrodynamic lubrication. *Appl. Eng. Sci.* **2021**, *7*, 100055. [CrossRef]
13. Grützmacher, P.G.; Profito, F.J.; Rosenkranz, A. Multi-scale surface texturing in tribology-current knowledge and future perspectives. *Lubricants* **2019**, *7*, 95. [CrossRef]
14. Spencer, A.; Almqvist, A.; Larsson, R. A semi-deterministic texture-roughness model of the piston ring-cylinder liner contact. *Proc. Inst. Mech. Eng. Part J J. Eng. Tribol.* **2011**, *225*, 325–333. [CrossRef]

15. Brunetière, N.; Tournerie, B. Numerical analysis of a surface-textured mechanical seal operating in mixed lubrication regime. *Tribol. Int.* **2012**, *49*, 80–89. [CrossRef]
16. Gu, C.; Meng, X.; Wang, S.; Ding, X. Study on the mutual influence of surface roughness and texture features of rough-textured surfaces on the tribological properties. *Proc. Inst. Mech. Eng. Part J J. Eng. Tribol.* **2021**, *235*, 256–273. [CrossRef]
17. Podgornik, B.; Vilhena, L.M.; Sedlacek, M.; Rek, Z.; Zun, I. Effectiveness and design of surface texturing for different lubrication regimes. *Meccanica* **2012**, *47*, 1613–1622. [CrossRef]
18. Sedlacek, M.; Podgornik, B.; Vizintin, J. Planning surface texturing for reduced friction in lubricated sliding using surface roughness parameters skewness and kurtosis. *Proc. Inst. Mech. Eng. Part J J. Eng. Tribol.* **2012**, *226*, 661–667. [CrossRef]
19. Sedlacek, M.; Gregorcic, P.; Podgornik, B. Use of the roughness parameters Ssk and Sku to control friction—A method for designing surface texturing. *Tribol. Trans.* **2017**, *60*, 260–266. [CrossRef]
20. Meylan, B.; Saeidi, F.; Wasmer, K. Effect of surface texturing on cast iron reciprocating against steel under cyclic loading in boundary and mixed lubrication conditions. *Lubricants* **2018**, *6*, 2. [CrossRef]
21. Yue, H.; Deng, J.; Zhang, Y.; Meng, Y.; Zou, X. Characterization of the textured surfaces under boundary lubrication. *Tribol. Int.* **2020**, *151*, 106359. [CrossRef]
22. Zum Gahr, K.H.; Wahl, R.; Wauthier, K. Experimental study of the effect of microtexturing on oil lubricated ceramic/steel friction pairs. *Wear* **2009**, *267*, 1241–1251. [CrossRef]
23. Wahl, R.; Schneider, J.; Gumbsch, P. Influence of the real geometry of the protrusions in micro textured surfaces on frictional behaviour. *Tribol. Lett.* **2012**, *47*, 447–453. [CrossRef]
24. Stephens, L.; Siripuram, R.; Hayden, M.; McCartt, B. Deterministic micro asperities on bearings and seals using a modified liga process. *J. Eng. Gas. Turbines Power* **2004**, *126*, 147–154. [CrossRef]
25. Vlădescu, S.C.; Olver, A.V.; Pegg, I.G.; Reddyhoff, T. Combined friction and wear reduction in a reciprocating contact through laser surface texturing. *Wear* **2016**, *358*, 51–61. [CrossRef]
26. Huang, J.; Guan, Y.; Ramakrishna, S. Tribological behavior of femtosecond laser-textured leaded brass. *Tribol. Int.* **2021**, *162*, 107115. [CrossRef]
27. Biboulet, N.; Bouassida, H.; Lubrecht, A. Cross hatched texture influence on the load carrying capacity of oil control rings. *Tribol. Int.* **2015**, *82*, 12–19. [CrossRef]
28. Pettersson, U.; Jacobson, S. Textured surfaces in sliding boundary lubricated contacts-mechanisms, possibilities and limitations. *Tribol.—Mater. Surf. Interfaces* **2007**, *1*, 181–189. [CrossRef]
29. Schneider, J.; Braun, D.; Greiner, C. Laser textured surfaces for mixed lubrication: Influence of aspect ratio, textured area and dimple arrangement. *Lubricants* **2017**, *5*, 32. [CrossRef]
30. Kovalchenko, A.; Ajayi, O.; Erdemir, A.; Fenske, G.; Etsion, I. The effect of laser surface texturing on transitions in lubrication regimes during unidirectional sliding contact. *Tribol. Int.* **2005**, *38*, 219–225. [CrossRef]
31. Wos, S.; Koszela, W.; Dzierwa, A.; Pawlus, P. Effects of operating conditions and pit area ratio on the coefficient of friction of textured assemblies in lubricated reciprocating sliding. *Materials* **2022**, *15*, 7199. [CrossRef] [PubMed]
32. Codrignani, A.; Frohnapfel, B.; Magagnato, F.; Schreiber, P.; Schneider, J.; Gumbsch, P. Numerical and experimental investigation of texture shape and position in the macroscopic contact. *Tribol. Int.* **2018**, *122*, 46–57. [CrossRef]
33. Bhushan, B. *Introduction to Tribology*; John Wiley & Sons: New York, NY, USA, 2013. [CrossRef]
34. Zhang, H.; Zhang, D.; Hua, M.; Dong, G.; Chin, K. A study on the tribological behavior of surface texturing on babbitt alloy under mixed or starved lubrication. *Tribol. Lett.* **2014**, *56*, 305–315. [CrossRef]
35. Saeidi, F.; Meylan, B.; Hoffmann, P.; Wasmer, K. Effect of surface texturing on cast iron reciprocating against steel under starved lubrication conditions: A parametric study. *Wear* **2016**, *348*, 17–26. [CrossRef]
36. Pawlus, P.; Reizer, R. Functional importance of honed cylinder liner surface texture: A review. *Tribol. Int.* **2022**, *167*, 107409. [CrossRef]

lubricants

MDPI

Article

Combined Effects of DLC Coating and Surface Texturing on Seizure and Friction in Reciprocating Sliding

Slawomir Wos *, Waldemar Koszela, Andrzej Dzierwa and Pawel Pawlus

Department of Manufacturing Technology and Production Engineering, Rzeszow University of Technology, al. Powstancow Warszawy 12, 35-959 Rzeszów, Poland; wkktmiop@prz.edu.pl (W.K.); adzierwa@prz.edu.pl (A.D.); ppawlus@prz.edu.pl (P.P.)

* Correspondence: wosslawomir@prz.edu.pl

Abstract: Surface texturing is designed to improve the functional properties of machine elements by generating dimples on the surface contacted. Friction and wear resistance can also be improved by creating diamond-like carbon (DLC) coatings. These two techniques were combined to extend the lifetime of the elements and minimise friction in reciprocating conformal sliding contact. This work is functionally important for assemblies operating under high normal loads. Experiments were carried out in initially lubricated reciprocating sliding contact using an Optimol SRV 5 tribotester in the flat-on-flat configuration. The disc samples were untextured, laser textured, and DLC-coated untextured and textured. The combination of DLC coating and surface texturing caused an enhancement of the tribological performance of the sliding pair compared to that of untextured discs with and without DLC coating and textured discs without DLC coating. The DLC coating of the untextured disc caused a growth in the lifetime of a friction pair by a factor of 2.4. Seizure resistance also increased due to surface texturing of the steel disc for pit area ratios of 9 and 13%. Combining surface texturing with pit area ratios of 3 and 9% and DLC coating led to a decrease in the coefficients of friction of sliding pairs compared to only textured and coated discs. The DLC coating caused a decrease in the wear of the disc sample and reduction in wear levels of the counter samples in comparison to those of textured discs without DLC coatings.

Keywords: DLC coating; reciprocating sliding; surface texturing; friction; wear; seizure resistance

1. Introduction

Well-lubricated and long-lasting surfaces are important in industry because they reduce friction and wear, which helps equipment run reliably, leading to increased productivity and efficiency. Improvements in tribological behaviour can be achieved by surface texturing and DLC coating.

Surface texturing relies on generating dimples (oil pockets) or grooves on machine elements in sliding and rolling contacts. This technique leads to an improvement in friction, especially under mixed or boundary lubrication [1,2]. Surface texturing can also result in wear improvement [3]. Lu and Wood [4] reviewed surface texturing applications in mechanical engineering (cutting tools, piston rings and cylinder liners, sealing, and journal bearings). Surface texturing can also be used to improve the quality and adhesion of anti-wear and low-friction coatings [5,6].

Surface texturing can also cause a decrease in oil demand, leading to an improvement in seizure resistance [7]. The growth of the oil film lifetime caused by surface texturing was

 https://doi.org/10.3390/lubricants13060244

previously studied. Andersson et al. [8] prolonged the life of the oil film by a factor of up to eight by combining the selection of pit area ratio (density of oil pockets) and depth of the dimple under conditions of mixed lubrication. Hu and Lu [9] achieved an increase in the lifetime of the oil film by a factor of nine. Duarte et al. [10] proved that a cross-like pattern caused an increase in the lifetime of the oil film by a factor of 16 under mixed lubrication. Rosenkranz et al. [11] created periodic cross-like patterns in stainless-steel samples. Due to surface texturing, the oil film lifetime increased by a factor of 130 compared to the untextured surface in the pin-on-disc configuration in rotational sliding during mixed lubrication. Koszela et al. obtained a growth in the lifetime of a steel–bronze assembly in the ring-on-block configuration by a factor of five [12].

Researchers carried out seizure resistance tests by increasing loads. Kuroiwa et al. [13] improved the seizure resistance of a lead-free copper alloy in contact with ductile iron in the ring-on-ring configuration under an incrementally increased load. The texture specimens presented superior anti-seizure properties in comparison to those of untextured surfaces. The influence of the array of dimples was substantial. Galda et al. [14] found that surface texturing of a steel surface in contact with cast-iron in the block-on-ring configuration with increasing normal load could lead to increased resistance to seizure. Oil pocket shape and orientation were very important factors. Fan and Zhong [15] improved the seizure resistance of cast-iron cylinder liners by creating dimples. The experiment was carried out by increasing the load under starved lubrication in reciprocating motion.

Creating wear-resistant coatings can lead to a reduction in tribology losses [16]. Diamond-like carbon (DLC) coatings are characterised by low wear and high frictional resistance [17]. The applications of DLC coatings frequently led to an improvement in seizure resistance. Iwata et al. [18,19] reduced friction and improved seizure resistance by applying DLC coating to the crank journal. Usui et al. [20] achieved a significant improvement in the anti-seizure characteristic of DLC-coated piston pins compared to uncoated pins. Wu et al. [21] obtained a substantial increase in the scuffing limit temperature by coating gears with DLC compared to the grinding state.

Surface texturing can be combined with DLC coating to improve tribological properties. Arslan et al. [22] achieved an improvement in the tribological performance of a DLC coating by surface texturing in boundary reciprocating sliding. This improvement was obtained for specific sizes of dimples (diameter and depth). Shum et al. [23] obtained decreases in friction and wear for the combination of DLC coating and surface texturing for appropriate dimple density and depth of the dimple under reciprocal lubricated conditions compared to untextured DLC coatings. Oil pockets enhanced lubricant retention and were traps for wear particles. Amanov et al. [24] achieved an improvement in the tribological behaviour of a Si–DLC coating by surface texturing under lubricated reciprocating sliding at several temperatures (from room temperature to 200 °C). Mishra and Penchaliah [25] found that, under all lubrication regimes, coated cylinder samples with oil pockets had better frictional behaviour than untextured coated samples in reciprocating motion. Koskinen et al. [26] experimentally studied the influence of DLC coatings and surface texturing in contact with a steel ball on the coefficient of friction (COF) in reciprocating lubricated sliding. Friction decreased as oil capacity increased. Ding et al. [27] improved the tribological behaviour of DLC coatings under water-lubricated conditions by surface texturing. The creation of dimples resulted in a reduction in the COF and an extension of coating lifetime. Dufils et al. [28] analysed the combined effects of surface texturing and DLC coatings on the tribological behaviour of the thermoplastic polymer PEEK in reciprocating sliding. Coated PEEK textures, characterised by shallow dimples and a low pit area ratio, caused reduced wear and friction in distilled water in comparison to smooth DLC-coated PEEK. Koszela et al. [29] studied the effects of cylinder surface texturing and DLC coating on the functional performances of internal combustion motorcycle engines. The best behaviour was achieved for textured cylinders with a DLC coating (an increase in maximum speed and effective power). Vallinot et al. [30] studied the individual and combined effects of DLC coating and surface texturing in the lubricated pin-on-disc configuration. The combination of DLC coating and surface texturing could cause a decrease in friction compared to the behaviours of only coated disc samples, especially under severe operating conditions.

It is obvious from the analysis of the literature that the combined effects of DLC coating and surface texturing on the seizure and the friction resistances under high normal loads and conformal contact under starved lubrication have not been experimentally investigated yet. However, similar instigations were performed for lower load conditions [25,29].

This study aims to answer the following question: Can the combination of DLC coating and surface texturing cause improvement in tribological behaviour of sliding elements under hard working conditions (high normal load, starved lubrication, and reciprocating motion)? The positive effect of surface texturing on friction reduction can disappear as a result of wear. The application of a DLC coating to textured surfaces can prolong this effect. The combined effect of surface texturing and DLC coating can also improve the seizure resistance of sliding elements in difficult working conditions. This technique can be used for internal combustion engine cylinder liners and valve guides.

2. Materials and Methods

The tests were performed in an initially lubricated reciprocating motion using an Optimol SRV 5 tribotester in the disc-on-disc configuration (Figure 1). A small disc (counter sample; object 2 in Figure 1) of 21 mm diameter and 5 mm thickness co-acted with a disc (sample; object 4 in Figure 1) of 25 mm diameter and 7.9 mm thickness. The contact was conformal with a ring shape, with an outer diameter of 20 mm and an inner diameter of 17 mm. All the counter samples (small discs) were untextured; however, the samples (discs) were in different configurations, such as textured and untextured, DLC-coated untextured, and DLC-coated textured. During the tests, 3 different samples were tested for each configuration. The roughness amplitude of the untextured disc samples and the counter sample, defined by the Ra parameter, was 0.09–0.14 μm. This parameter also characterises the roughness of the base surface of samples free of dimples (for both DLC-coated and uncoated samples).

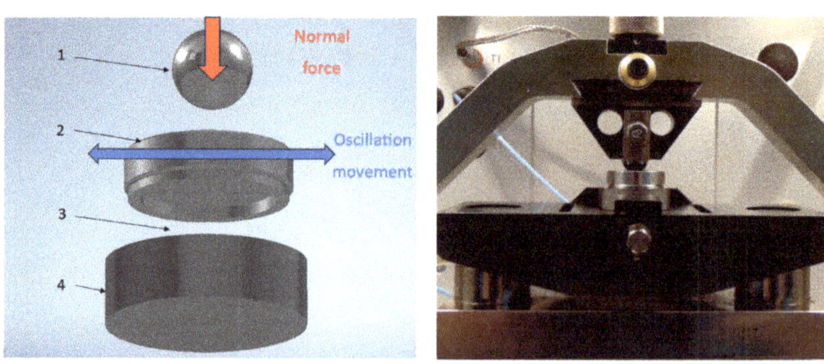

(a) (b)

Figure 1. (**a**) Scheme and (**b**) photo of sliding pair: 1—ball, 2—oscillating counter sample, 3—lubricant, 4—disc sample.

The dimples were created using a SpeedMarker 300 laser engraver (Trotec, Marchtrenk, Austria). Table 1 presents the texturing parameters. The engraver was equipped with an ytterbium fibre laser. Due to the laser characteristics, the beam parameters were set to minimalise material buildups around the oil pocket. No additional polishing was needed before coating with DLC or before the tribological tests with the uncoated samples. The assumed pit area ratios were achieved by changing the diameters of the circular dimples.

Table 1. Texturing parameters.

Parameter	Values
Laser power	20 W (100%)
Focal length	254 mm
Diameter of the beam focusing area	64 μm
Pulse duration	1.5 ns
Pulse repetition rate	820 kHz
Feed speed	200 mm/s
Laser path determination type	Wobble (spiral path around designated shape)
Laser power	20 W (100%)

Laser surface texturing was performed after grinding and before DLC coating. BALINIT® DLC coatings were applied by the Oerlikon Balzers firm (Balzers, Liechtenstein). Material a-C:H was used for coatings with PACVD coating process technology at a temperature smaller than 250 °C. The hardness of the coating according to ISO 14577 [31] was 15–24 GPa. The maximum available working temperature was 300 °C. The application of a DLC coating is ideal for extreme operating conditions, such as high loads, high and very low sliding speeds under starved lubrication, and, occasionally, dry contact conditions. The coated surface protects against excessive abrasion and adhesive wear under high loads.

The displacement frequency was 20 Hz, the stroke was set to 3 mm, and the normal load was set to 1000 N. The pressure in the contact area was 11.47 MPa. Discs and small discs from steel 42CrMo4 with a hardness of 44 ± 2 HRC were tested. The initial temperature was 30 °C. Before each test, a drop of mineral oil L-AN-46 (0.08 ± 0.01 mL) was placed on the inner side of the contact zone. The amount of oil was sufficient to fill all the lubrication pockets in the contact zone. The kinematic viscosity of this oil at 40 °C was 46.0 mm^2/s, the kinematic viscosity at 100 °C was 6.66 mm^2/s, and the viscosity index was 96. The combined impacts of DLC coating and surface texturing were previously studied in reciprocating motion [22–30]. Therefore, the experiment was carried out in this configuration. The conformal contact was chosen because of the possible application of the results of this research (cylinder liners and valve guides). Testing parameters were selected to obtain difficult working conditions leading to seizures.

The duration of the experiment was 2 h when the seizure did not occur and was stopped when the seizure occurred (COF reached 0.3 value). COF was calculated as the average value of the maximum friction coefficients (in reversal points) of each cycle after every second. The friction coefficients under mixed lubrication [32] were analysed. The surface textures of the disc samples before and after the tribologic tests were measured using a Talysurf CCI Lite white light interferometer, with an objective 5×. The measuring area of 3.29 mm × 3.29 mm contained 1024 × 1024 data points. The measured surfaces were levelled only; digital filtration was not used. Worn surfaces were also analysed by Phenom ProX scanning electron microscopy (SEM) (Phenom-World BV, Eindhoven, The Netherlands).

Circular oil pockets were created on the surfaces of the disc samples. Three types of textured surfaces were formed. The pit area ratios were 3, 9, and 13%, which correspond to dimple diameters of 0.2, 0.3, and 0.4 mm and to maximum depths of 30, 20, and 15 μm, respectively. Differences in depth and diameter resulted from different volumes of single oil pockets, which were 0.0005, 0.001, and 0.0015 mm^3, respectively. The distances between the centres of each dimple were the same in all cases. The pit area ratio was calculated as the ratio of the area occupied by the dimples to the entire surface area. Figure 2 presents pseudo-colour views and extracted profiles of textured disc surfaces.

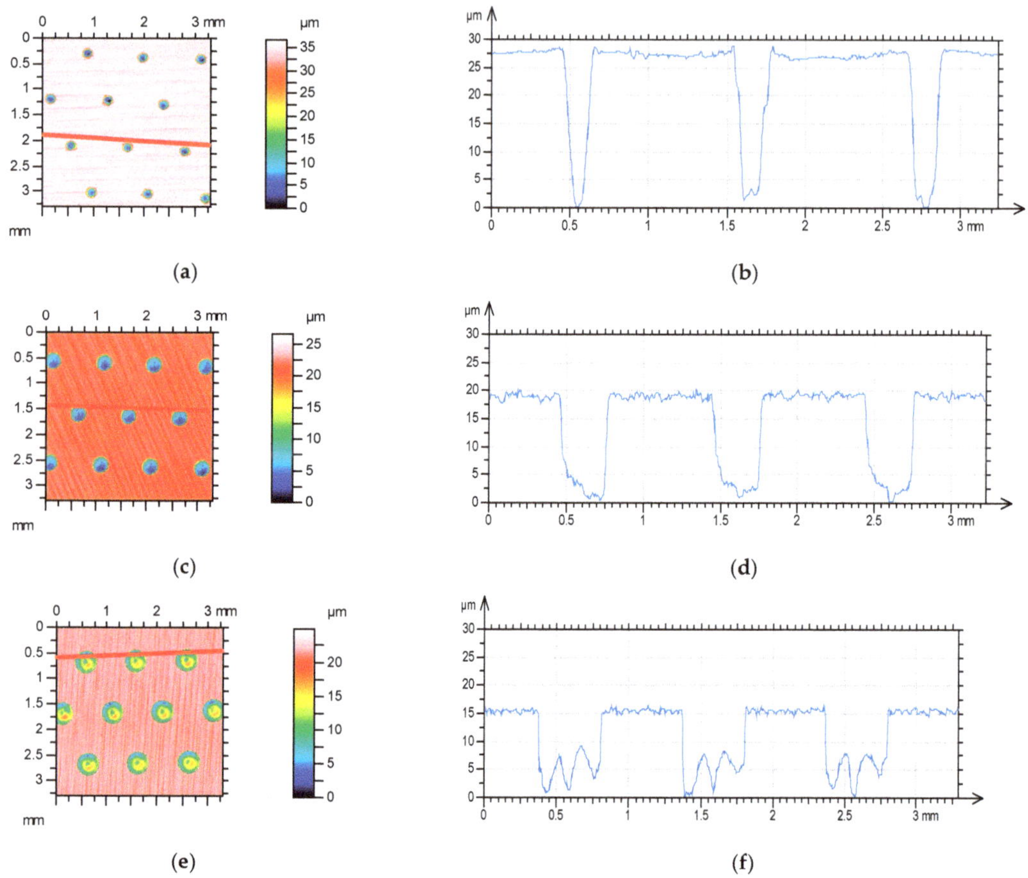

Figure 2. (**a**,**c**,**e**) Pseudo-colour views and (**b**,**d**,**f**) profiles of surfaces with oil pocket densities of (**a**,**b**) 3, (**c**,**d**) 9, and (**e**,**f**) 13%.

Figure 3 presents photos of untextured and selected textured coated and uncoated surfaces. One can see that the DLC coating made the colours of the samples darker. However, the DLC coating had negligible effects on surface textures.

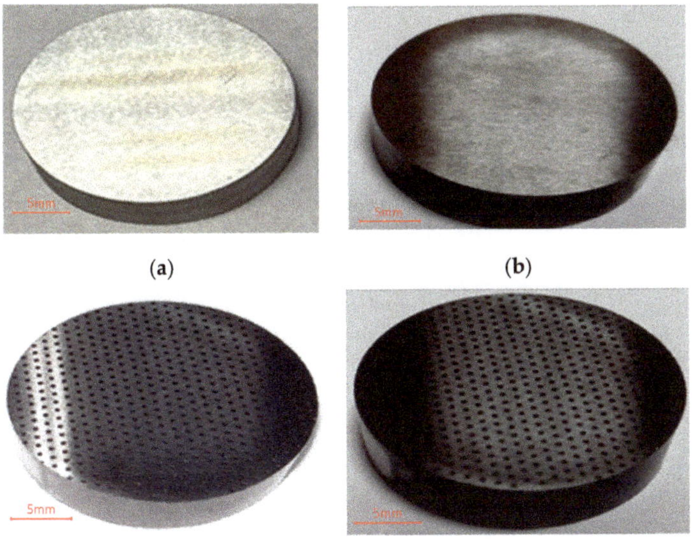

(a) (b) (c) (d)

Figure 3. Photos of untextured (**a**) uncoated and (**b**) coated surfaces, (**c**) uncoated textured surface with a pit area ratio of 13%, and (**d**) coated textured surface with a pit area ratio of 13%.

3. Results and Discussion

Figure 4 shows the friction coefficients versus time for assemblies having untextured samples without and with DLC coatings. Non-coated disc surfaces had a high tendency to seizure. The COF increased to values near 0.2, and then for all samples, it increased considerably. For two repetitions of tests, the coefficients of friction had values higher than 0.3, and the tests were stopped. For the third case, the COF increased to 0.28 and then decreased to 0.16, so the test was not completed.

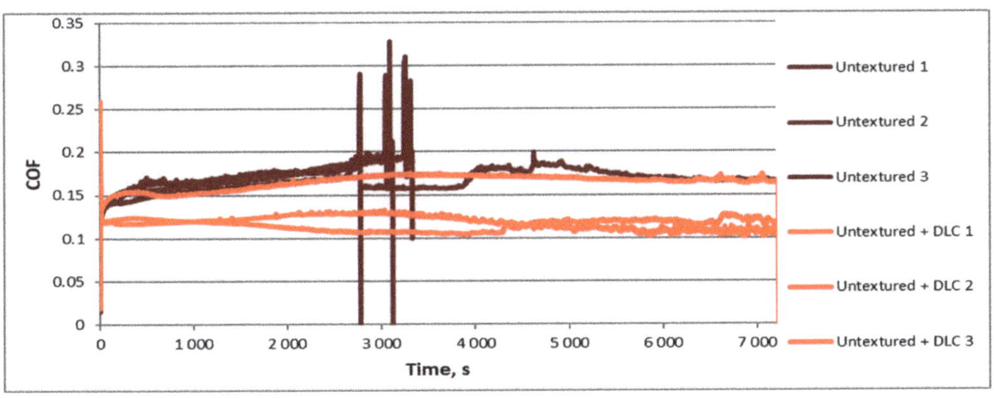

Figure 4. The friction coefficient against time for assemblies having untextured disc samples without and with DLC coatings.

The introduction of a DLC coating improved seizure resistance. For two test repetitions, the friction coefficients were between 0.1 and 0.13; in the remaining case, the friction coefficients were between 0.15 and, finally, 0.17. Generally, DLC coating of the disc caused a growth of the friction pair lifetime by a factor of 2.4 (ratio of the test duration to mean time of seizure point) and caused a reduction in the coefficient of friction compared to the assembly with the uncoated disc.

Figure 5 presents the friction coefficient versus time for the textured disc sample with a pit area ratio of 3% without and with DLC coatings. For two uncoated samples, the coefficients of friction after initial variations gained a value of 0.15 after 300 s. The seizure occurred for one sliding pair with an uncoated disc surface after 2300 s. For the second repetition, after 400 s, the coefficient of friction was rather stable. For the third sliding pair containing textured surfaces without DLC coating, the coefficient of friction reached 0.13 after 200 s and then decreased to a value close to 0.1.

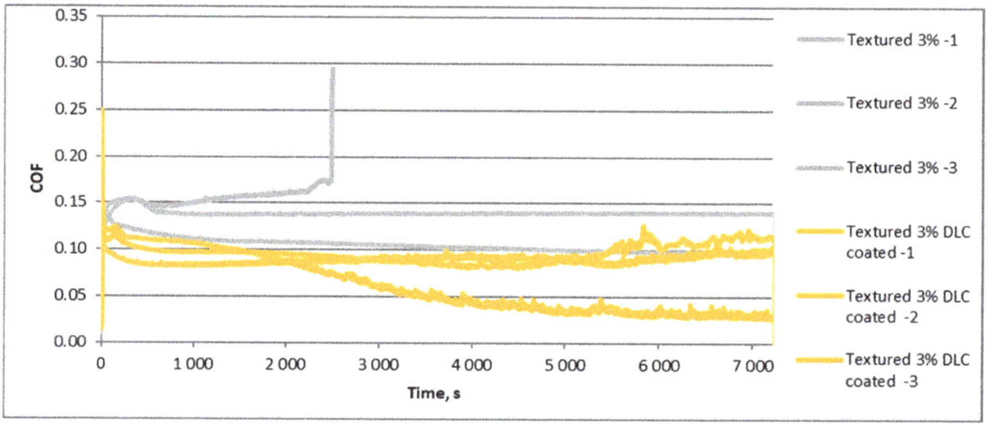

Figure 5. The friction coefficient against time for assemblies containing textured disc samples with oil pocket density of 3% without and with DLC coatings.

Sliding pairs with coated textured discs after initial fluctuation obtained smaller values compared to sliding pairs with uncoated surfaces up to a test duration of 5300 s. During sliding, for two test repetitions, the coefficients of friction were rather stable. For the third repetition of the test, the COF, after reaching a value of 0.12 in 200 s, decreased as the test progressed. The final friction coefficients for the assemblies having coated textured disc samples were between 0.03 and 0.12.

Figure 6 presents a friction coefficient as a function of time for the assemblies that had textured disc samples with a pit area ratio of 9%. Surface texturing without DLC coating did not cause a seizure. For two test repetitions, the COF, after initial fluctuation, obtained a value of 0.155 after 1000 s and then slowly increased. For the third repetition, the COF obtained a value of 0.15 after 300 s and slowly decreased.

Different results were obtained for the textured disc surfaces with DLC coatings. After initial fluctuations, the friction coefficients stabilised after 200 s and obtained values of about 0.11. Then, after 1500 s, they decreased and obtained final values between 0.05 and 0.075.

Figure 7 presents a friction coefficient as a function of time for the assemblies with textured disc samples for a pit area ratio of 13%. For two repetitions of tests, the assemblies with textured discs, after initial instabilities, obtained a friction coefficient of 0.165 after 300 s, and from this time, the friction coefficients slowly decreased. For the other repetition, the friction coefficients, after 200 s, obtained a value of 0.15 and then decreased.

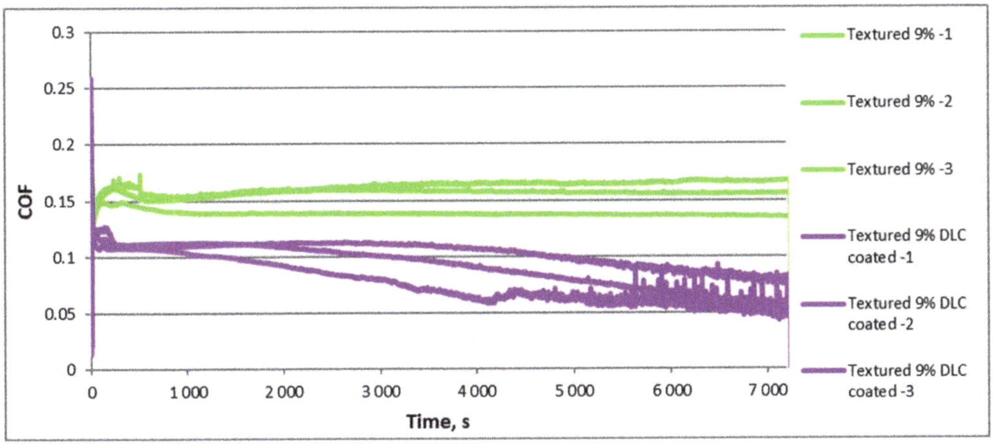

Figure 6. The friction coefficient versus time for assemblies having textured disc samples with a pit area ratio of 9% without and with DLC coatings.

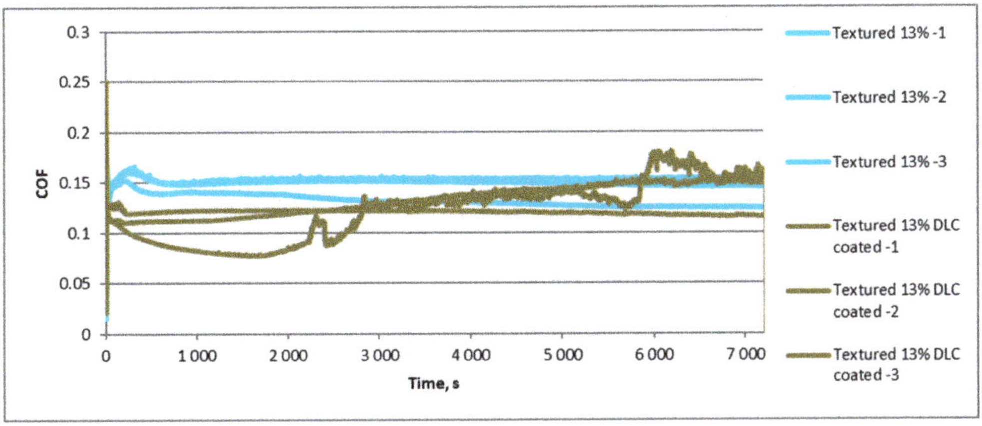

Figure 7. The friction coefficient against time for assemblies containing textured disc samples with dimple density of 13% without and with DLC coatings.

Different situations occurred for the textured disc surfaces with DLC coatings. Before 2800 s, similar to the textured surfaces with other pit area ratios, the friction coefficient was smaller compared to only textured disc surfaces. For one repetition of the test, from this time, the friction coefficient slowly decreased. For two other repetitions, the friction coefficients increased and gained final values between 0.15 and 0.165.

Figure 8 presents the mean values and error bars of the COF in the initial period after friction stabilisation (between 1000 and 2000 s), in the final test part (the last 100 s), and in the entire tests (after 1000 s); in the last two cases when a seizure occurred, the coefficients of friction before sharp increases were taken into account. The effects of surface texturing and DLC coatings on the final values of the coefficient of friction were assumed to be the most important; therefore, the initial 1000 s was not analysed. The behaviours of the friction pairs after friction stabilisation (between 1000 and 2000 s) were studied because of friction reduction due to DLC coatings obtained for all used samples.

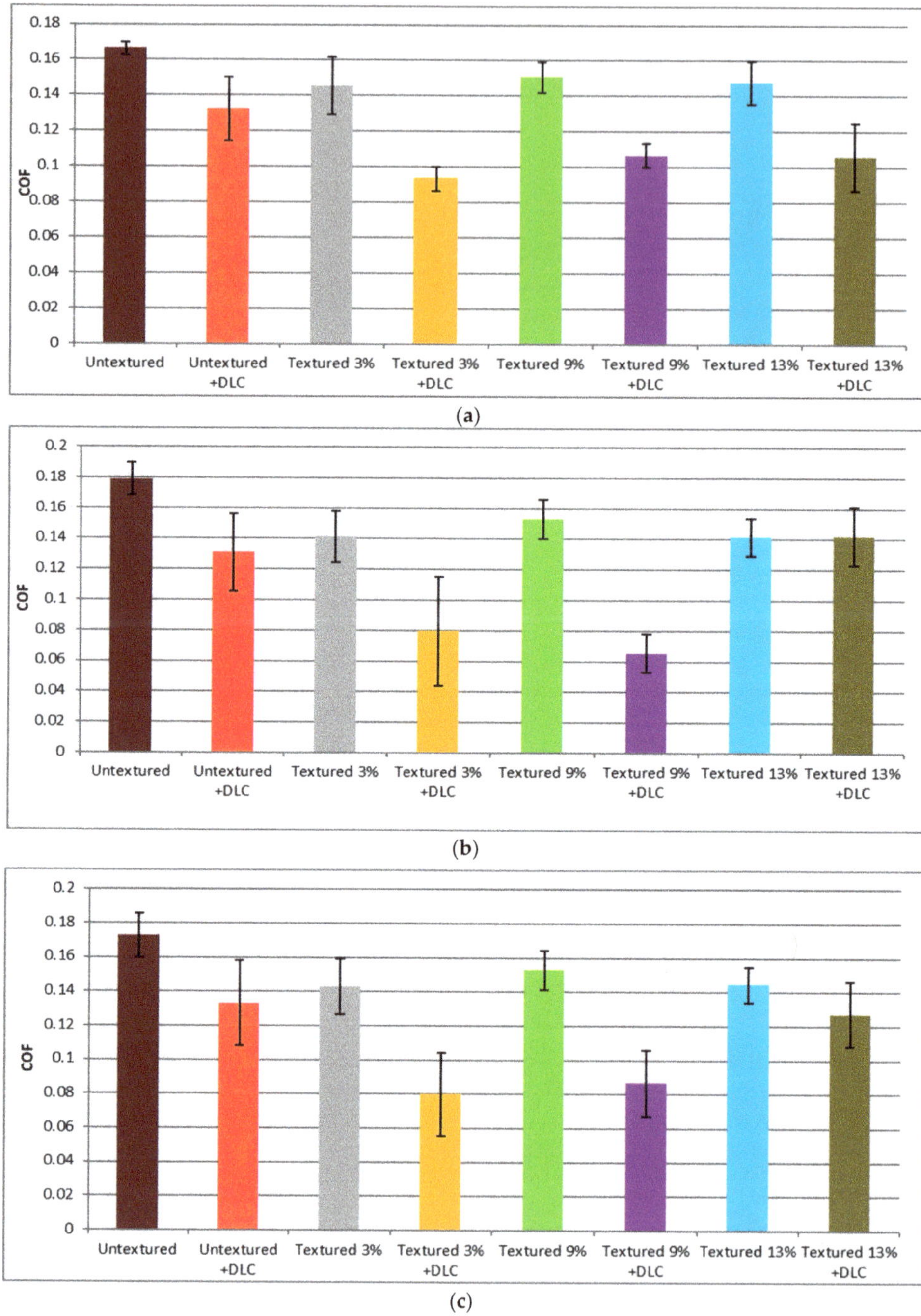

Figure 8. Average values and error bars of the COF (**a**) in the early test part—between 1000 and 2000 s, (**b**) in the final test part—last 1000 s, and (**c**) in all tests—after 1000 s.

In the initial test part (Figure 8a), surface texturing in all analysed cases caused decreases in COF; however, these reductions were low (up to 7%) compared to the untextured samples. The DLC coating caused a decrease in the COF of approximately 20% compared to the untextured sample without the DLC coating. A higher friction reduction was achieved by the combination of surface texturing and DLC coating compared to only surface texturing, and the highest decrease was 38%. This performance was obtained for the textured samples with all pit area ratios analysed.

A decrease in the COF due to surface texturing was also achieved in the final parts of the tests (Figure 8b) for all textured sliding pairs of approximately 20%. Application of the DLC coating to an untextured disc sample caused a reduction in the COF of 28%. In contrast to the initial test part, the application of a DLC coating to textured surfaces caused a decrease in the COF only for samples with pit area ratios of 3 and 9%. The highest decrease in the final friction coefficients was near 60% for a pit area ratio of 9%. The non-beneficial impact of surface texturing on resistance to motion was caused by an increase in friction for two repetitions of the test after 1600 s (Figure 7).

The application of surface texturing also caused a decrease in the friction coefficient of approximately 20% in all tests (Figure 8c). The use of DLC coating for an untextured surface also caused a decrease in the COF of 20%. Similar to the final test part (Figure 8b), the DLC coating did not cause a substantial reduction in the COF for the sliding pair containing a textured disc sample with a dimple density of 13%. For the other textured assemblies, the reduction in the COF was about 50%.

Figure 9 presents photos of sliding pairs after the seizure process, with marked corresponding areas where the seizure occurred. In all the cases analysed, the seizure occurred on one side of the disc samples. Tracks of seizure are visible both on the disc sample and counter sample. The smallest scars of the textured sample and counter sample can be seen (Figure 9e,f).

Figure 10 presents examples of non-seizured sliding pairs after the tribologic tests. Wear tracks are not visible in the untextured disc sample with the DLC coating (Figure 10a). However, one can see wear traces on the steel counter sample (Figure 10b). The textured disc without the DLC coating with a pit area ratio of 3% (Figure 10c) showed wear tracks, which were also visible on the counter sample (Figure 10d). For the textured disc sample without the DLC coating with an oil pocket density of 13% (Figure 10e), the wear tracks were more uniform compared to those obtained for the pit area ratio of 3% (Figure 10c). However, the counter samples shown in Figure 10d,f look similar. Wear tracks are not visible in the textured disc samples for a pit area ratio of 3% with the DLC coating shown in Figure 10g. A similar view occurred on a textured disc with a DLC coating for a pit area ratio of 9%. However, wear tracks are visible on the DLC-coated textured disc with a pit area ratio of 13% (Figure 10i). It is difficult to quantitatively compare the wear volumes of the disc surfaces; the wear depended only on the truncation of the highest roughness peaks. The DLC coating caused a decrease in the wear of the disc samples. Unlike the counter sample that co-acted with an untextured disc with the DLC coating (Figure 10a), the counter samples that co-acted with textured discs with pit area ratios of 3 and 13% with DLC coatings do not show wear tracks (Figure 10h,j). In the case of the non-textured samples covered with DLC, the high hardness of the applied DLC layer caused increased wear of the counter sample. The use of surface texturing without the DLC coating resulted in wear traces that corresponded to the array of lubrication pockets. However, the combination of the DLC layer and laser texturing did not cause such effects.

Figure 11 presents isometric views of the disc surfaces after seizure and views of disc surfaces not subjected to seizure after the tests. Figure 11a,c,e,g show surfaces without the DLC coating after the test; the other graphs presented in Figure 11 are surfaces that are DLC coated. The untextured surface without DLC coatings presented in Figure 11a corresponds to the disc shown in Figure 9a. The seizure process resulted in the formation of valleys and peaks created by adhesion, and consequently, the height of the roughness increased compared to the ground surfaces, for which the maximum amplitude was approximately 2 μm. The textured surface with a pit area ratio of 3% without the DLC coating presented in Figure 11c corresponds to the disc shown in Figure 9e. In this case, the seizure resulted in the formation of new peaks and an increase in the height of roughness, which was lower compared to the untextured surface presented in Figure 11a. Among the surfaces shown in Figure 11a,c, the seizure area was the smallest for the last-mentioned surface. For the surfaces not subjected to seizure, wear tracks were visible on the textured surface with a pit area ratio of 9% without the DLC coating (Figure 11e) and not visible on the textured surface with an oil pocket density of 13% without the DLC coating (Figure 11g). The destruction of the DLC coating was visible on the textured surface with the highest pit area ratio of the 13% around the edge of oil pocket (arrow in Figure 11f). A similar removal of the DLC layer was observed in other parts of the coated disc sample with a pit area ratio of 9%. In all cases, use of a DLC-coated surface caused low non-measurable wear, except for DLC coating destruction around the biggest oil pockets. The use of the surface with the highest pit area ratio of 13% also resulted in non-measurable wear. However, the COF was higher in this case. The low wear was probably due to the largest volume of oil pockets with a pit ratio of 13% and the ability to capture the wear products efficiently.

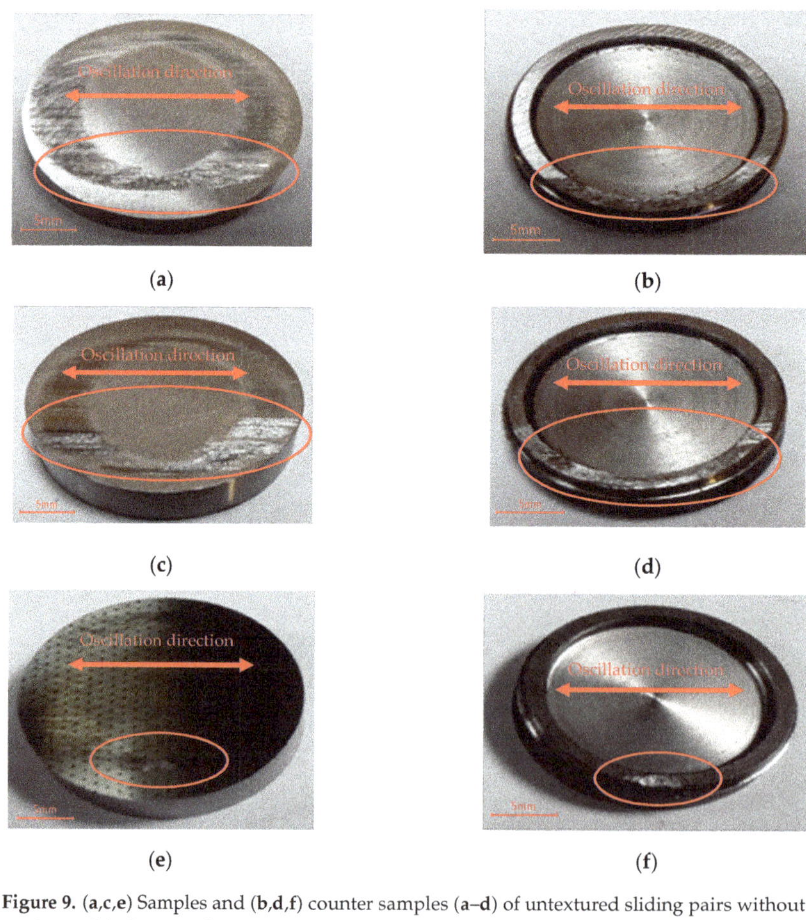

(a) (b) (c) (d) (e) (f)

Figure 9. (**a**,**c**,**e**) Samples and (**b**,**d**,**f**) counter samples (**a**–**d**) of untextured sliding pairs without DLC coating and (**e**,**f**) textured assembly with oil pocket density of 3% without DLC coating. With marked wear areas.

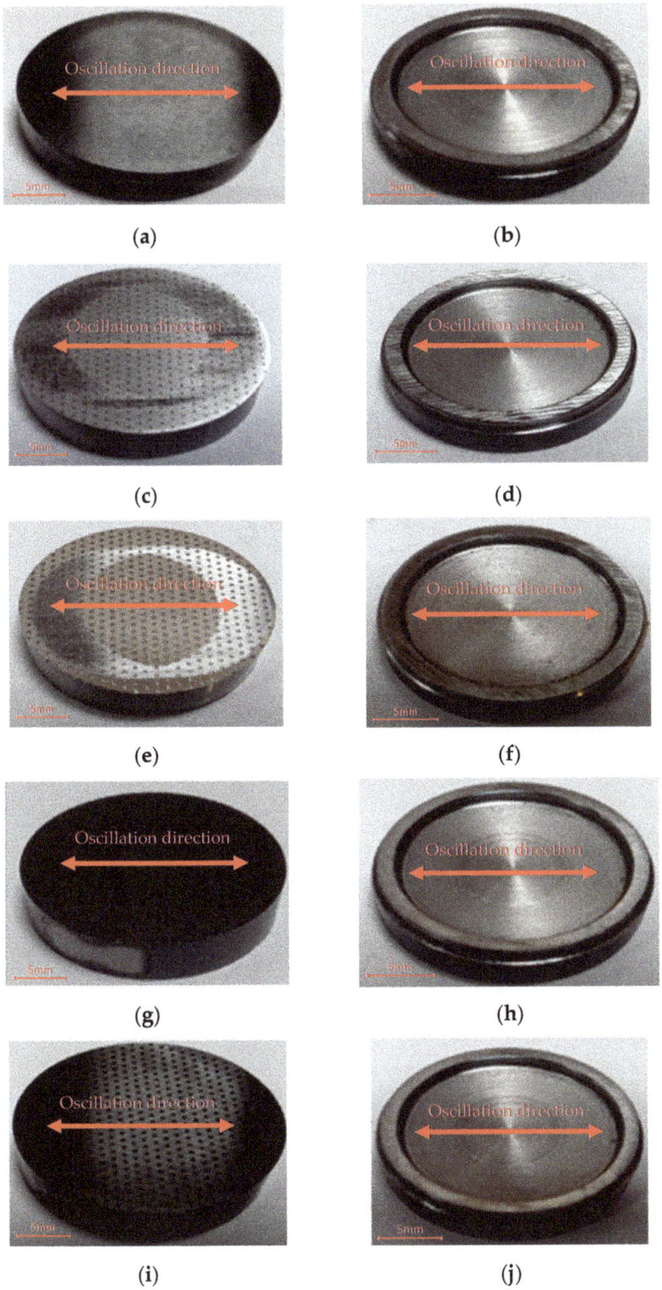

Figure 10. (**a**,**c**,**e**,**g**,**i**) Samples and (**b**,**d**,**f**,**h**,**j**) counter samples of sliding pairs with (**a**,**b**) untextured DLC coated disc, assemblies with textured discs without DLC coatings with dimple density of (**c**,**d**) 3% and (**e**,**f**) 13%, and sliding pairs with DLC-coated textured discs with dimple density of (**g**,**h**) 3% and (**i**,**j**) 13%.

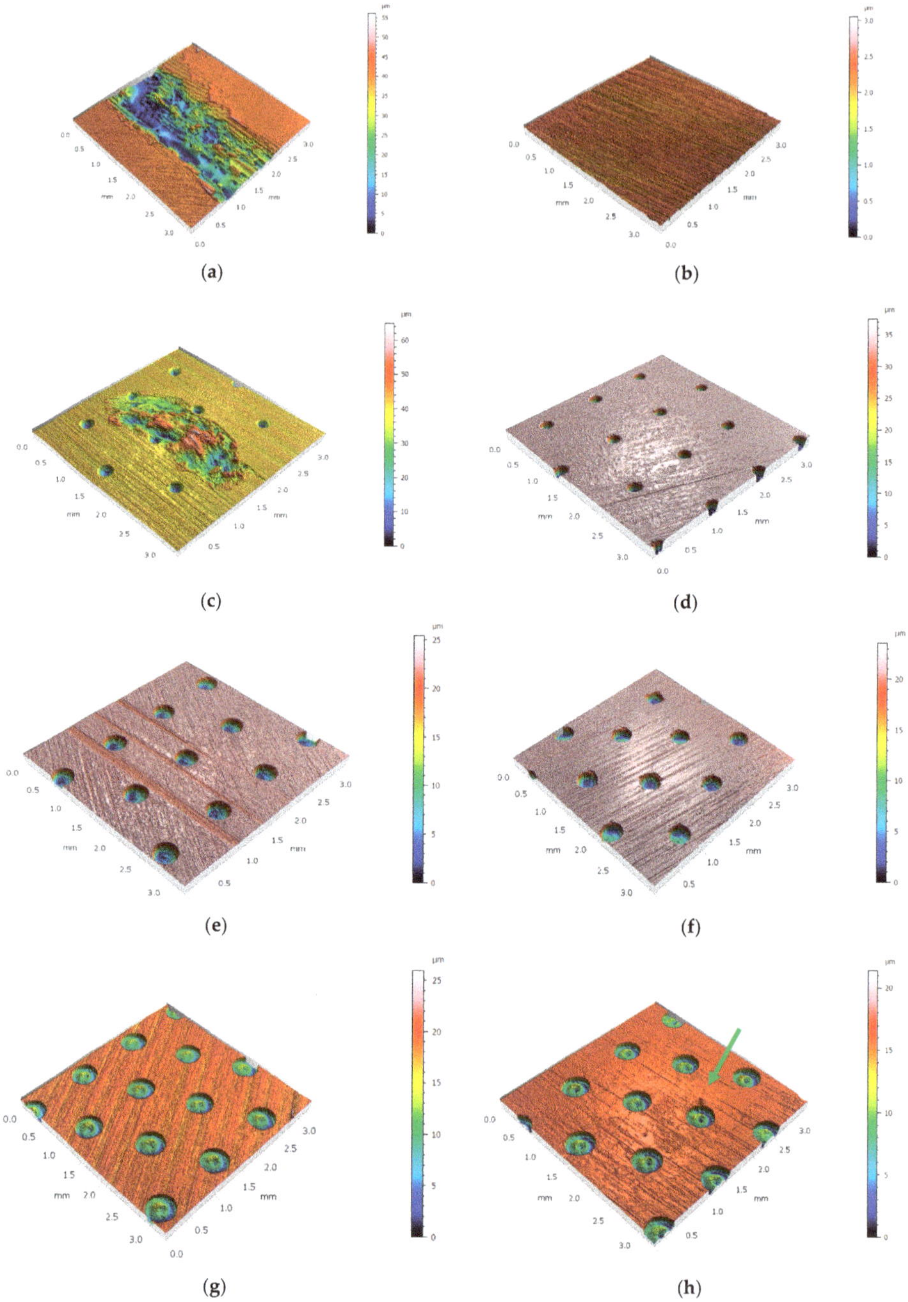

Figure 11. Isometric views of untextured disc surfaces (**a**) without DLC coatings after seizure and (**b**) with DLC coating after test, (**c**) of the surface of the textured disc with dimple density of 3% without DLC coatings after seizure (**d**) and corresponding disc with DLC coating, (**e**) of the textured disc surface with pit area ratio of 9% without DLC coating without seizure (**f**) and with DLC coating, and of the textured disc surface with dimple density of 13% (**g**) without DLC coating and with (**h**) DLC coating without seizure.

Figure 12 presents views of the disc sample surfaces before and after the tribological tests obtained using the Phenom ProX SEM microscope.

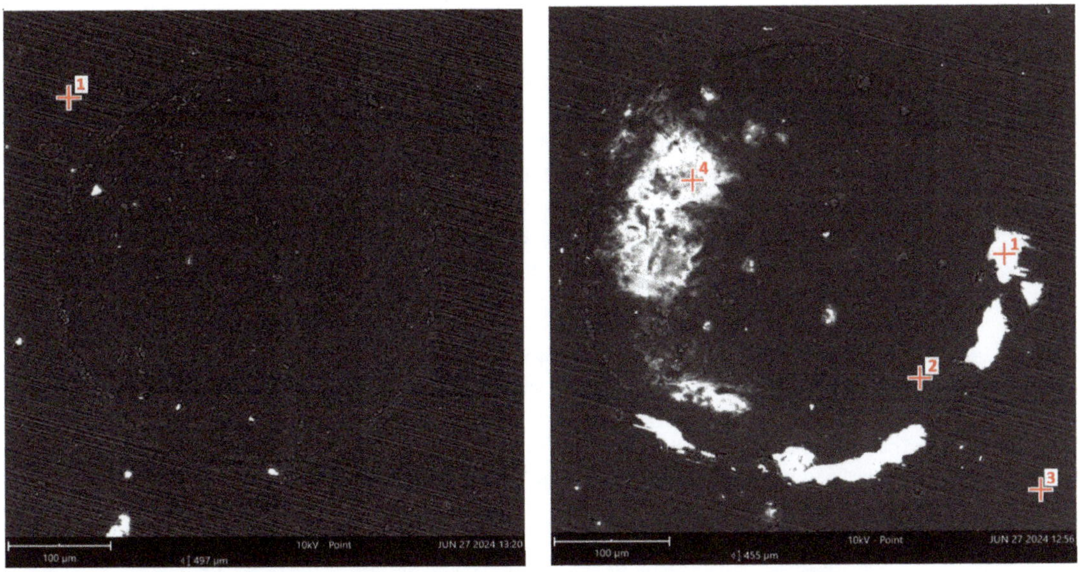

(**a**) (**b**)

Figure 12. (**a**) View of the dimple on the DLC-coated surface with oil pocket density of 13% before the tribological test; (**b**) view of the dimple on the DLC-coated surface with oil pocket density of 13% after the tribologic test, with marked regions of elemental concentration analysis.

The results obtained with the Phenom ProX SEM microscope show that wear of DLC-coated textured surfaces occurred at the outer area close to the oil pocket. Typical elemental analysis on the textured DLC-coated surfaces before the tribologic tests showed more than 96% of the carbon weight concentration, as in point 1 in Figure 12a. The scars of the greatest wear were visible on the textured DLC-coated surface with the highest dimple density of 13% (and of the highest dimple diameter of 0.4 mm) in regions that corresponded to the zone affected by heat by laser texturing outside of the oil pocket. The behaviour of the sample with a pit area ratio of 13% was shown because of the highest wear of the DLC-coated disc. Elemental analysis performed with a SEM microscope shows that the base material was at point 1 in Figure 12b (79% Fe; 14% C; 6% O of the weight concentration) with some oxidations. The regions at point 2 (inside oil pocket) and point 3 (outside oil pocket) show similar and high carbon concentrations (around 99% of weight) typical for regions coated with DLC. In some cases, regions, as in measurement point 4, were visible with a highly oxidised area inside the oil pocket. In this region, the elemental concentration was around 69% C, 17% O, 13% Fe of the weight concentration. This suggests that, during wear, regions that were highly oxidised had low adherence to the DLC layer. In places of wear around the oil pockets, the DLC coating was removed. This phenomenon could occur due to two mechanisms. Perhaps the cohesion between the heat-affected zone and the DLC layer was lower than that in other regions. As a result of the large pit area ratio and the high load, stress was probably concentrated on the edges of the oil pockets. The impact of limited lubricant in the contact region due to the highest oil pocket volumes was probably smaller because the amount of oil was sufficient to fill all dimples in the contact zone, and the effect of pit area ratio on the coefficient of friction of only textured surfaces was marginal.

Working conditions near reversal points under oscillation, high load, and the limited oil volume may lead to starved lubrication conditions. Laser texturing of the disc surface caused growth in the seizure resistance in the steel–steel initially lubricated sliding pair in reciprocating motion in such difficult circumstances. The lifetime of the friction pair increased 2.4 times due to DLC coating and surface texturing. However, this increase could probably be longer (tests were stopped after a duration of 2 h). This effect was evident mainly for textured discs of pit area ratios of 9 and 13%, and for a dimple density of 3%, seizure occurred for one sliding pair. The increase in friction pair lifetime by surface texturing was also found in works [5–12]. Surface texturing also caused an increase in seizure resistance in [13–15]. However, the authors of the cited articles applied dimples with pit area ratios greater than 3%. The tendency to seizure can be reduced by breaking the contact between two surfaces. This behaviour is hardly possible using a textured surface with a very small pit area ratio of 3%.

The application of the DLC coating on the disc surface led to a significant increase in the seizure resistance of the sliding assembly compared to that of the ground disc. DLC coating of the textured disc surface with the smallest ratio also led to an improvement in the anti-seizure characteristic. The seizure did not occur for sliding pairs with DLC-coated discs. The authors of articles [18–21] also found that the use of DLC coating caused growth of seizure resistance of machine elements. The DLC coating of the untextured disc surface also led to a reduction in the coefficient of friction compared to the uncoated disc. Friction reduction was also achieved by DLC coating of the crank journal [18,19].

Combining surface texturing with DLC coating could also cause decreases in the COF of sliding assemblies compared to only textured discs. This impact was visible in the initial portions of the tests for all textured assemblies, for the lowest oil pocket density of 3% before 5300 s, for the highest oil pocket density of 13% before 2800 s, and for the medium oil pocket density of 9% for the entire test. The further frictional behaviour was influenced by the pit area ratios of the disc samples. For the textured discs without the DLC coating, the friction force after initial fluctuation slowly decreased or obtained stable values for the smallest and highest pit area ratios. When the density of the dimples was 9%, the COF slowly increased with time for two repetitions or was constant. For the same pit area ratio of 9% combined with the DLC coating, the friction coefficient decreased as the test progressed. Therefore, a considerable decrease in the COF was obtained in the final test portion—up to 60%. When the pit area ratio of 3% was combined with the DLC coating for one test repetition, the COF decreased with time, while for other courses, the friction coefficient increased after 5000 s. Therefore, the final COF was smaller for the combination of surface texturing with DLC coating compared with only surface texturing; however, the friction reduction for the pit area ratio of 3% was lower compared to the oil pocket density of 9%. The beneficial effect of surface texturing and DLC coating on friction reduction was presumably caused by negligible wear of the co-acting pair. For a combination of a 13% dimple density and DLC coating, the friction coefficient increased after 1800 s for two test repetitions. This increase was probably related to higher disc wear compared to other DLC-coated textured disc samples. The removal of the DLC coating around the dimples with the largest diameter for a very high normal load was presumably caused by the high load stresses and low cohesion between the heat-affected zone and the DLC layer. This behaviour led to similar values of the final friction coefficients for textured sliding pairs with and without DLC coatings, and beneficial effects of DLC coating on the frictional behaviour of textured assemblies were found only for oil pocket densities of 3 and 9%. For pit area ratios of 3 and 9%, applications of DLC coating and surface texturing caused a friction decrease compared to only DLC coating. Friction reductions due to surface texturing combined with DLC coating in relation to DLC coating were also achieved in works [22–25,27–29]; however, the normal loads in the cited studies were smaller than in this research.

DLC coatings are applied to increase wear resistance [17]. In this research, wear levels of co-acting parts were qualitatively assessed. DLC coatings caused a reduction in disc wear for textured and untextured sliding parts. For the untextured assembly, wear of the only counter sample was visible. Similarly to [25], untextured and textured DLC samples typically did not show wear marks. In [30], no deformation of oil pockets occurred for textured DLC samples. Among the DLC coated textured discs, wear tracks and the removal of the DLC layer around the edges of the dimples were only found for a pit area ratio of 13% (Figures 11e and 12b). DLC coatings led to a reduction in the wear levels of the counter samples compared to those acted with textured discs without DLC coatings. The authors of References [22–24,27,28] also obtained improvement in wear resistance by combining DLC coating and surface texturing compared to DLC coating. On the basis of the analysis of surfaces using a white light interferometer and scanning electron microscope, it was found that the wear of the samples and the counter samples not subjected to seizure had an abrasive character.

4. Conclusions

This experiment was carried out to study the effects of laser texturing and DLC coating on seizure and friction of the sliding pair under conformal reciprocating conditions with a high normal load. The following conclusions were obtained:

1. The combination of surface texturing and DLC coating led to improved tribological performance of the lubricated assembly in reciprocating conformal sliding contact in comparison to untextured discs with and without DLC coating and textured discs without DLC coating.
2. The DLC coating of the disc surface caused an improvement in seizure resistance in the steel–steel initially lubricated sliding pair in reciprocating motion. The lifetime of the friction pair increased at least 2.4 times due to DLC coating.
3. The seizure resistance also increased due to the surface texturing of the steel discs. This effect was evident mainly for pit area ratios of 9 and 13%.
4. Combining surface texturing and DLC coating could cause decreases in the coefficients of friction of sliding assemblies compared to only textured discs and to only DLC-coated samples. This effect was visible for disc pit area ratios of 3 and 9%.
5. The DLC coating caused disc wear reductions for textured and untextured sliding parts. For the untextured assembly, wear of the only counter sample was visible. Wear of the disc sample and removal of the DLC layer around the edges of the dimples were found for the largest pit area ratio of 13%. DLC coatings caused reduction in wear levels of the counter samples compared to those co-acted with textured discs without DLC coatings.
6. The small pit area ratio should be avoided under hard working conditions in steel-to-steel contact due to seizure because of the low possibility of breaking the contact between the sliding surfaces.
7. Removal of the DLC layer around the large-diameter dimples, caused by the high load stresses and low cohesion between the heat-affected zone and the DLC layer, can lead to friction growth. Use of smaller types of oil pockets is recommended in combination with DLC coating.
8. This work is functionally important, especially for assemblies operated under high normal loads. Research results can be used for various applications, such as cylinder liners and valve guides from internal combustion engines. Further research will be focused on tribological research in conditions similar to those of industrial applications.

Author Contributions: Conceptualisation, W.K., S.W. and P.P.; methodology, S.W. and A.D.; validation, W.K., A.D. and P.P.; formal analysis, A.D. and W.K.; investigation, S.W. and P.P.; resources, W.K. and A.D.; data curation, S.W. and P.P.; writing—original draft preparation, S.W. and W.K.; writing—review and editing, A.D. and P.P.; visualisation, S.W.; supervision, W.K.; project administration, A.D.; funding acquisition, P.P. All authors have read and agreed to the published version of the manuscript.

Funding: This work was supported by the National Science Center; Kraków, Poland (Decision No. 2018/31/B/ST8/02946) "The effect of disc surface texturing on the tribological properties of pin-on-disc assembly".

Data Availability Statement: The raw data supporting the conclusions of this article will be made available by the authors on request.

Conflicts of Interest: The authors declare no conflicts of interest.

References

1. Etsion, I. State of the Art in Laser Surface Texturing. *J. Tribol.* **2005**, *127*, 761–762. [CrossRef]
2. Rosenkranz, A.; Grützmacher, P.G.; Gachot, C.; Costa, H.L. Surface Texturing in Machine Elements—A Critical Discussion for Rolling and Sliding Contacts. *Adv. Eng. Mater.* **2019**, *21*, 1900194. [CrossRef]
3. Gachot, C.; Rosenkranz, A.; Hsu, S.M.; Costa, H.L. A critical assessment of surface texturing for friction and wear improvement. *Wear* **2017**, *372–373*, 21–41. [CrossRef]
4. Lu, P.; Wood, R.J.K. Tribological performance of surface texturing in mechanical applications—A review. *Surf. Topogr. Metrol. Prop.* **2020**, *8*, 043001. [CrossRef]
5. Baiocco, G.; Genna, S.; Salvi, D.; Ucciardello, N. Laser Texturing to Increase the Wear Resistance of an Electrophoretic Graphene Coating on Copper Substrates. *Materials* **2023**, *16*, 5359. [CrossRef]
6. Arenas, M.A.; Ahuir-Torres, J.I.; García, I.; Carvajal, H.; De Damborenea, J. Tribological behaviour of laser textured Ti6Al4V alloy coated with MoS2 and graphene. *Tribol. Int.* **2018**, *128*, 240–247. [CrossRef]
7. Wos, S.; Koszela, W.; Pawlus, P. Determination of oil demand for textured surfaces under conformal contact conditions. *Tribol. Int.* **2016**, *93*, 602–613. [CrossRef]
8. Andersson, P.; Koskinen, J.; Varjus, S.; Gerbig, Y.; Haefke, H.; Georgiou, S.; Zhmud, B.W. Buss Microlubrication effect by laser-textured steel surfaces. *Wear* **2007**, *262*, 369–379. [CrossRef]
9. Hu, T.; Hu, L. The study of tribological properties of laser-textured surface of 2024 aluminum alloy under boundary lubrication. *Lubr. Sci.* **2012**, *24*, 84–93. [CrossRef]
10. Duarte, M.; Lasagni, A.; Giovanelli, R.; Narciso, J.; Louis, E.; Mücklich, F. Increasing lubricant film lifetime by grooving periodical patterns using laser interference metallurgy. *Adv. Eng. Mater.* **2008**, *10*, 554–555. [CrossRef]
11. Rosenkranz, A.; Heib, T.; Gachot, C.; Mücklich, F. Oil film lifetime and wear particle analysis of laser-patterned stainless steel surfaces. *Wear* **2015**, *334-335*, 1–12. [CrossRef]
12. Koszela, W.; Galda, L.; Dzierwa, A.; Pawlus, P. The effect of surface texturing on seizure resistance of a steel-bronze assembly. *Tribol. Int.* **2010**, *43*, 1933–1942. [CrossRef]
13. Kuroiwa, Y.; Amanov, A.; Tsuboi, R.; Sasaki, S.; Kato, S. Effect of surface texturing arrangement for improving anti-seizure property of lead-free copper alloy. *J. Teknol.* **2014**, *66*, 81–89. [CrossRef]
14. Galda, L.; Dzierwa, A.; Sep, J.; Pawlus, P. The effect of oil pockets shape and distribution on seizure resistance in lubricated sliding. *Tribol. Lett.* **2010**, *37*, 301–311. [CrossRef]
15. Fan, P.; Zhong, M. Improving the Seizure Resistance of Automobile Engine Liners By Cost Effective ns Fiber Laser Texturing Approach. In *International Summer Session: Lasers and Their Applications*; Optica: New York, NY, USA, 2011. [CrossRef]
16. Holmberg, K.; Erdemir, A. Influence of tribology on global energy consumption, costs and emissions. *Friction* **2017**, *5*, 263–284. [CrossRef]
17. Sharifahmadian, O.; Pakseresht, A.; Amirtharaj Mosas, K.K.; Galusek, D. Doping effects on the tribological performance of diamond-like carbon coatings: A review. *J. Mater. Res. Technol.* **2013**, *27*, 7748–7765. [CrossRef]
18. Iwata, T.; Oikawa, M.; Chida, R.; Ishii, D.; Ogihara, H.; Mihara, Y.; Kano, M. Excellent Seizure and Friction Properties Achieved with a Combination of an a-C:H:Si DLC-Coated Journal and an aluminum Alloy Plain Bearing. *Coatings* **2021**, *11*, 1055. [CrossRef]
19. Ogihara, H.; Iwata, T.; Mihara, Y.; Kano, M. The effects of DLC-coated journal on improving seizure limit and reducing friction under engine oil lubrication. *Int. J. Engine Res.* **2021**, *23*, 1267–1274. [CrossRef]
20. Usui, S.; Kato, H.; Nakajima, H.; Shimokawa, K. *Effect of Diamond-Like Carbon Coating on Anti-Scuffing Characteristics of Piston Pins*; SAE Technical Paper; SAE: Warrendale, PA, USA, 2019. [CrossRef]

21. Wu, J.; Wei, P.; Liu, G.; Chen, D.; Zhang, X.; Chen, T.; Liu, H. A comprehensive evaluation of DLC coating on gear bending fatigue, contact fatigue, and scuffing performance. *Wear* **2024**, *536–537*, 205177. [CrossRef]
22. Arslan, A.; Masjuki, H.H.; Varman, M.; Kalam, M.A.; Quazi, M.M.; Al Mahmud, K.; Gulzar, M.; Habibullah, M. Effects of texture diameter and depth on the tribological performance of DLC coating under lubricated sliding condition. *Appl. Surf. Sci.* **2015**, *356*, 1135–1149. [CrossRef]
23. Shum, P.W.; Zhou, Z.F.; Li, K.Y. Investigation of the tribological properties of the different textured DLC coatings under reciprocating lubricated conditions. *Tribol. Int.* **2013**, *65*, 259–264. [CrossRef]
24. Amanov, A.; Watabe, T.; Tsuboi, R.; Sasaki, S. Improvement in the tribological characteristics of Si-DLC coating by laser surface texturing under oil-lubricated point contacts at various temperatures. *Surf. Coat. Technol.* **2013**, *232*, 549–560. [CrossRef]
25. Mishra, P.; Penchaliah, R. Synergistic effect of surface texturing and coating on the friction between piston ring and cylinder liner contact. *Proc. Inst. Mech. Engineers. J. Eng. Tribol.* **2020**, *235*, 1298–1311. [CrossRef]
26. Koskinen, J.; Tapper, U.; Andersson, P.; Varjus, S.; Kolehmainen, J.; Tervakangas, S.; Buss, W. Friction reduction by texturing of DLC coatings sliding against steel under oil lubrication. *Surf. Coat. Technol.* **2010**, *204*, 3794–3797. [CrossRef]
27. Ding, Q.; Wang, L.; Wang, Y.; Wang, S.C.; Hu, L.; Xue, Q. Improved Tribological Behavior of DLC Films Under Water Lubrication by Surface Texturing. *Tribol. Lett.* **2010**, *41*, 439–449. [CrossRef]
28. Dufils, J.; Faverjon, F.; Héau, C.; Donnet, C.; Benayoun, S.; Valette, S. Combination of laser surface texturing and DLC coating on PEEK for enhanced tribological properties. *Surf. Coat. Technol.* **2017**, *329*, 29–41. [CrossRef]
29. Koszela, W.; Pawlus, P.; Reizer, R.; Liskiewicz, T. The combined effect of surface texturing and DLC coating on the functional properties of internal combustion engines. *Tribol. Int.* **2018**, *127*, 470–477. [CrossRef]
30. Vallinot, I.B.; Ochoa, E.d.l.G.; Otero, J.E.; Tanarro, E.C.; Martínez, I.F.; Varela, J.A.S. Individual and combined effects of introducing DLC coating and textured surfaces in lubricated contacts. *Tribol. Int.* **2020**, *151*, 106440. [CrossRef]
31. *ISO 14577-1:2015*; Metallic Materials—Instrumented Indentation Test for Hardness and Materials Parameters. Part 1: Test Method; ISO: Geneva, Switzerland, 2015.
32. Kelley, J.; Poll, G.; Pape, F. Investigation of the possible applications for microtextured rolling bearings. *Front. Manuf. Technol.* **2022**, *2*, 1012343. [CrossRef]

lubricants

MDPI

Article

Predicting Low Sliding Friction in Al-Steel Reciprocating Sliding Experiment after a Controlled Grinding of the Steel Counterface

Gopakumar Parameswaran *, Vikram Jayaram and Satish V. Kailas

Indian Institute of Science, Bangalore 560012, Karnataka, India; qjayaram@iisc.ac.in (V.J.); satvk@iisc.ac.in (S.V.K.)
* Correspondence: gopa99@gmail.com

Abstract: The aim of this study was to identify the areal surface parameters that correlated with lowering of sliding friction. Different ground surfaces were created on stainless steel and the lubricated sliding friction generated at the contact interface with a flat-faced aluminum pin was studied. The frictional force encountered is an order of magnitude lower for a P1200-finished surface than the other ground surfaces. Using 3D surface profilometry, a unique surface parameter ratio "Spk/Sk" was found to predict the frictional performance of these surfaces. When this surface parameter ratio was less than 1, average sliding friction was close to 0.1. When this ratio was greater than 1, the coefficient was an order of magnitude lower. Using energy dispersive spectrometry, such surfaces after wear showed the presence of a uniform dispersed layer of iron oxide on the surface of the pin. This was absent on the surfaces having high friction, indicating the role of the steel counter surface in building this beneficial transfer layer. Scanning electron microscopy provided topography images to visualize the surface wear. The motivation for the authors was to use a commercially scaled process like precision grinding for the surface modifications on stainless steel.

Keywords: friction; back-scattered imaging; 3D surface profilometry; sliding; surface texturing; energy dispersive spectroscopy; lubricated sliding; reciprocated sliding; surface modification; topography

1. Introduction

The manufacturing industry, and conventionally the automotive industry, rely on grinding as a cost-efficient and scalable process to modify surface roughness with a view to improving dimension, fit, and appearance, apart from meeting performance requirements [1], including sliding friction and wear [2], where mating surfaces are involved. Although surfaces and interfaces are multifold, those that are in contact and move relative to each other gather special attention with regards to surface modification and refinement. Such functional surfaces need to be studied, analyzed, and controlled as part of ongoing research and while setting design specifications needed for manufacturing, inspection, and quality control stages. The parameters thus defined [3,4] will have a key relationship with the outgoing quality of products manufactured, the part and system durability requirements, and the functional requirements that the part is required to meet. Karkalos et al. [5], in their studies on slot milling, found that the influences of the coolant and depth of cut were key factors in milling for reducing surface roughness.

Burnishing as a finishing operation was examined by Tayeb et al. [6], where machining parameters like burnishing speed, force, and ball diameters influenced the surface roughness. These machined surfaces determined the friction at contact, the quantum of wear [7], and the ability to retain or repel fluids. Jeng [8] conducted surface roughness experiments using riblet tapes to generate repeatable surfaces in the early 1990s; using the contact-based characterization techniques, the study proposed reducing roughness heights to enhance sliding friction and reducing wear where mating surfaces are involved.

Texturing has always been a favorite topic of research, especially to those working in the fields of tribology, surface coatings, and surface modification. Laser surface texturing is widely accepted, and a popular fabrication technique used by these researchers [9,10]. Many of these studied were targeted to understand the effect of surface topography on reducing friction [11,12]. The authors here, in a previous publication [13], discussed their observations on the shape of the frictional force profiles using different surface topographies machined on a stainless steel counterface. There, it was noted that the shape of the frictional force profiles changed with changes in the surface topography of the stainless steel countersurface. It was also observed that the shape of the plots changed with the change in the speed of the reciprocating table and with the viscosity of the lubricant used. It was thus pertinent to further investigate these different topographies to find the reason for the change in the shape of the frictional force profiles. The techniques of SEM, SEM-EDS, BSE- SEM, and three-dimensional (3D) optical profilometry were used effectively to understand the mechanisms in greater detail.

In modern times, optical interference microscopy is used as a major noncontact characterization strategy towards developing new surface modification processes and for quality control. Scanning surfaces to obtain information on three-dimensional surface parameters [14] helps standardize and ensure the functionality of manufactured parts. There are more than 40 areal surface parameters defined to characterize a surface [15]. Areal surface parameters represent more information and data points about a surface than the conventionally used 2D surface parameters, and hence are more statistically significant. Three-dimensional profilometry is hence being widely used by researchers to understand the surface topography and predict sliding friction based on the influence of some key surface parameters. Comparison between the two commonly used dimensional parameters in the industry and the current 3D surface parameters indicates the wealth of information contained with them. Dong at al. [16,17] recognized that for control of surface roughness, the profiling techniques and two-dimensional (2D) parameters were inadequate. They defined an effective and widely accepted 3D parameter set for mass adoption. Functional indices were proposed to describe some fundamental functional properties of bearing and fluid retention from the characteristics of the surface bearing area ratio curve. Sedlacek [18] et al. used the surface roughness parameters S_{sk} and S_{ku} to predict the tribological behavior of contact surfaces; they found that higher values of kurtosis and lower skewness lead to lower friction. Turned surfaces which have positive skewness display higher sliding friction than surfaces with porosities which have a negative value of skewness. Surface with porosities act like a micro-reservoir and help retain the lubricant and wear particles; thus, negatively skewed surfaces display lower sliding friction. A still lower skewness meant that friction was still lower. This is true for surfaces which have almost the same S_a and S_z, but different values of S_{sk} and a kurtosis of almost 3.

Contact profilometers continue to still be used for their economy and ease of usage. However, where the design requires tighter control of surface morphology, for example like machining of fuel injectors, gears, and bearing races [19], the use of dedicated 3D optical profilometers in process control is necessary. In the present study, the topography measurements were carried out by 3D vertical interferometry equipment, whose reliability is higher than that of the conventional stylus measurements.

W.P. Dong et al. [20] investigated the 3D properties of a surface, mainly the amplitude and hybrid parameters. They postulated that the bearing area ratio curve can be used to calculate the material volume of the surface topography and the void volume enclosed in the interfacial area of the material. These parameters are closely interrelated to the functional properties of surfaces, such as the bearing, wear, running-in, and fluid retention. The surface bearing ratio, obtained from the area material ratio curve, is related to the ratio of the contacting area obtained by truncating the surface summits by a plane parallel to the mean plane at a given truncation level over the sampling area.

Grzesik et al. studied [21] the areal surface parameters of a machined surface to conclude that the increased wear resistance came from an increased valley void volume and a reduced material volume in the peaks. V_{mp} and V_{vv} are terms used for quantifying the volume in the peaks and valleys, respectively, where a lower V_{mp} indicated higher wear resistance, and vice versa, while a higher V_{vv} indicated a higher fluid retention tendency.

Pradeep L Menezes et al. [22,23] elaborated oil retention properties while observing that random smoother surfaces could retain more oil and could reduce the coefficient of friction during sliding as compared to surfaces prepared by unidirectional sliding of abrasives and an "8" pattern of abrading and surface preparation. They concluded that the parameter "Del a", which is an average of the slope of the surface profile, influences the ploughing component of friction and increases with the increase in "Del a". The transfer layer formation of the soft surface on the hard counter surface was found to the highest where the "Sda" or "Del a" parameter was the highest. The ploughing component was associated with the plastic deformation of the asperities, and hence was highest on surfaces with the highest Sda, whereas the adhesive component of friction dominated where oil lubrication was absent.

Studies by Prajapati and Tiwari [24] on pin-on-disk dry sliding tests found a correlation between the coefficient of friction (COF) and areal surface parameters for AISI 304 steel. They reported a negative correlation between COF and the areal surface parameters autocorrelation length Sal and skewness Sk, and a positive correlation between RMS roughness Sq and mean summit curvature Ssc.

Sedlacek et al. [25] used the information from the areal surface analysis to investigate the correlation between commonly used parameters like average surface roughness Sa, skewness Ssk, and kurtosis Ssk and the sliding friction during boundary lubricated sliding. They demonstrated that the standard roughness parameters Sa and Sq are not sufficient to determine the tribological properties of contact surfaces. They showed that surfaces with higher Sku and negative Ssk values tend to reduce friction even with different and higher average surface roughness values. They also found that under boundary lubrication, Ssk and Svk also showed a good correlation with the coefficient of friction and could be used for planning surfaces towards achieving a desired tribological behavior.

Reddy et al. [26] investigated the effect of areal surface parameters according to ISO 25178 [14] on the tribological behavior during lubricated sliding at varied weights and sliding speeds. They found that V_{vv}, S_{vk}, S_{pk}, S_k, S_{ku}, S_k, and S_q showed a strong correlation with friction and wear depth. They concluded that V_{vv} and S_{vk} retained the lubricant and increased the life of the component. Pettersson et al. [27,28] found that by introducing controlled peaks and valleys in a relatively flat surface, the frictional performance can be improved. Lubricant can remain trapped inside these mini reservoirs, lowering friction and increasing the durability of the system. Ibatan et al. [29] stated that in the hydrodynamic and mixed lubrication regime, the microcavity in the surface acts as a reservoir for fluid lubricant, while in the boundary lubrication regime, it traps wear particles to reduce the third body abrasion.

The authors had previously discussed [13] the shape of the frictional force curves for the different ground surfaces on the steel countersurface. The P1200-finished surface yielded extremely low sliding friction and generated almost a flat frictional force curve when compared with P320- and mirror-finished surfaces. Here, the steel flat was in reciprocated sliding against an aluminum flat pin, in the presence of a lubricant. The authors wanted to explain why the shape of these force curves were so different from each other. In one of their experiments here, the authors could generate different P320-finished surfaces by varying the quantity of grinding fluid. These surfaces had almost similar root mean square surface roughness but showed widely different frictional force responses. The aim of this study was to understand how the areal surface parameters in each of the topographies could relate to the obtained coefficient of COF. In this paper, the authors intend to discuss the 3D surface parameters using vertical scanning interferometry and predict the magnitude of sliding friction that was obtained. The reasons for lowering the sliding friction for the P1200 surface were already discussed in the previous paper; the authors further substantiate it with the aid of scanning electron microscopy and energy dispersive spectrometry.

2. Materials and Methods

Flat specimens of grade 304 stainless steel with Rockwell hardness B98 of 30 mm × 30 mm were laser cut from a single stock having a thickness of 4 mm; the faces were ground and lapped to a mirror finish using a table grinder, as shown in Figure 1. This was achieved by grinding the surface with a series of coated abrasives on the table grinder, starting from a coarse P320 followed by P400, P600, P800, P1000, P1200, P1500, and finally P2000 grit sizes.

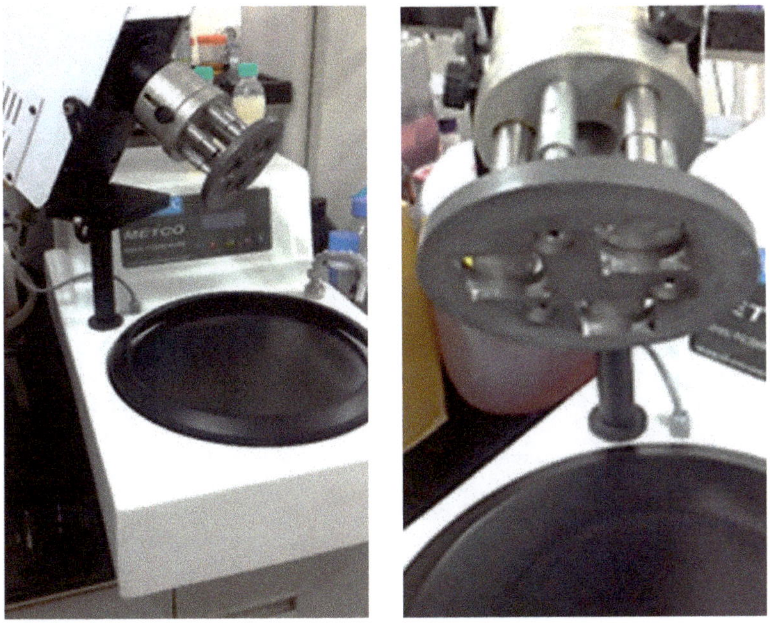

Figure 1. The table grinder with the custom-made square specimen holder [13].

After each step, the specimens were cleaned in an ultrasonic bath with ethanol followed by hexane. The surfaces were lapped with a ceramic polishing slurry to achieve the mirror finish. To prepare a specimen with the ground finish, sandpaper of the select grit was used to grind the surface through a controlled and repeatable process; these specimens are referred to by the grit size used for preparation, viz., P1200 and P320. Variations in areal surface parameters were also achieved by varying the quantity of grinding fluid used for wetting the surfaces; hence, this quantity was controlled to generate repeatable surface topographies in this study.

The stainless steel flat and aluminum pin specimens were then tested for friction on a pin-on-flat reciprocating tribometer driven by a Scotch yoke mechanism, as shown in Figure 2. The Scotch yoke mechanism was responsible for the reciprocation, and hence the velocity profile was sinusoidal. Aluminum 6061 T6 pins of hardness level HRB 61 with a diameter of 3 mm were used in the test. These pins were run-in suitably while the run-in debris was intermittently cleaned from the contact zone.

Once the run-in was completed, the pins were tested in the tribometer set up to a vertical Z load of 50 N and 20 mm stroke at preset frequencies of reciprocation. The areal surface parameters were measured using vertical scanning interferometry. A Gaussian filter was applied with a S-filter cutoff of 8 μm and L-filter cutoff of 0.08 mm, and the objective magnification used was 10X. The total sliding distance was 72 m, while the frequency of sliding was maintained at 1 Hz for 1800 reciprocating cycles.

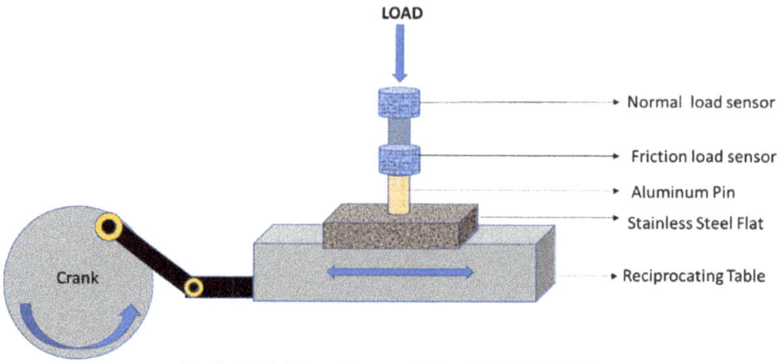

Figure 2. Schematic of the tribometer and the loading arrangement [13].

Group 1 base oil of API Group 1 SN500, which had a kinematic viscosity of 100 cSt at 40 °C, was used in this study. The secondary emission mode and back-scattering mode of a scanning electron microscope were used to scan the surface composition and image the surface topography; an energy dispersive spectrometer quantified the composition of the surface and tribolayers.

During boundary lubrication, as defined by the Stribeck curve there is an inverse relationship between the sliding velocity and the frictional force. The frictional force reaches a maximum at velocities approaching zero. At maximum velocities, the friction force is the lowest. Figure 3 below shows the velocity profile and the instantaneous frictional force plotted against the crank angle. An alternating "M" and inverted "M" shape was seen, which conforms to the classical Stribeck behavior. This plot was used as a reference to compare the actual frictional force plots observed in the sliding experiments that followed.

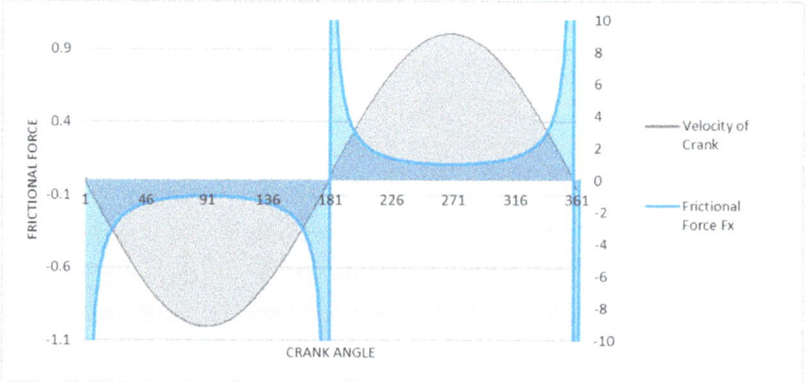

Figure 3. The velocity profile of the reciprocating table and the theoretical frictional force profile [13].

3. Background

The authors in their previous work [11] observed that the sliding friction was extremely low for a P1200finished surface as compared to other similarly prepared surfaces such as mirror, P2000, P800, and P320-finished surfaces. In their work, they varied different components of the tribo-system, like oil viscosity, speed of reciprocation, and the contact pressures; they found that in all the combinations, the P1200-finished surfaces exhibited extremely low friction, while the mirror and P320 exhibited higher friction. This phenomenon was widely pronounced at lower sliding velocities and where viscosity of the oil was relatively higher. The authors decided to investigate further using these three surfaces in this study. Figure 4 shows the shape of the frictional profiles on mirror-, P1200-, and P320-finished surfaces.

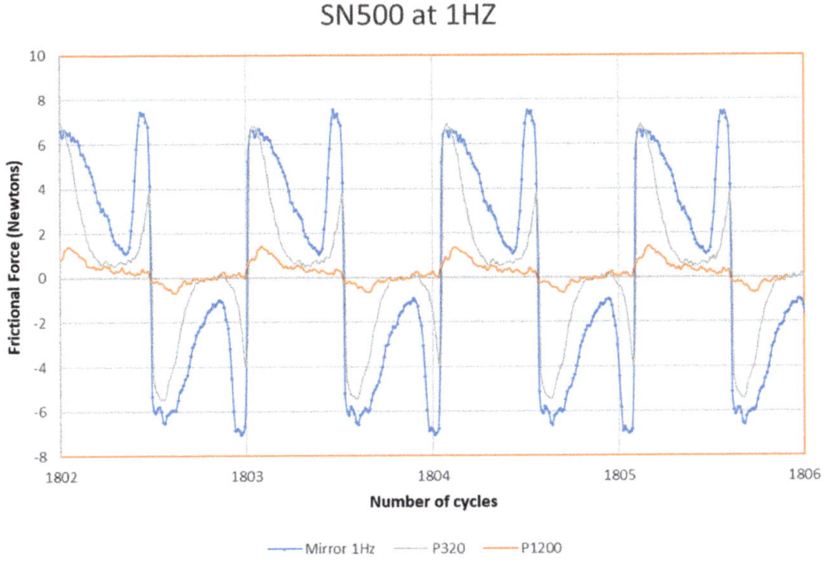

Figure 4. Plot of frictional force vs. number of reciprocating cycles for different finished surface.

The pin-on-flat tribo-system accelerates from zero velocity in the boundary lubrication zone to a maximum velocity at the middle of the stroke corresponding to the mixed lubrication zone. It later returns to the boundary lubrication zone as the system decelerates to zero velocity, before the direction reversal occurs and this cycle repeats until the end of the test. Neglecting any surface effects and using the Stribeck phenomena governing the relationship between the sliding friction with other tribo-system parameters, the authors could theoretically predict the shape (Figure 3) of the frictional force curve for a complete reciprocating cycle.

In these sliding experiments, the friction force curves did not obey the classical Stribeck behavior when the topography was changed from a mirror-finished surface to a P1200-finished surface. The force profiles had almost disappeared to indicate minimal friction, as seen in Figure 4. For a P320-finished surface, the acceleration peaks were very prominent while the deceleration peaks were smaller.

It was believed that P1200 surfaces in combination with SN500 virgin base oil could entrap enough oil to ensure a continuous oil film was formed and retained even when the surface contact zone was decelerating to zero velocities. Sudeep et al. [30] stated that the wettability of textured surfaces increased as compared with smooth and lapped surfaces. While accelerating from zero velocities, the friction peaks were minimal, which would have resulted from the minimal surface peak heights resulting from the topography generated by the P1200 abrasive finishing process.

In this study, the contact surfaces of the stainless steel flat and the pin were observed under a 3D optical profilometer to understand what changes were occurring on the surface topography as the system reached steady state, while each test was run for 1800 reciprocating cycles with a stroke length of 20 mm. Surfaces were characterized using three-dimensional surface parameters and correlated with friction results obtained from sliding.

Further examination of the P1200- and P320-finished surfaces was conducted by comparing the wear surfaces on secondary emission (SE) imaging and back-scattered emission (BSE) imaging on a scanning electron microscope. The wear scars were also tested under energy dispersive spectrometry (EDS) to identify and characterize the tribo-layer formed on the contact surfaces.

4. Results

Using back-scattered imaging on a scanning electron microscope, the surface was scanned after completing 1800 reciprocating cycles. The images taken for the mirror-, P1200-, and P320-finished surfaces are shown in Figure 5. The frictional forces as recorded on the tribometer for these corresponding surfaces are plotted in Figure 4.

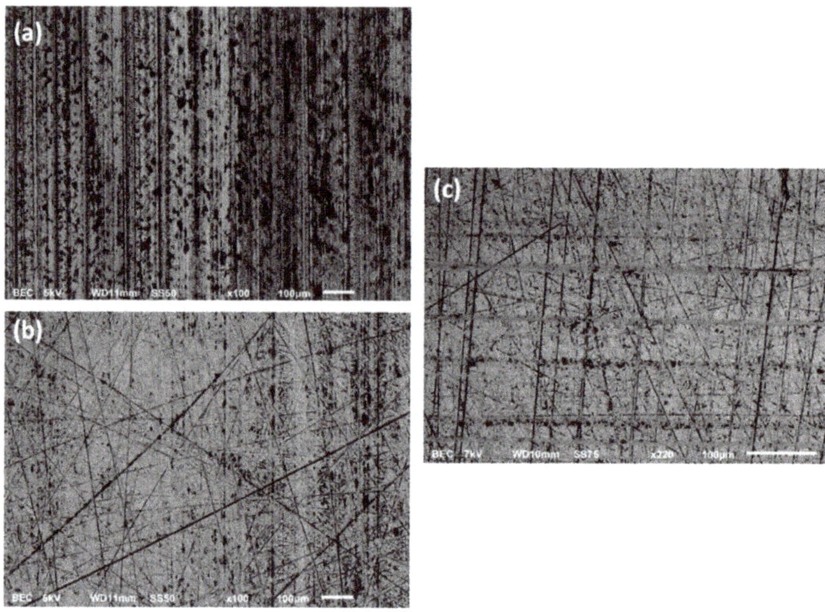

Figure 5. BSE SEM micrographs after sliding for (**a**) mirror-finished surface, (**b**) P1200-finished surface, and (**c**) P320-finished surface.

The material removal is visible from the deep and parallel wear lines on the surface; this is positioned along the direction of the sliding of the pin. The dark spots seen are the aluminum wear particles from the pin transferred onto the steel counterface; see Figure 6.

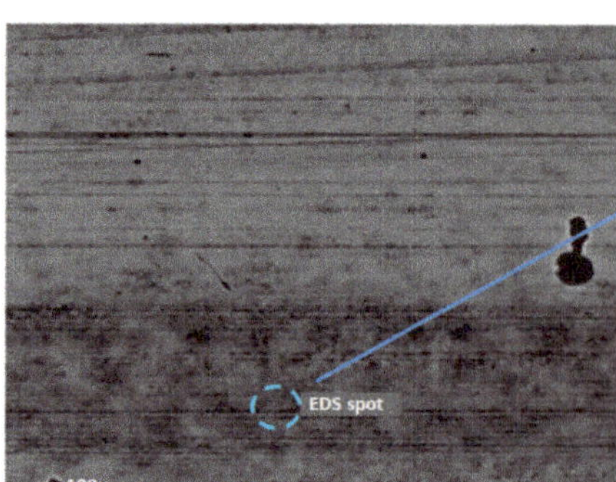

eZAF Smart Quant Results

Element	Weight %	Atomic %	Net Int.
AlK	15.43	27.44	14012.66
CrK	13.44	12.4	17061.58
MnK	7.12	6.22	7145.21
FeK	47.35	40.67	40712.25
NiK	11.18	9.14	7153.98
CuK	5.47	4.13	2990.32

Figure 6. BSE micrograph and EDS scan on black wear marks after sliding on P1200-finished stainless steel flat. Aluminum is present in the scan, indicating transfer of aluminum from the pin.

We can observe from Figure 5a that the mirror-finished surface after the end of test had deep wear marks resembling delamination and adhesive wear; dark lines are visible in the BSE image, indicating the wear of the aluminum pin and subsequent material transfer on the steel surface. From Figure 5a,b, it is observed that the wear was comparatively lower for P1200- and P320-finished surfaces. The black wear tracks did appear; however, they were significantly less with much lower intensities.

The wear marks and dark spots indicating material transfer are far less as compared to Figure 5a, indicating less wear of the aluminum pin. The EDS scan in Figure 6 indicates the presence of aluminum.

The BSE images, however, could not explain why P1200 surfaces had extremely low friction as compared to P320 surfaces, and why for P320-finished surfaces the Fx plots had significantly milder deceleration peaks as compared to the acceleration peaks seen in Figure 4.

The surfaces of the stainless steel flat were scanned before the sliding experiment using a 3D optical profilometer. The three different machined surfaces, viz., mirror-, P1200-, and P320-finished surfaces, were tabulated using the 3D surface parameters, which quantified the volume of peaks, valley roughness, and core roughness volumes; this is shown in Table 1.

Table 1. Shows the different areal surface parameters analyzed on the optical surface profiler for the different surfaces.

Surface	S_{Pk}	S_k	S_{Vk}	S_{pk}/S_k	S_{Vk}/S_k	S_{bi}	S_{ci}	S_{vi}	S_q	COF
Mirror	0.05	0.07	0.1	0.71	1.43	0.031	1.186	0.094	0.005	0.12
Mirror	0.96	1.56	1.56	0.62	1.00	0.66	1.411	0.176	0.9	0.12
Mirror	0.8	1.51	1.34	0.53	0.89	0.491	1.349	0.166	0.79	0.12
Mirror	0.62	1.77	2.08	0.35	1.18	0.841	1.082	0.199	1.04	0.12
Mirror	1.27	1.66	2.49	0.77	1.50	0.483	1.221	0.211	1.17	0.12
Mirror	1.83	2.02	1.59	0.91	0.79	0.569	1.691	0.132	1.23	0.12
P1200	0.12	0.09	0.18	1.33	2.00	0.076	0.627	0.166	0.08	0.02
P1200	0.18	0.1	0.24	1.80	2.40	0.107	0.67	0.19	0.1	0.02
P1200	0.2	0.11	0.28	1.82	2.55	0.163	0.701	0.196	0.12	0.02
P1200	0.14	0.1	0.22	1.40	2.20	0.135	0.729	0.174	0.09	0.02
P1200	0.16	0.1	0.25	1.60	2.50	0.156	0.616	0.189	0.1	0.02
P1200	0.19	0.1	0.29	1.90	2.90	0.117	0.639	0.206	0.11	0.02
P320	0.16	0.27	0.34	0.59	1.26	0.111	0.93	0.2	0.17	0.06
P320	0.21	0.22	0.39	0.95	1.77	0.215	0.873	0.231	0.17	0.06
P320	0.16	0.21	0.35	0.76	1.67	0.205	0.854	0.222	0.16	0.06
P320	0.14	0.23	0.32	0.61	1.39	0.173	0.891	0.2	0.16	0.06
P320	0.24	0.3	0.47	0.80	1.57	0.078	0.993	0.21	0.22	0.06
P320	0.26	0.35	0.52	0.74	1.49	0.123	0.989	0.207	0.26	0.06
P320-smooth	0.25	0.25	0.37	1.00	1.48	0.185	1.03	0.207	0.18	0.02
P320-smooth	0.26	0.23	0.41	1.13	1.78	0.211	1.059	0.224	0.19	0.02
P320-smooth	0.2	0.17	0.4	1.18	2.35	0.159	0.819	0.239	0.16	0.02
P320-smooth	0.22	0.18	0.43	1.22	2.39	0.121	0.745	0.241	0.17	0.02
P320-smooth	0.22	0.17	0.38	1.29	2.24	0.263	0.67	0.223	0.17	0.02
P320-smooth	0.22	0.14	0.33	1.57	2.36	0.118	0.775	0.24	0.13	0.02

In the analysis shown in Table 1, the surface parameter S_{pk}/S_k happens to be less than 1 for the mirror-finished surface, where frictional curves followed the Stribeck curve relationship. For a P1200-finished surface, the surface parameter ratio S_{pk}/S_k was noted to be greater than 1. It was found that the frictional force was exceptionally low, and the frictional force plot was subdued in the friction scale domain used for these tests. This was explained using the shape of the frictional force profiles in Figure 4.

By increasing the quantity of grinding fluid, the 'P320-smooth'surface was prepared using P320 abrasive. This had a similar magnitude of root mean square (RMS) roughness as the P320 finished surface Zhou et al. [31] showed that by varying the grinding fluid parameters, the grinding forces changed, which further influenced the surface parameters and surface residual forces. The tabulation in Table 1 depicts the three surfaces using 3D surface parameters quantifying the volume of peak, valley roughness, and core roughness.

As seen in Table 1, the P1200-finished surfaces and "P320-smooth" surfaces had a surface parameter ratio S_{pk}/S_k greater than 1. For the mirror-finished surface and P320-finished surface, the surface parameter ratio S_{pk}/S_k was less than 1. The "P320-smooth" surfaces had very similar magnitudes and shapes of the frictional force plots as observed in the P1200-finished surfaces. Exceptionally low frictional forces were noted, and the frictional force plot was almost absent in the friction scale domain used for these tests. P320-finished surfaces showed higher sliding friction and the frictional curves followed the principles of the Stribeck curve.

Figure 7 shows that the surface of the pin is highly non-uniform, indicating severe sliding wear during running-in. This is also expected as the running-in procedure does not employ a viscous lubricant, as the objective is to ensure maximum surface-to-surface contact at the start of the test.

S_{pk}/S_k	0.774
S_{pk}	0.687μ
S_k	0.888μ

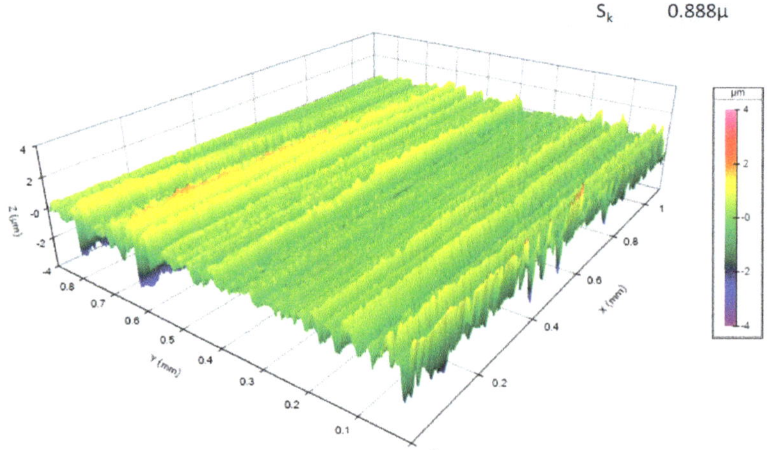

Figure 7. Three-dimensional profile of the pin after "running-in" under 10X objective. The topography looked rough with deep ploughing marks and the surface parameter ratio S_{pk}/S_k of this surface was less than 1.

Figures 7 and 8 show the condition of the pin surface before and after the sliding test on a P1200-finished stainless steel surface. The measured surface parameter ratio on the pin surface, S_{pk}/S_k was greater than 1 at the end of the test. A notable transformation in the surface topography was observed on the pin after the sliding test. The change in topography is evident, as the deep sliding wear marks present at the start of the test from the running-in have completely disappeared, to give a smooth topography at the end of the test.

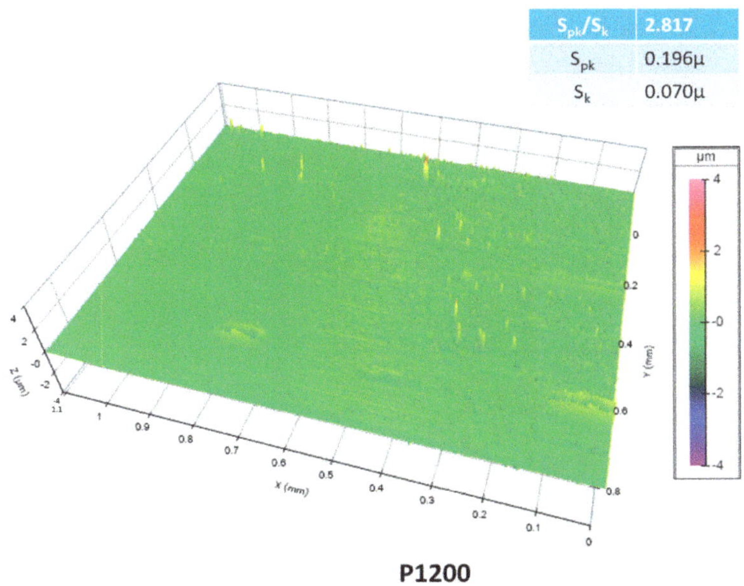

Figure 8. Three-dimensional pin profile after 1800 cycles on P1200-finished surface. The topography looked smooth and the surface parameter ratio S_{pk}/S_k of this surface was greater than 1.

Figure 9a,b show the back-scattered images of the pin surface at different magnifications, captured after the reciprocating sliding experiment on a P1200-finished steel surface.

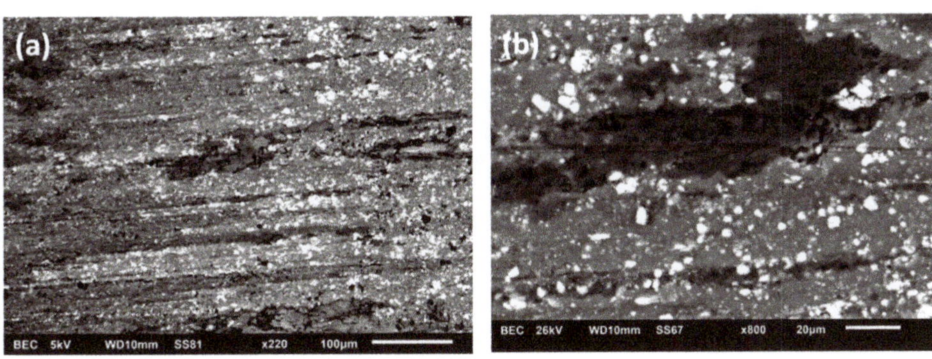

Figure 9. (**a**). Back-scattered emission image of the pin after P1200 sliding test (at 220×). Observe the even layer of bright spots, identified later from EDS as an iron oxide layer. (**b**). Back-scattered emission image of the pin after P1200 sliding test (at 800×). The bright transfer layer is even more evident in this picture at higher magnification.

A fine dispersion of particles embedded on the pin surface was of the order of a micron in diameter. They are seen as evenly dispersed bright spots in the back-scattered image and were formed during the sliding experiment.

The EDS scan as shown in Figure 10 confirms the presence of an iron oxide layer in the bright spots seen on the surface. The scan in Figure 11 looks at the black spots found on the pin surface, as well as a point in the bulk matrix on the pin surface. The black spots correspond to carbon and oxygen either from the lubricant or from oxidation, while the bulk corresponds to the surface aluminum oxide layer.

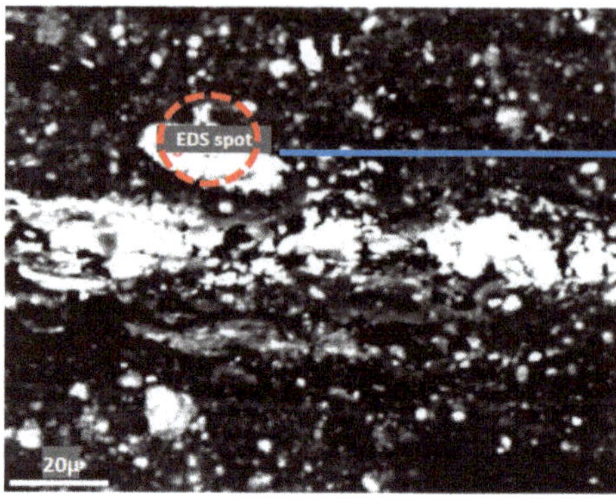

BRIGHT SPOT

eZAF Smart Quant Results

Element	Weight %	Atomic %	Net Int.
C K	20.38	40.46	2623.93
O K	25.11	37.42	3153.16
FeL	29.51	12.6	1226.74
AlK	4.81	4.25	304.01
ZrL	20.18	5.27	357.56

Figure 10. EDS scan of back-scattered emission image of the pin after P1200 sliding test.

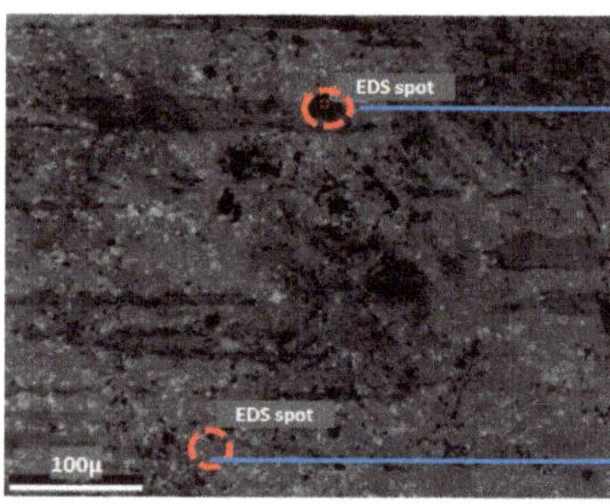

DARK SPOT

eZAF Smart Quant Results

Element	Weight %	Atomic %	Net Int.
C K	60.24	65.45	1443.58
N K	18.23	16.98	175.08
O K	21.54	17.57	289.54

MATRIX

eZAF Smart Quant Results

Element	Weight %	Atomic %	Net Int.
C K	5.13	8.28	85.55
O K	47.55	57.68	996.75
AlK	47.33	34.04	444.46

Figure 11. EDS scan of back-scattered emission image of the pin in after P1200 sliding test.

It was seen that this tribo-layer formed on the pin surface was unique to the P1200 topography. An energy dispersive spectrometry of these bright spots confirmed iron oxide to be present in significant proportions.

Three-dimensional surface image of pins from the sliding experiment on P320-finished surfaces, including the "P320-smooth" surfaces is shown in Figure 12. The pin surface was scanned on the 3D optical profilometer before and after completing the sliding on the different topographies discussed above.

The pin surface appears smooth after the test; this is a drastic change from the start of the test, where the pin running-in surface was replete with heavy sliding wear marks.

Figure 13a shows a similar surface micrograph as already seen in Figure 11; a huge transformation in the surface topography of the pin is seen after the sliding test. The deep sliding wear marks present at the start of the test have completely disappeared to give a smooth topography.

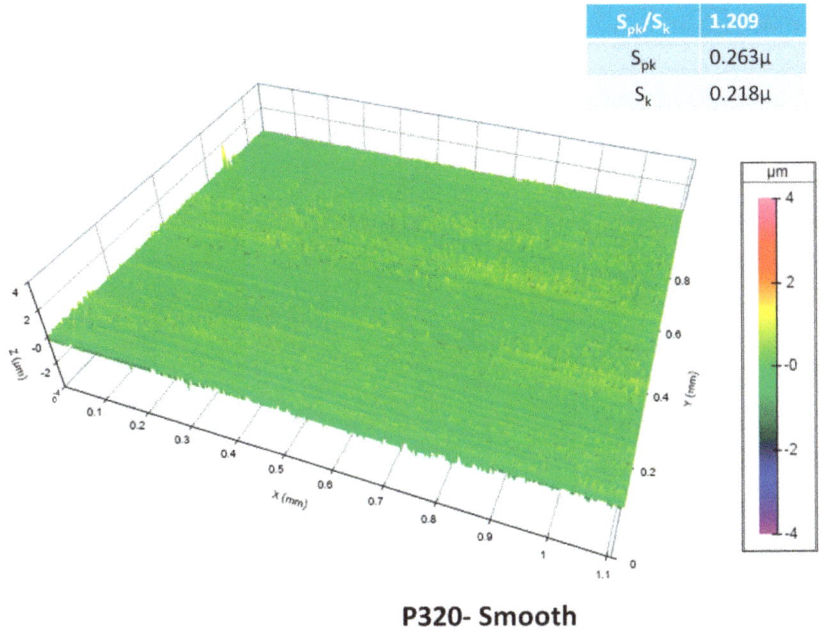

Figure 12. Three-dimensional profile of the pin after 1800-cycle sliding test on "P320-smooth" flat. The topography looked smooth and the surface parameter ratio S_{pk}/S_k of this surface was greater than 1.

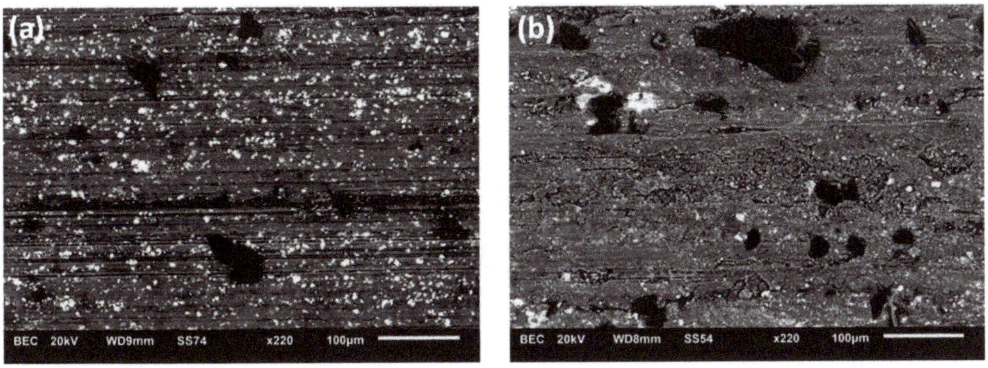

Figure 13. (**a**). Back-scattered emission image of the pin after 1800-cycle sliding test on "P320-smooth" flat (at 220×). (**b**). Back-scattered emission image of the ping after sliding on P320, the rougher surface; an absence of the iron oxide transfer layer is noted.

As observed with P1200, the pin surface shows a bright uniformly spread transfer layer.

With P320-finished surfaces, it was observed that the aluminum pin surface was worn during the sliding. The 3D micrograph of the pin after sliding on a P320-finished surface is shown in Figure 14. There was no improvement in the surface topography of the pin surface, as was noted with the P320-smooth surface.

From the back-scattered image shown in Figure 13b, there were no sharp and bright contrasts as observed earlier with P1200 and P320-smooth specimens. The micrograph had a uniform dark matrix which resembled that of the aluminum surface of the pin.

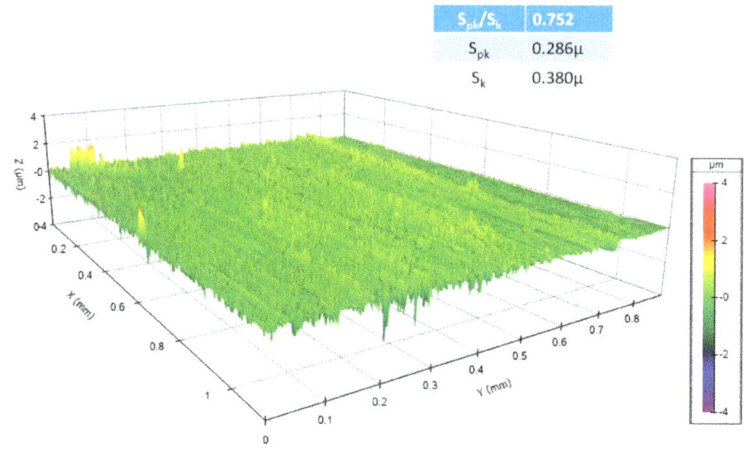

P320- Rough

Figure 14. Three-dimensional profile of the pin after sliding on P320 surface. Unlike "P320-smooth" or "P1200" pin samples, the pin topography here shows deep sliding wear marks and evidence of heavy wear. The topography looked rough with deep ploughing marks and the surface parameter ratio S_{pk}/S_k of this surface was less than 1.

Regression:

A polynomial regression analysis was performed on the COF response using the S_{pk}/S_k value as a continuous variable for the data set listed in Table 1. The plot of the variable and response parameter and the regression line is shown in Figure 15. The authors performed transformations on the parameters in the x and y axis to obtain a meaningful analysis. The x axis was transformed to the square of the variable, and for the y axis the logarithm of the response was taken; this operation reduced the randomness of the residuals and ensured a normal distribution.

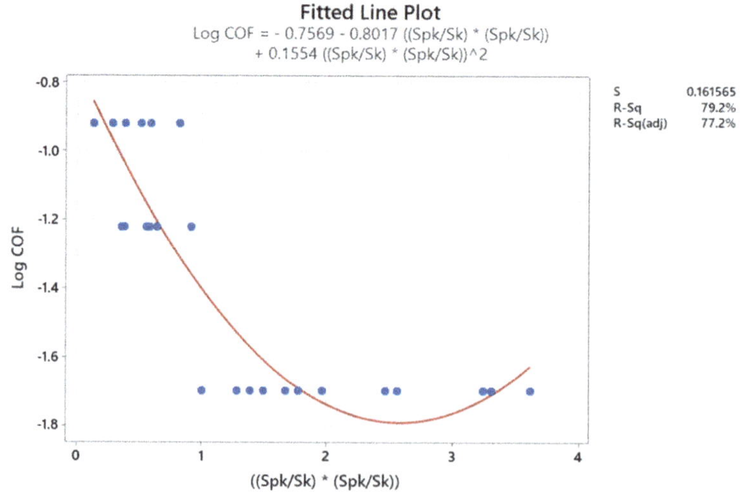

Figure 15. Plot of $(Spk/Sk)^2$ v/s Log (COF) and the regression line. The blue dots represent the transformed responses, and the red plot is the regression line.

A coefficient of determination R^2 of 79% was obtained. This is shown in Table 2. Minitab® 20.2 software was used for this analysis. In these results, the function of S_{pk}/S_k of the topography explains approximately 79% of the variation in the average COF obtained in the sliding tests. The R^2 value indicates that the model fits the data well. The regression coefficients in the model are statistically significant as the p value is <0.05.

Table 2. Polynomial regression analysis of variables having "Spk/Sk" against COF response.

	Sequential Analysis of Variance			
Source	**DF**	**SS**	**F**	***p***
Linear	1	1.56984	32.41	0
Quadratic	1	0.51742	19.82	0
	Regression Equation			
$\text{Log (COF)} = -0.7569 - 0.8017\,((\text{Spk/Sk}) * (\text{Spk/Sk})) + 0.1554\,((\text{Spk/Sk}) * (\text{Spk/Sk}))^2$				
	Model Summary			
S		**R-sq**		**R-sq(adj)**
0.161565		79.20%		77.22%

The regression results indicate that the ratio parameter S_{pk}/S_k is significant. In Figure 16, the residuals from the regression equation are plotted in a normal probability distribution, and the scatter shows them to be normally distributed. The order of observations shows randomness with respect to the residuals, and hence the model is a good fit.

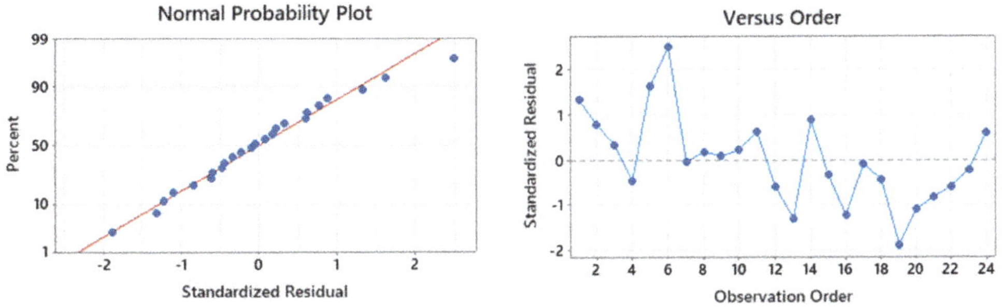

Figure 16. Plot of the residuals from the regression and residuals based on the order of observations. The scatter points are in blue dots and the best fit straight line is in red. Residuals follow the normal probability distribution well and are randomly distributed among the multiple observations.

5. Discussion

In Figure 5, the authors observed the wear tracks after completion of 1800 cycles under the back-scattered emission field of a scanning electron microscope. What emerged from the wear track images was a correlation between the frictional force plots and intensity of the wear tracks. The wear tracks having characteristic dark wear lines suggested that the aluminum from the pin had transferred onto the stainless steel surface; the frictional force plots showed the characteristic alternating "M" and "W" shapes adhering to the Stribeck behavior seen in Figure 3. On the contrary, where the wear track was light or almost invisible, the frictional force plots did not follow the shape of the idealized Stribeck curve. For the P1200-finished surface, the frictional force plots looked nearly flat when compared with other specimens. These observations suggest that a critical film thickness of the lubricant at the interface exists, above which the surface-to-surface contact tends to reduce. When the sliding velocities increase, the oil film thickness increases at the contact zone and the intensity of the wear tracks tones down and the coefficient of friction drops.

From the SEI image of the steel counterface in Figure 17, considerable wear on the steel counterface was observed, which was originally mirror-finished. Because aluminum is a softer metal compared to stainless steel, the wear of aluminum will be more, and hence the contrast on the EDS scan of a black transfer layer identified as aluminum was seen in Figure 6 on the steel counterface.

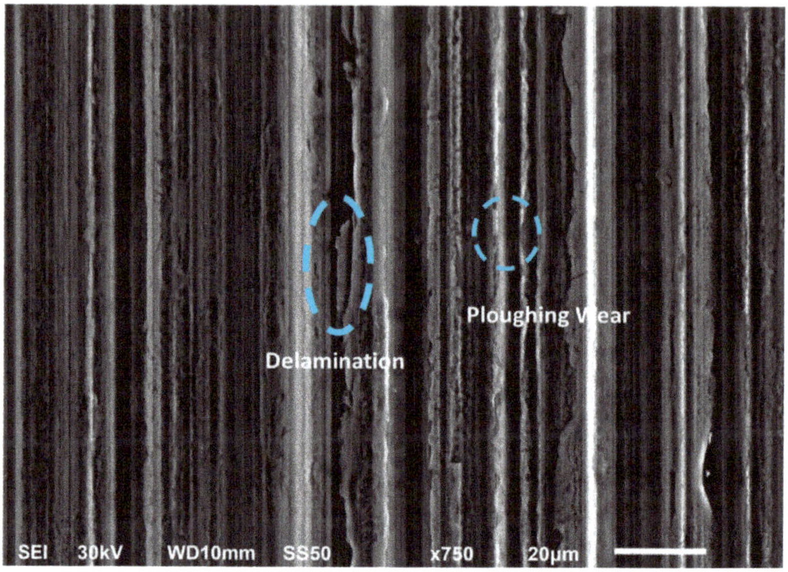

Figure 17. Secondary emission image on the mirror-finished surface after the completion of sliding. Observe the delamination and ploughing wear as indicated in the image.

Westlund et al. [32] worked on a series of microscale sliding experiments involving a polished stainless steel flat and aluminum 6082 pin. The stainless steel flat was polished using diamond particles to a mirror finish. They found that the relationship between friction and transfer of materials between the sliding surfaces strongly depends on the initial roughness of the harder countersurface, material composition, presence of oxygen, and lubricants. They observed that on the hard steel flat, the smooth surface is compromised by transferred aluminum from the pin or due to third-body abrasion effects. They also concluded that using lubricants with tribofilm-forming additives reduces material transfer from the aluminum pin, which makes the sliding surface rough. They found that the aluminum pin was heavily loaded with oxidized layer including steel fragments. Similar observations of the transfer from hard steel to the softer aluminum were reported by Menezes et al. [33] and Bowden and Tabor [34], where they investigated the transfer of steel fragments to copper and the immediate subsurface plastic deformation of the steel surface.

The authors observed high friction in experiments with mirror-finished stainless steel, which is associated with the severe deformation seen in Figure 17; this caused wear of stainless steel, including observations of ploughing wear and delamination wear. The surface oxide layer eventually wears off in high friction, forming a thick hard layer which could wear off the steel. Westlund et al. [32] found that the soft pin surfaces are modified with these steel particles; these are capable of being a part of a strongly modified tip that can scratch the hard steel surface, which was precisely seen in these experiments.

Krupka et al. [35,36] mentioned in their work that surface texturing, for a mixed film regime, provided an increase in the average film thickness and the local film thickness across the topographic features forming the texture. In such a tribo-system, when all other variables like oil viscosity, reciprocating speed, and contact pressure were kept constant, the topography modification at the contact zone and a tribo-transfer layer formation on the interacting surfaces seemed responsible for the tribo-system reaching a steady state, and in some favorable instances this mechanism brought the frictional forces to a minimum. For P320-finished surfaces, the total of the adhesive and abrasive component of friction [37], representing the net frictional force, was lower during deceleration than acceleration. This indicated that the steel topography was aiding in fluid retention and viscous flow of the lubricant, and consequently a leading trail of the lubricant film was maintained ahead of the pin which, while decelerating, reduced the sliding friction. This film was, however, absent when direction reversals occurred, explaining the higher friction seen while the pin is accelerating.

The wear track micrographs helped us conclude that mirror-finished surfaces had the maximum share of adhesion forces, and hence the highest adhesive delamination wear, irrespective of the sliding velocity. Here, the wear patterns observed in the SEM showed material delamination, with adhesive wear being the most prominent wear mechanism. As we move to a coarse surface like P320, the share of abrasive forces on the net friction force was higher. The abrasive forces reduced with the number of cycles. This was also evidenced by the stratified presence of aluminum over the surface of the stainless steel flat, as observed in the back-scattered image of the wear surface in Figure 5c.

As defined in ISO 25,178 Part II [38], the surface parameter ratio S_{pk}/S_k is defined as the ratio of the reduced peak height to the core roughness depth. The magnitude of this ratio was found to be synonymous with the sliding friction observed in the experiments. The authors introduced this areal surface parameter ratio after scanning and compiling the observations and findings for the different stainless steel specimens prepared. The regression analysis could relate the frictional performance with the surface topography and demonstrate the relevance. As seen in Table 1, this ratio could predict the observations of near-zero frictional force plots for P1200-finished and "P320-smooth" surfaces.

Luiz Albero Franco et al. [39], in their study on 3D surface parameters, tried to find the actual meaning of S_{pk} and its relation to the volume of the hill or peaks, V_{mp}. They stated that the parameter S_{pk}, as defined in the ISO 25,178-2 standard [14] as the average height of the peaks protruding over the core surface, is inadequate in its definition. Rather, S_{pk} is a height parameter that is closely related to "V_{mp}", the volume occupied by the peaks on a surface. They derived that, for random surfaces, the average height of the hills that extends above the core region is one half of the reduced peak height "S_{pk}".

The aluminum pin surface was found to evolve as the test progressed and was influenced by the topography of the steel countersurface. Where friction was extremely low, the pin surface was modified and became more even and smooth; this is shown in Figures 7, 8 and 12, which had a S_{pk}/S_k ratio greater than 1. The pin, when observed under a BSE scan, as seen in Figure 9a,b and Figure 13a, had a bright layer evenly spread on its surface. This layer was identified as iron oxide transferred from the stainless steel flat surface during sliding. This layer was found to form preferentially with the P1200-finished and "P320-smooth" surfaces only, and was absent on the mirror-, P320-, and "P320-rough"-finished surfaces. This indicated that this iron oxide tribo-layer had an impact on the shape of the frictional plots and was responsible for the extremely low values of friction observed. For surfaces where friction was higher, the pin surfaces remained rough, as seen in Figure 14, and the S_{pk}/S_k ratio was less than 1.

For surfaces where S_{pk}/S_k is greater than 1, the volume of the material occupied in the peaks is much more than for surfaces where S_{pk}/S_k is less than 1. When the ratio of S_{pk}/S_k is related to the texture amplitude normalized by the overall roughness magnitude, it gives an insight into the texture symmetry, which plays a distinct role in maximizing the S_{pk}/S_k ratio. Das et al. [40] stated that when S_{pk}/S_k is high, there is a greater amount of texture amplitude symmetry, indicating the presence of a greater volume of material occupied in symmetrical peaks than for other surfaces where S_{pk}/S_k is low. The S_{pk}/S_k ratio highlights the importance of peak structures in a topography. The volume occupied by the peaks in relation to the core roughness and the symmetry in the texture amplitude increased with increase in this surface parameter ratio. A S_{pk}/S_k ratio greater than 1 represents a larger volume of symmetrically sized asperities supporting the load at the contact zone, enabling a uniform transfer layer of iron oxide to develop on the aluminum countersurface. These topographies, apart from reducing surface wear caused by surface adhesion, also allow the valley structures to hold the oil efficiently. Erdemir [41] inferred that the lubrication regimes in classical Stribeck diagrams could be modified by surface modifications. Controlling specific surface parameters resulted in modifying both the width and the height of the boundary lubrication regimes, and thus achieving lower friction and wear at sliding and rotating contact interfaces.

In industry, precision grinding techniques are being used as a method of surface finishing, which is often the penultimate step in the material removal cycle of a machining operation. Scenarios exist for such parts where the end application might have critical performance requirements involving sliding in wet or lubricated conditions. A good example is of the diesel injectors used in modern-day diesel engines, where lubricated sliding minimizing friction is a key performance requirement [42,43]. The authors believe that the approach highlighted in this study of evaluating the surfaces from critical areal surface parameters can prove to be an effective method of quality control for limiting the variabilities during the metal finishing operation. This can act as a boon to the energy balance by conserving useful energy and reducing friction.

6. Conclusions

Certain topographies achieved extremely low sliding friction in the lubricated regime. The critical three-dimensional areal surface parameter S_{pk}/S_k with respect to the stainless steel surface could fairly predict whether the tribo-system can achieve a minimum sliding friction. The stainless steel flat ground using a P1200 coated abrasive showed the lowest average coefficient of friction, of the order of 0.02. This surface had an average root mean square roughness value of 0.15 microns and a S_{pk}/S_k value above 1.

It was found that stainless steel topographies where S_{pk}/S_k value was above 1 had a beneficial transfer layer of iron oxide gradually depositing on the softer counterface. This might lower the sliding friction to such low levels. Surfaces having an S_{pk}/S_k value below 1 experienced higher friction. The S_{pk}/S_k ratio highlights the importance of peak structures in a topography. These features on the surface helped reduce the sliding friction and consequent wear found on a stainless steel countersurface. The topography of the stainless steel surface had an influence on the evolution of the aluminum pin topography during sliding. For topographies having an S_{pk}/S_k value below 1, the transfer layer was not found on the pin surface.

The beneficial changes noted on the pin surface were responsible for the extraordinary reduction in the sliding friction for the tribo-system. The authors expect this study to be relevant to the metal finishing industry, especially where applications focus on lubricated sliding. Where friction reduction is a design criterion, controlling critical surface parameters can be important in improving manufacturing and system reliability, enhancing system response and reducing energy consumption.

Author Contributions: Conceptualization, G.P., V.J. and S.V.K.; methodology, G.P.; formal analysis, G.P., V.J. and S.V.K.; investigation, G.P.; resources, V.J. and S.V.K.; data curation, G.P.; writing—original draft preparation, G.P.; writing—review and editing, V.J. and S.V.K. All authors have read and agreed to the published version of the manuscript.

Funding: This research received no external funding.

Data Availability Statement: All relevant data are presented in the paper.

Conflicts of Interest: The authors declare no conflict of interest.

Abbreviations

F_x	Frictional force
COF	Coefficient of friction
P2000, P1200, P320	PSA liner paper backed coated abrasives as per FEPA (P- Scale) based on mean grain size of abrasive mineral ds-50 particle size.
Hz	Unit of reciprocating frequency expressed as Hertz
MPa	S.I. Unit of pressure, abbreviated as Mega Pascals
S_a	Average areal surface roughness
S_{da}	Average dale area
S_{dq}	Root mean square gradient
S_{dr}	Surface area ratio
S_{ds}	Density of summits
SE	Secondary emissions
SEM	Scanning electron microscope
SI	International system of units
SiC	Silicon carbide
S_k	Reduced core roughness
S_{ku}	Surface kurtosis
SN150	Mineral Base oil belonging to the Group 1 American Petroleum Institute classification, refined using Solvent Neutral processing methods having a target viscosity of 30 cSt at 40 deg C.
SN500	Mineral Base oil belonging to the Group 1 American Petroleum Institute classification, refined using Solvent Neutral processing methods having a target viscosity of 100 cSt at 40 deg C.
SN60	Mineral Base oil belonging to the Group 1 American Petroleum Institute classification, refined using Solvent Neutral processing methods having a viscosity of 12 cSt at 40 deg C.
S_p	Max peak height
S_{pk}	Reduced summit height
S_q	Root mean square roughness
SS	Stainless steel
S_{sc}	Mean summit curvature
S_{sk}	Surface skewness
S_{td}	Texture direction
S_{tdi}	Texture direction index
S_{ti}	Texture index
S_v	Max valley depth
S_{vi}	Valley fluid retention index
S_{vk}	Reduced valley height
3D	Three dimensional

References

1. Molnar, V. Tribological Properties and 3D Topographic Parameters of Hard Turned and Ground Surfaces. *Materials* **2022**, *15*, 2505. [CrossRef] [PubMed]
2. Ahmed, M.M.I.; Wei, L.; Abdel-Hamid, I.M.; Mohamed, A.E.O.; Fadl, A.E.; Ahmed, M.A.M.; Muhammad, F.E.; Ammar, E. Cooling and lubrication techniques in grinding: A state-of-the-art review, applications, and sustainability assessment. *Chin. J. Aeronaut.* **2023**, *36*, 76–113.
3. An, B.; Wang, X.; Xu, Y.; Jackson, R.L. Deterministic elastic-plastic modelling of rough surface contact including spectral interpolation and comparison to theoretical models. *Tribol. Int.* **2019**, *135*, 246–258. [CrossRef]
4. Öztürk, S.; Kahraman, M.F. Modeling and optimization of machining parameters during grinding of flat glass using response surface methodology and probabilistic uncertainty analysis based on Monte Carlo simulation. *Measurement* **2019**, *145*, 274–291. [CrossRef]
5. Karkalos, N.E.; Karmiris-Obratanski, P.; Kurpiel, S.; Zagorski, K.; Markopoulos, A.P. Investigation on the surface quality obtained during trochoidal milling of 6082 aluminum alloy. *Machines* **2021**, *9*, 75. [CrossRef]
6. El-Tayeb, N.S.M.; Low, K.O.; Brevern, P.V. On the surface and tribological characteristics of burnished cylindrical Al-6061. *Tribol. Int.* **2009**, *42*, 320–326. [CrossRef]

7. Whittaker, R.K.; Hothi, H.S.; Eskelinen, A.; Blunn, G.W.; Skinner, J.A.; Hart, A.J. Variation in Taper Surface Roughness for a Single Design Effects the Wear Rate in Total Hip Arthroplasty. *J. Orthop. Res.* **2017**, *35*, 1784–1792. [CrossRef] [PubMed]
8. Jeng, Y.R. Experimental Study of the Effects of Surface Roughness on Friction. *Tribol. Trans.* **1990**, *33*, 402–410. [CrossRef]
9. Kligerman-Ryk, G.; Kligerman, Y.; Etsion, I. Experimental investigation of laser surface texturing for reciprocating automotive components. *Tribol Trans.* **2002**, *45*, 444–449.
10. Kovalchenko, A.; Ajayi, O.; Erdemir, A.; Fenske, G.; Etsion, I. The effect of laser surface texturing on transitions in lubrication regimes during unidirectional sliding contact. *Tribol. Int.* **2005**, *38*, 219–225. [CrossRef]
11. Ronen, A.; Etsion, I.; Kligerman, Y. Friction-reducing surface texturing in reciprocating automotive components. *Tribol. Trans.* **2001**, *44*, 359–366. [CrossRef]
12. Marian, V.G.; Kilian, M.; Scholz, W. Theoretical experimental analysis of a partially textured thrust bearing with square dimples. *Proc. Inst. Mech. Eng. Part J J. Eng. Tribol.* **2007**, *221*, 771–778. [CrossRef]
13. Parameswaran, G.; Jayaram, V.; Kailas, S.V. Frictional Characteristic Curves of Ground Surfaces in Lubricated Sliding. *Lubricants* **2023**, *11*, 354. [CrossRef]
14. *ISO 25178-2*; 2012 Geometrical Product Specifications (GPS)—Surface Texture: Areal—Part 2: Terms, Definitions and Surface Texture Parameters. ISO: Geneva, Switzerland, 2012.
15. Stout, K.; Blunt, L. *Three-Dimensional Surface Topography*, 2nd ed.; Penton Press: London, UK, 2000; ISBN 9781857180268.
16. Dong, W.P.; Sullivan, P.J.; Stout, K.J. Comprehensive study of parameters for characterizing three-dimensional surface topography. Part III. Parameters for characterizing amplitude and some functional properties. *Wear* **1994**, *178*, 29–43. [CrossRef]
17. Dong, W.P.; Sullivan, P.J.; Stout, K.J. Stout, Comprehensive study of parameters for characterizing three-dimensional surface topography I: Some inherent properties of parameter variation. *Wear* **1992**, *159*, 161–171. [CrossRef]
18. Sedlaček, M.; Gregorčič, P.; Podgornik, B. Use of the Roughness Parameters Ssk and Sku to Control Friction—A Method for Designing Surface Texturing. *Tribol. Trans.* **2016**, *60*, 260–266. [CrossRef]
19. de Groot, P. Principles of interference microscopy for the measurement of surface topography. *Adv. Opt. Photon.* **2015**, *7*, 1–65. [CrossRef]
20. Dong, W.P.; Stout, K.J. An integrated approach to characterization of surface wear: Qualitative characterization. *Wear* **1995**, *181–183*, 700–716. [CrossRef]
21. Grzesik, W.; Zak, K.; Kiszka, P. Comparison of Surface Textures Generated in Hard Turning and Grinding Operations. *Procedia CIRP* **2014**, *13*, 84–89. [CrossRef]
22. Menezes, P.L.; Kailas, S.V. Effect of surface roughness parameters and surface texture on friction and transfer layer formation in tin-steel tribo-system. *J. Mater. Process. Technol.* **2008**, *208*, 372–382. [CrossRef]
23. Menezes, P.L.; Kailas, S.V. Effect of surface topography on friction and transfer layer during sliding. *Tribol. Online* **2008**, *3*, 25–30. [CrossRef]
24. Prajapati, D.K.; Tiwari, M. The correlation between friction coefficient and areal topography parameters for AISI 304 steel sliding against AISI 52100 steel. *Friction* **2021**, *9*, 41–60. [CrossRef]
25. Sedlaček, M.; Podgornik, B.; Vižintin, J. Correlation between standard roughness parameters skewness and kurtosis and tribological behaviour of contact surfaces. *Tribol. Int.* **2012**, *48*, 102–112. [CrossRef]
26. Kuriachen, B.; Joshy, J.; Joy, M.L. Influence of areal surface parameters on the tribological behavior of Ti6Al4V under lubricated condition. *Tribol. Int.* **2023**, *179*, 108147. [CrossRef]
27. Pettersson, U.; Jacobson, S. Textured surfaces in sliding boundary lubricated contacts—Mechanisms, possibilities and limitations. *Tribol. -Mater. Surf. Interfaces* **2007**, *1*, 181–189. [CrossRef]
28. Pettersson, U.; Jacobson, S. Influence of surface texture on boundary lubricated sliding contacts. *Tribol. Int.* **2003**, *36*, 857–864. [CrossRef]
29. Ibatan, T.; Uddin, M.S.; Chowdhury, M.A.K. Recent development on surface texturing in enhancing tribological performance of bearing sliders. *Surf. Coat. Technol.* **2015**, *272*, 102–120. [CrossRef]
30. Sudeep, U.; Tandon, N.; Pandey, R.K. Tribological Studies of Lubricated Laser-Textured Point Contacts in Rolling/Sliding Reciprocating Motion with Investigations of Wettability and Nano hardness. *Tribol. Trans.* **2015**, *58*, 625–634. [CrossRef]
31. Zhou, N.; Peng, R.L.; Pettersson, R. Surface integrity of 2304 duplex stainless steel after different grinding operations. *J. Mater. Process. Technol.* **2016**, *229*, 294–304. [CrossRef]
32. Westlund, V.; Heinrichs, J.; Jacobson, S. On the Role of Material Transfer in Friction Between Metals: Initial Phenomena and Effects of Roughness and Boundary Lubrication in Sliding Between Aluminium and Tool Steels. *Tribol. Lett.* **2018**, *66*, 97. [CrossRef]
33. Menezes, P.L.; Kishore; Kailas, S.V.; Lovell, M.R. Tribological response of soft materials sliding against hard surface textures at various numbers of cycles. *Lubr. Sci.* **2013**, *25*, 79–99. [CrossRef]
34. Bowden, F.P.; Tabor, D. *The Friction and Lubrication of Solids. Part 1*; Clarendon Press: Oxford, UK, 1954.
35. Křupka, I.; Vrbka, M.; Hartl, M. Effect of surface texturing on mixed lubricated non-conformal contacts. *Tribol. Int.* **2008**, *41*, 1063–1073. [CrossRef]
36. Křupka, I.; Hartl, M. The effect of surface texturing on thin EHD lubrication films. *Tribol. Int.* **2007**, *40*, 1100–1110. [CrossRef]
37. Hutchings, I.; Shipway, P. Friction. In *Tribology: Friction and Wear of Engineering Materials*, 2nd ed.; Section 3.4.2; Elsevier: Amsterdam, The Netherlands, 2017; pp. 41–45.

38. Whitehouse, D.J. *Handbook of Surface and Nanometrology*, 2nd ed.; CRC Press: Boca Raton, FL, USA; Taylor and Francis Group: Abingdon, UK, 2011.
39. *ISO 25178-2*; Franco, L.A.; Sinatora, A. 3D Surface Parameters Actual Meaning of Spk and Its Relationship to Vmp. Precision Engineering: Monmouth, UK, 2015; pp. 106–111.
40. Das, J.; Linke, B. Evaluation and systematic selection of significant multi-scale surface roughness parameters (SRPs) as process monitoring index. *J. Mater. Process. Technol.* **2017**, *244*, 157–165. [CrossRef]
41. Erdemir, A. Review of engineered tribological interfaces for improved boundary lubrication. *Trib. Int.* **2005**, *38*, 249–256. [CrossRef]
42. Oğuz, H.; Düzcükoğlu, H.; Ekinci, Ş. The Investigation of Lubrication Properties Performance of Euro-Diesel and Biodiesel. *Tribol. Trans.* **2011**, *54*, 449–456. [CrossRef]
43. Xiao, H.; Liu, S. Tribological properties of biodiesel: A review of recent advances. In *Energy Sources, Part A: Recovery, Utilization, and Environmental Effects*; Taylor and Francis Group: Abingdon, UK, 2020; pp. 1–20.

lubricants

MDPI

Article

Multiscale Texture Features to Enhance Lubricant Film Thickness for Prosthetic Hip Implant Bearing Surfaces

Fitsum Berhe Tewelde [1], Quentin Allen [2] and Tianfeng Zhou [1,3,*]

[1] School of Mechanical Engineering, Beijing Institute of Technology, Beijing 100081, China; fitsum.berhe@bit.edu.cn
[2] Department of Manufacturing Engineering, Brigham Young University, Provo, UT 84602, USA; quentin_allen@byu.edu
[3] Chongqing Innovation Center, Beijing Institute of Technology, Chongqing 401120, China
* Correspondence: zhoutf@bit.edu.cn

Abstract: The longevity of prosthetic hip implants is significantly influenced by wear. Surface textures of various length scales can reduce the friction coefficient and wear of lubricated bearing surfaces. The optimization of multiscale texture parameters, aimed at maximizing lubricant film thickness, was achieved through hydrodynamic lubrication simulations that solve the Reynolds equation with a mass-conserving cavitation model under various operating conditions. The outcomes indicate that adding "interstitial" texture features to a pattern of microscale texture features can further increase the lubricant film thickness. Additionally, the lubricant film thickness increases as the interstitial texture feature aspect ratio and texture density decrease. Pin-on-disc experiments align with simulation findings, demonstrating that multiscale texturing with ultra-fast laser ablation on Ti6Al4V discs significantly improves wettability and reduces the friction coefficient of ultra-high molecular weight polyethylene pins when compared to untextured and microscale textured surfaces. The multiscale surface texturing also changes the evident wear mechanisms on the pins, reducing the incidence of abrasive scratches and adhesive wear compared to both untextured and just microscale textured surfaces.

Keywords: artificial joints; bio-tribology; laser surface texturing; film thickness; hydrodynamic lubrication

1. Introduction

Total hip arthroplasty (THA) surgery is a globally performed procedure that involves the removal of a damaged or diseased hip joint, replaced by a prosthetic implant fixed in the bone to restore anatomy and mobility [1]. Despite its success, the statistical survival of prosthetic hip implants decreases over time, resulting in complicated revision surgeries that can give rise to economic and health challenges [2]. Studies reveal that the typical lifespan of implants ranges from 15 to 25 years, a duration considered unacceptable especially for young patients [3]. Common causes of prosthetic hip implant failures, such as osteolysis, loosening, and mechanical complications [4], are often linked to wear debris from the implant's bearing surfaces [5]. Consequently, addressing the challenge of reducing wear on these bearing surfaces remains a crucial focus of ongoing research.

Minimizing the friction and wear in prosthetic hip implants encompasses both material enhancement and surface engineering approaches. Improving the materials and mechanical properties involves optimizing the selection, manufacturing, and characteristics of materials to enhance durability. Specifically, vitamin E-infused and highly cross-linked ultra-high molecular weight polyethylene (UHMWPE) exhibits lower wear rates than traditional UHMWPE material, due to the increased strength and toughness caused by the cross-linking [6], and the enhanced fatigue resistance caused by the antioxidant vitamin E [7]. Improved ceramics manufacturing has reduced the incidence of fracture in the ceramic bearing surfaces of prosthetic hip implants from as high as 13% in some reports to less than 0.004% [8].

Surface engineering methods include tribology-inspired solutions such as coating the bearing surfaces with hard and smooth material such as diamond-like carbon intended to improve wear resistance [9]. Nevertheless, the issue of delamination poses a constraint on its utilization [10]. A patterned surface texture has proven effective in reducing friction and wear by incorporating various mechanisms [3,11]. Large and deep texture features serve as wear particle traps, thereby minimizing abrasive interactions and promoting a stable friction coefficient [12,13], and function as lubricant reservoirs, facilitating a consistent and readily available lubrication supply during sliding [14,15]. The textured surface contributes to a reduction in the nominal contact area, diminishing direct surface contact and friction [16]. Furthermore, shallow texture features aid in increasing hydrodynamic pressure, promoting a load-carrying lubricating film between the contacting surfaces [17].

Numerous studies show that textured surfaces reduce friction and wear between prosthetic hip implant bearing surfaces. Ito et al. experimentally showed that large texture features (500 μm diameter, 100 μm depth) on a cobalt–chromium femoral head produced 36% less wear in a UHMWPE counter face than a smooth femoral head because of the lubricant reservoirs and trapped wear debris [18]. Tarabolsi et al. demonstrated large accumulation of wear debris in texture features on cobalt–chromium- and ceramic-bearing surfaces and measured a significant decrease in the total wear volume compared to smooth bearing surfaces [12]. Dong et al. experimentally measured that shallow texture features (<10 μm deep) increased the lubricant film thickness compared to smooth bearing surfaces [19]. Borjali et al. showed that shallow texture features on cobalt–chromium discs reduced the friction coefficient and wear rate of UHMWPE pins by inducing the hydrodynamic lubrication regime under conditions that caused boundary/mixed regime in smooth bearing surfaces [20]. Allen and Raeymaekers used elastohydrodynamic lubrication simulations to model how the lubricant film thickness changes with texture design parameters and bearing operating conditions, indicating that optimal texture designs can be determined for individual operating conditions [21,22].

Multiscale texture features commonly occur on natural surfaces and can provide enhanced tribological performance by combining various functions on the same surface. For instance, the microstructures on a lotus leaf make the surface hydrophobic and allow for self-cleaning [23]. On shark skin, textured scales reduce drag [24] and hierarchical structures on gecko feet facilitate extraordinary adhesive force [25]. Articular cartilage surfaces naturally possess "dimple" surface texture features with diameters ranging from 20 to 50 μm and depths of 0.5 to 2 μm. These dimple texture features play a crucial role in enhancing lubrication for the joint bearing surface [26]. Thus, multiscale textures exist to optimize surface physical properties, allowing adaptation to a variety of natural environments. Furthermore, these textures have the potential to reduce friction and enhance wear resistance [27,28], thereby increasing the efficiency and longevity of tribo-pairs.

The influence of multiscale textures in diverse tribological applications has been studied by several research groups. Segu et al. examined the effect of multiscale/shape texture features on lubricant regimes using laser-textured round and ellipse dimples on a steel surface and showed a consistently low friction coefficient attributed to hydrodynamic lubrication effects [29]. Resendiz et al. demonstrated that multiscale texturing on Al6061-T6 surfaces by inclined milling and micro shot blasting process reduced friction and wear compared to untextured and single-scale textured surfaces, attributed to increased lubricant fluid pressure [30]. Zhang et al. studied hierarchical structures created by an ion beam etching of a laser-textured Co–Cr–Mo alloy surface and observed friction and wear performance improvement, attributed to a decreased contact area, stored lubricant, and increased lubricant film thickness [31]. Hsu et al. investigated how sliding orientation and specific texture parameters are crucial for achieving effective friction reduction in each lubrication regime [32]. In a subsequent experimental investigation, the same research group compared single-scale, mixture (alternating between two shapes), and overlapped (placing one texture feature on top of another) multiscale texture designs under a wide range of loads and speeds. The results indicated that multiscale surface textures reduce friction more effectively across various lubrication regimes compared to single-scale dimples [33].

Therefore, multiscale textures on bearing surfaces offer a distinct advantage over single-scale textures, particularly in dynamic loading and varying speed conditions like those experienced by prosthetic hip implants. Multiscale textures can enhance performance by accommodating varying speeds and forces, improving lubrication, reducing wear, and extending the lifespan of prosthetic hip implants through shortening the time working under the boundary and/or mixed lubrication regime during the gait cycle.

Previous studies have primarily focused on single-scale textures to enhance tribological properties. However, this paper aims to determine the influence of multiscale textures on lubricant film thickness in the context of prosthetic hip implant applications. Various multiscale texture design parameters were investigated within reasonable bearing operating conditions to identify the optimum surface parameters. The objective is to develop a bio-inspired solution to decrease the friction coefficient and wear of prosthetic hip implants by increasing the lubricant film thickness between articulating surfaces.

2. Materials and Methods

2.1. Optimization of Multiscale Texture Design Parameters

We developed a hydrodynamic lubrication (HL) model of a bearing surface with a repeating pattern of multiscale texture features in lubricated sliding parallel with a flat counter face as shown in Figure 1. The multiscale texture pattern occurs as a combination of the microscale texture features, with smaller "interstitial" texture features in between, and the pattern repeats in the x- and y-directions. The cylindrical shaped texture features (flat texture floor shape) are fully described by the non-dimensional parameters of texture density, $S_p = \pi {r_p}^2/4{r_1}^2$, and texture aspect ratio, $\varepsilon = h_p/2r_p$, where r_p is the texture feature radius, h_p is the depth of the texture feature, and r_1 is the half-length of a unit square that surrounds each texture feature. The smaller, interstitial texture features are similarly defined with texture density and aspect ratio, i.e., $S_{pi} = \pi {r_{pi}}^2/4{r_1}^2$ and $\varepsilon_i = h_{pi}/2r_{pi}$, where r_{pi} is the radius and h_{pi} is the depth of the interstitial texture features.

We solve the Reynolds equation for the lubricant film pressure using finite differencing. The two-dimensional, steady-state, isothermal, and iso-viscous Reynolds equation is given in Equation (1),

$$\frac{\partial}{\partial x}\left(h^3\frac{\partial p}{\partial x}\right)+\frac{\partial}{\partial y}\left(h^3\frac{\partial p}{\partial y}\right)=6\mu U\frac{\partial h}{\partial x}, \tag{1}$$

where x and y are Cartesian coordinates, h and p are the thickness and pressure of the lubricant film, respectively, μ is the lubricant dynamic viscosity, and U is the relative surface velocity in the x-direction. We introduce non-dimensional parameters so the simulation results can be applied to any bearing system, independent of texture dimensions and lubricant properties. Specifically, $X = x/r_p$, $Y = y/r_p$, $H = h/c$, and $P = p/p_0$, where p_0 is atmospheric pressure (101,325 Pa) and c is the separation between bearing surfaces. Furthermore, the parameter $\lambda = 3\mu U/2r_p p_0$ is known as the flow factor, as it incorporates the lubricant viscosity and the relative bearing surface velocity. The parameter $\delta = c/2r_p$ is the non-dimensional bearing spacing. The Reynolds equation expressed with non-dimensional parameters is given in Equation (2).

$$\frac{\partial}{\partial X}\left(H^3\frac{\partial P}{\partial X}\right)+\frac{\partial}{\partial Y}\left(H^3\frac{\partial P}{\partial Y}\right)=\frac{\lambda}{\delta^2}\frac{\partial H}{\partial X}, \tag{2}$$

The simulation domain consists of half of an array of 5 texture features with symmetry boundary conditions on the centerline and the edge of the array to reduce the computation time. The inlet and outlet of the texture feature array are kept at atmospheric pressure.

We employ an Elrod algorithm to enforce mass-conserving Jakobson–Floberg–Olsson (JFO) cavitation boundary conditions. Specifically, we include a fractional film content parameter, θ, in the non-dimensional Reynolds equation.

$$\frac{\partial}{\partial X}\left(H^3\frac{\partial P}{\partial X}\right)+\frac{\partial}{\partial Y}\left(H^3\frac{\partial P}{\partial Y}\right)=\frac{\lambda}{\delta^2}\frac{\partial}{\partial X}(\theta H), \tag{3}$$

Thus, each point in the solution domain can revert between the cavitation and full-film regimes. In the cavitation regime, the pressure remains at the cavitation threshold (chosen as the vapor pressure of water at 5630 Pa for a worst-case conservative estimate). The center of a texture feature is positioned at the bearing inlet, as the JFO cavitation condition does not produce positive lubricant film pressure if the bearing inlet is untextured [34,35]. We discretize the solution domain with a matrix of nodes, equally spaced in the x- and y-directions. We set an initial non-dimensional bearing spacing, δ_0, and iterate on solutions of Equation (3) until successive iterations of the lubricant film pressure are different by less than a critical convergence criterion. Convergence studies show that n = 201 nodes across the length of a unit cell ($2r_1$) and a convergence criterion of 0.0001 are sufficient to achieve accurate simulation results of the lubricant film pressure. We compare the simulated lubricant film pressure to the desired bearing load and change the value of δ until the converged solution carries the applied load.

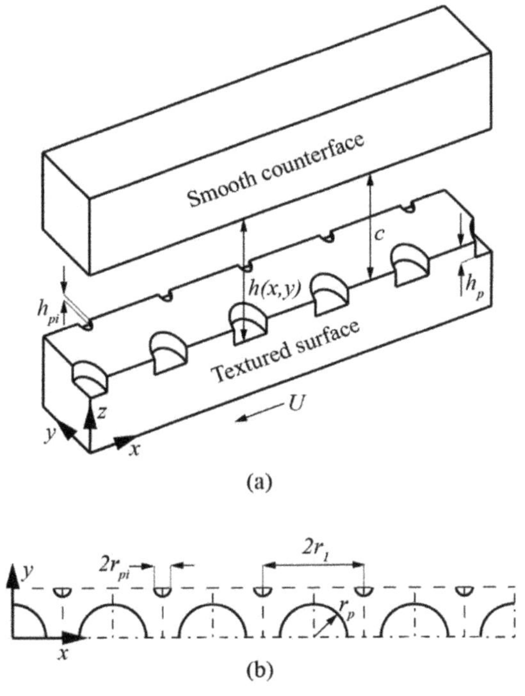

Figure 1. Schematic illustration of the hydrodynamic lubrication (HL) model domain: (**a**) isometric view; (**b**) top view.

We test different multiscale texture design parameters to identify the most advantageous combination for maximizing the lubricant film thickness. Table 1 shows the maximum and minimum values of the texture design parameters and bearing operating conditions considered in these simulations, as well as the nominal values that were held constant when varying other parameters.

Table 1. Operating parameters varied for the study.

Parameter	Minimum Value	Maximum Value	Nominal Value
Load-carrying capacity, W	2.467	9.869	4.935
Flow factor, λ	0.075	0.300	0.300
Micro texture density, S_p	0.100	0.400	0.200
Micro texture aspect ratio, ε	0.020	0.090	0.050
Interstitial texture density, S_{pi}	0.005	0.030	0.020
Interstitial texture aspect ratio, ε_i	0.0025	0.030	0.020

2.2. Sample Preparation and Laser Surface Texturing

We conducted an experimental procedure with the aim of validating the simulation outcomes, emphasizing the achievement of maximum lubricant film thickness. The critical texture parameter selection criterion depends on feasible minimum depth of the smallest texture feature, which is set at 0.73 µm for effective direct laser texturing. Depths below this threshold posed a risk of introducing roughness and waviness to the texture feature bottom surface, potentially exceeding the top flat surface. We selected three samples for multiscale textures with the same microscale texture features and the same interstitial texture aspect ratio but varying in interstitial texture density. Additionally, we selected two samples for comparison with just microscale texturing and an untextured surface.

The widely used biomaterial, CoCrMo, releases harmful metal ions, resulting in tissue reactions and a decreased lifespan. Thus, Ti6Al4V implants with enhanced surface engineering are being studied as an alternative due to their biocompatibility and strength. Biomedical-grade Ti6Al4V alloy discs (grade 5), with a diameter of 30 mm and a thickness of 6 mm, were polished to achieve a mirror finish surface roughness Ra below 50 nm. The surface roughness was measured with laser scanning confocal microscope (KEYENCE VK-100, Keyence Co., Osaka, Japan). The polished samples were then subjected to 15 min of ultrasonic cleaning with acetone. To ensure complete drying and prevent the introduction of moisture-related issues or contaminants, the heat drying process followed at 70 °C for 30 min. For the friction test, the Ti6Al4V disc was paired with flat biomedical-grade UHMWPE (GUR 1020) pin with a diameter of 4 mm and a length of 10 mm. The molecular weight of the UHMWPE is around 5.5 million g/mol. The material characteristics of UHMWPE and Ti6Al4V are provided in Table 2.

Table 2. Material properties of UHMWPE and Ti6Al4V.

	Young's Modulus (GPa)	Poisson's Ratio	Hardness	Surface Roughness (Ra (µm))
UHMWPE	0.68	0.45	63 (Shore D)	0.45
Ti6Al4V	120	0.36	414 (HV)	0.05

The design for just microscale texture features was adapted from a prior numerical study [22], aiming for maximum film thickness, comprising an aspect ratio of 0.022 and a texture area density of 0.2. Interstitial textures, designed with a constant aspect ratio of 0.025 and a varying texture area density ranging from 0.0176 to 0.0706, were selected for the manufacturing process. Table 3 presents geometrical parameters of the texture features on Ti6Al4V surface.

Laser surface texturing was carried out using a commercially available picosecond laser system, (diode-pumped solid-state laser, Advanced Optowave, Ronkonkoma, NY, USA). The system generates τL = 10 ps laser pulse at wavelength λ = 1064 nm and we employed pulse energy Ep = 250 µJ at a repetition rate of fp = 100 kHz to produce the microtextures. The picosecond laser produces pulses of linearly polarized light with the Gaussian intensity distribution. The diameter of laser spot was 14.4 µm at the focal point on the sample surface. A spiral laser path approach was used to remove material inside the circular shape micro textures [36]. The fabrication of various dimple sizes was accomplished using a single-step process [28], although various process repetitions were necessary to achieve the required dimple depths. Before surface characterization, the samples underwent a 30-min ultrasonic cleaning in acetone. A confocal microscope (Keyence VK-100, Keyence Co., Osaka, Japan) was used for surface characterization of the fabricated microtextures.

Table 3. Surface texture design parameters.

Samples	Texture Type	Width ø (μm)	Depth (μm)	Aspect Ratio	Area Density
D000d00	Untextured	-	-	-	-
D100d00	Micro texture	100	2.2	0.022	0.2
D100d30	Micro texture	100	2.2	0.022	0.2
	Interstitial	30	0.73	0.022	0.017
D100d45	Micro texture	100	2.2	0.002	0.2
	Interstitial	45	1	0.022	0.039
D100d60	Micro texture	100	2.2	0.022	0.2
	Interstitial	60	1.3	0.022	0.072

2.3. *Contact Angle Measurement*

The wettability of the sample surfaces was characterized by the contact angle (CA) measurement with water droplets (OCA25, Dataphysics Instruments GmbH, Filderstadt, Germany). This instrument is capable of measuring a contact angle with a resolution of 0.1°. Measurements were conducted at a room temperature of approximately 25 ± 1% °C and a relative humidity of around 50 ± 5%. Contact angle values were obtained through the direct observation, analysis, and calculation of the shape of liquid droplets (sessile drop) formed on horizontally placed solid surfaces. The deposition of drops was detected using an automatic recognition technique integrated within the interface measurement and analysis software of the system. The droplet size was regulated utilizing an electric injection device with a liquid addition and return rate (LAR) set at 0.5 μL/s. Initially, 4 μL of liquid was dispensed, followed by inserting the needle into the liquid for continuous addition of another 4 μL. The system remained stable for 100 s before retracting the 4 μL of liquid back into the needle. We conducted five repetitions of the contact angle (CA) test at different locations on each sample to ensure reproducibility and precision. The average values were then calculated and presented, with the maximum standard deviation observed being 3.1%, assuring that the recorded values in each area of the sample were similar, thus justifying their averaging.

Water was selected as the test fluid for contact angle (CA) measurements due to its versatility and validity. A study by Zhang et al. [31] revealed similar wetting trends between water and bovine serum on textured surfaces. Notably, bovine serum consistently demonstrated lower contact angles than water, implying that using water could yield more conservative results.

The advancing contact angle is determined as the solid/liquid contact area expands, while the receding contact angle is measured as the contact area decreases on the surface. The difference between advancing and receding angle is known as contact angle hysteresis (CAH) in the liquid–solid interface. A large CAH (i.e., significant differences between advancing and receding contact angles) indicates strong surface interactions that enhance the wetting behavior.

2.4. *Pin-on-Disc Friction and Wear Measurement*

The friction and wear performance of just microscale textured, multiscale textured, and untextured surfaces were evaluated with a pin-on-disc setup by a reciprocating sliding test conducted on a UMT-2 tribometer under lubricated conditions. Figure 2a illustrates the schematic diagram of the pin-on-disc setup, and Figure 2b shows a photograph of the setup. The upper sample, featuring a cylindrical shape and a flat-ended UHMWPE pin, slides over Ti6Al4V discs. All tests were conducted at a constant frequency of 1 Hz, a sliding distance of 15 mm for 1000 cycles, and a load of 6.28 N, creating a contact pressure of 0.5 MPa. This contact pressure is within the expected range experienced by real hip implants, and the choice of these parameters was in accordance with the parameters used for the simulations.

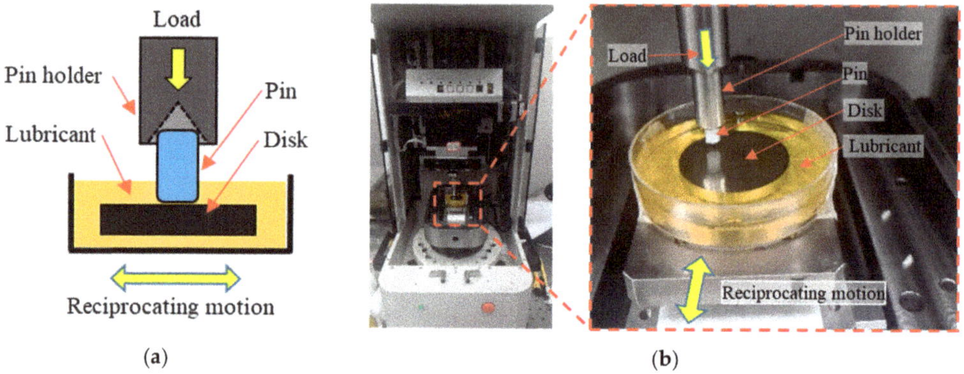

Figure 2. Pin-on-disc apparatus for flat-end pin to self-align pin on flat disc: (**a**) schematic diagram; (**b**) photograph image of pin-on-disc arrangement.

Bovine serum containing a protein concentration of 15 mg/mL served as the lubricant. We maintained lubricating liquid viscosity by controlling temperature and minimizing contaminants. We ensured consistent lubrication conditions and prevented disintegration by uniformly applying the lubricant, inspecting for degradation, implementing quality control measures and maintaining a stable testing environment.

For the self-alignment of the flat pin surface with the flat disc surface, a conical-shaped pin holder was utilized to hold and apply pressure to the top side of the pin, as depicted in Figure 2a. To ensure the repeatability of friction measurements, a minimum of three measurements were conducted for each sample, and the average values were subsequently reported.

3. Results

3.1. Typical HL Solution for Each Scale of Texture Features

The simulation results include the lubricant film pressure as well as the lubricant film thickness at each point in the domain. Figure 3a shows a typical three-dimensional result of the lubricant film pressure across the solution domain with W = 4.935, λ = 0.300, S_p = 0.200, ε = 0.050, S_{pi} = 0.010, and ε_i = 0.020. Figure 3b shows the contour of lubricant film pressure along the edge of the texture feature array for simulations with the combined multiscale texture features (black solid line) and for just the microscale texture features (red dashed line). Figure 3c shows the contour of lubricant film pressure along the centerline of the microscale texture feature array. We see from Figure 3c that the lubricant film begins at atmospheric pressure at the bearing inlet and increases to a maximum value at the outlet of the first texture feature because the texture outlet creates a converging flow. Each texture inlet causes a diverging flow that reduces the lubricant film pressure to the cavitation limit. The simulations with just microscale texture features and the combined multiscale texture simulations are virtually identical at this centerline.

The edge contour in Figure 3b passes between microscale texture features and through the center of the interstitial texture features. From this figure, we observe that the solution with only microscale texture features shows a repeating increase and decrease in the lubricant film pressure, corresponding to the locations of the microscale texture features in the X-direction. The combined multiscale texture solution shows an additional slight decrease followed by a slight increase in pressure due to the extra diverging and converging channels from the interstitial texture features. Figure 3d shows a magnified view of the lubricant film pressure contour for the centermost texture features at the edge of the array for the microscale texture features (black solid line) and three different sizes of interstitial texture features in a multiscale array (blue, red, and green lines). Increasing the size of the interstitial texture features leads to a greater disruption of the lubricant film pressure in this region of the array, but the deviations are similar in both the positive and negative directions. The interstitial texture features do not cause cavitation at the conditions simulated here, and the net influence of the interstitial texture features on the lubricant film pressure is small compared to the microscale texture features. Both texturing conditions have the same load-carrying capacity, $W = 4.935$, but the combined multiscale texture features resulted in a slightly thicker lubricant film ($\delta = 0.0488$) compared to the solely microscale texture features ($\delta = 0.0484$). This is due to the interstitial texture features increasing the volume of lubricant between the bearing surfaces and acting as additional micro-hydrodynamic bearings.

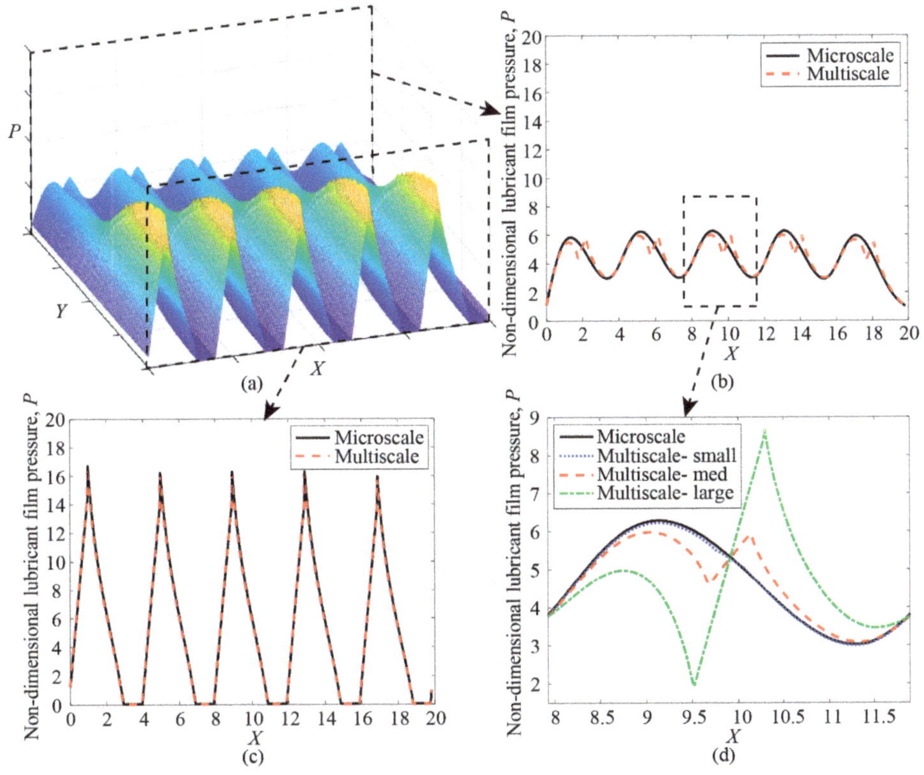

Figure 3. Typical results of the HL simulations with $W = 4.935$ and $\lambda = 0.300$: (**a**) 3D view of the non-dimensional lubricant film pressure for the combined multiscale texture features; (**b**) pressure contour at the edge of the array; (**c**) pressure contour at the centerline of the array; and (**d**) magnified view of the pressure contour of the middle texture from (**b**), with just microscale textures (black), and multiscale textures with small ($\varepsilon_i = 0.0025$, $Sp_i = 0.005$—blue), medium ($\varepsilon_i = 0.020$, $S_{pi} = 0.010$—red), and large ($\varepsilon_i = 0.030$, $S_{pi} = 0.030$—green) interstitial texture features.

3.2. *Effect of the Microscale Texture Features*

To determine the effect of changing the microscale texture features on a multiscale textured surface, we perform simulations with varied texture feature design parameters (S_p and ε), keeping the interstitial texture features constant ($S_{pi} = 0.01$, $\varepsilon_i = 0.02$). In each simulation, we calculate the non-dimensional lubricant film thickness (i.e., non-dimensional bearing spacing, δ) required to generate sufficient lubricant film pressure to carry the applied load. Figure 4 plots this film thickness as a function of the microscale texture aspect ratio. Figure 4a shows different values of the load-carrying capacity, W, in different colors and Figure 4b shows a magnified view of the simulation results with $W = 2.467$. Figure 4c shows different values of the flow factor, λ, in different colors and Figure 4d shows a magnified view of the simulation results with $\lambda = 0.075$. Increasing the load-carrying capacity and decreasing the flow factor reduce the lubricant film thickness because the bearing surfaces are compressed closer together, or because the lubricant viscosity or the bearing surface velocity has decreased, which reduces the hydrodynamic pressure generation. We also see that a texture density of $S_p = 0.2$ results in the maximum lubricant film thickness for each simulated condition. There is an optimal texture aspect ratio that maximizes the lubricant film thickness, however, the value of the optimal texture aspect ratio changes with the bearing operating conditions. Specifically, the optimal texture aspect ratio increases with operating conditions that increase the lubricant film thickness (i.e., increasing λ and decreasing W), similar to the results reported in [22].

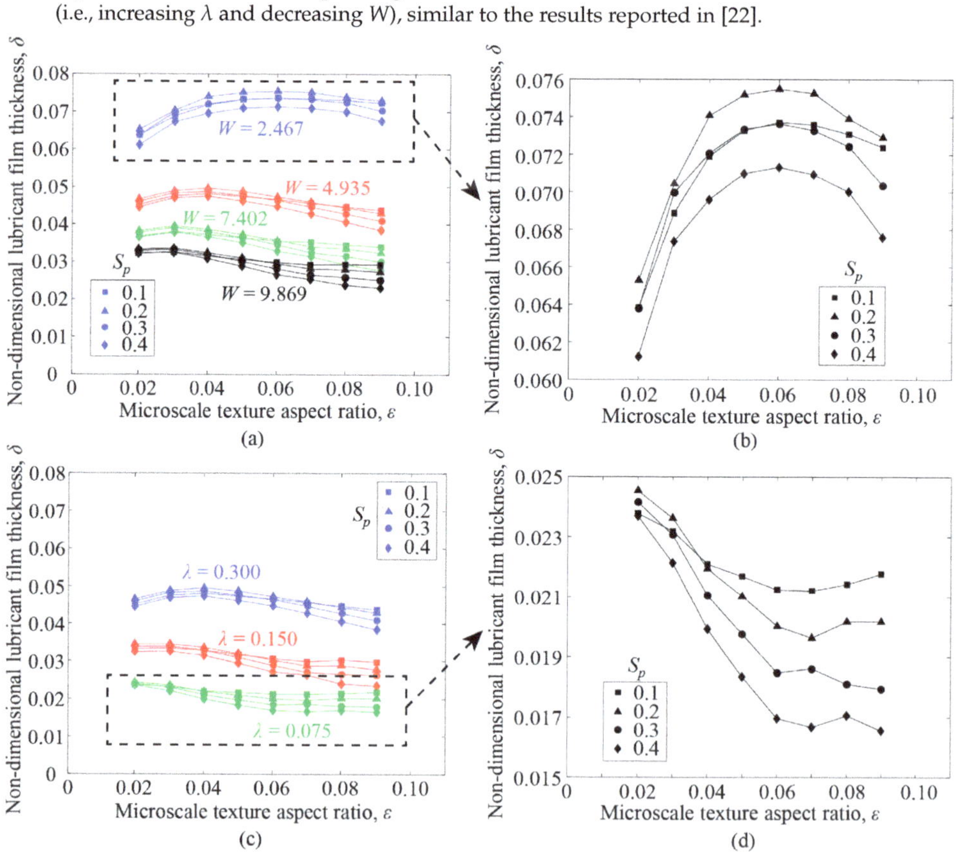

Figure 4. Non-dimensional lubricant film thickness plotted against microscale texture aspect ratio for: (**a**) different load carrying capacity; (**b**) a magnified view of the results with $W = 2.467$; (**c**) different flow factor; and (**d**) a magnified view of the results with $\lambda = 0.075$.

3.3. Effect of the Interstitial Texture Features

To determine the effect of changing the interstitial texture features on a multiscale textured surface, we perform simulations with varied interstitial texture feature design parameters (S_{pi} and ε_i), keeping the microscale texture features constant ($S_p = 0.02$, $\varepsilon = 0.05$).

Figure 5 shows the non-dimensional lubricant film thickness as a function of the interstitial texture aspect ratio. Figure 5a includes results for $\lambda = 0.300$ and different values of the load-carrying capacity ($W = 2.467$ in blue, $W = 4.935$ in red, $W = 7.402$ in green, and $W = 9.869$ in black), and Figure 5b shows a zoomed-in view of the simulations with $W = 4.935$. A dotted line shows the lubricant film thickness for a bearing surface with only microscale texture features under the same bearing operating conditions. Figure 5c plots the non-dimensional lubricant film thickness as a function of the interstitial texture aspect ratio for different values of the flow factor ($\lambda = 0.300$ in blue, $\lambda = 0.150$ in red, and $\lambda = 0.075$ in green). Figure 5d shows a magnified view of the results with $\lambda = 0.075$. Again, we observe that the lubricant film thickness decreases as the load-carrying capacity increases and as the flow factor decreases, because the increased loads push the bearing surfaces closer together and the lubricant viscosity or bearing surface velocity are decreased. At the conditions simulated here, the interstitial texture features have a slight effect on the overall lubricant film thickness. However, from the zoomed-in view in Figure 5b, we see that the lubricant film thickness increases as the interstitial texture density and aspect ratio decrease. We also observe that the interstitial texture features increase the lubricant film thickness compared to the microtextured surface at these simulation conditions ($\lambda = 0.300$ and $W = 4935$).

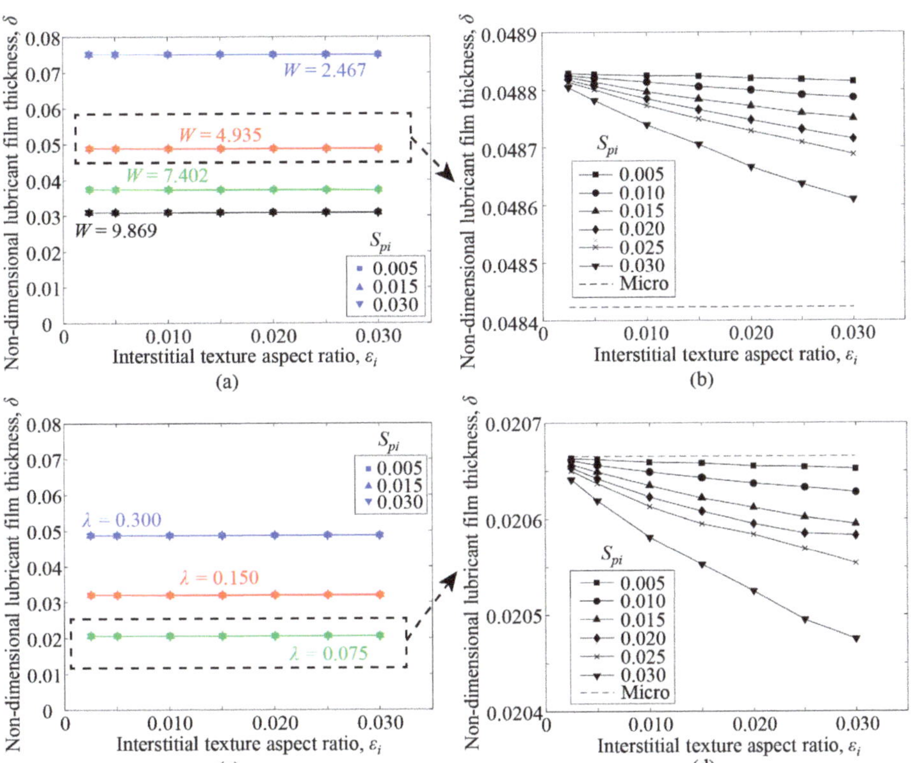

Figure 5. Non-dimensional lubricant film thickness as a function of interstitial texture feature aspect ratio plotted for: (**a**) constant flow factor $\lambda = 0.075$ and different values of the load carrying capacity indicated with different colors, and (**c**) constant load $W = 4.935$ and different values of the flow factor indicated with different colors (**b**,**d**) are zoomed in plots to show the detailed data points of different interstitial texture densities for $W = 4.935$ and $\lambda = 0.075$, respectively.

In Figure 5d, we see the same trend repeated, namely that the lubricant film thickness increases as the interstitial texture feature density and aspect ratio decrease. However, at these simulated conditions, the interstitial texture features resulted in a very slight decrease in lubricant film thickness compared to the microscale texture features, and the multiscale texture features approach the microscale solution as the interstitial texture aspect ratio decreases.

The interstitial texture features provide a slight increase in the lubricant film thickness compared to the microscale textures for the simulations with high load (W = 4.935, 7.402, and 9.869) and cause a slight decrease in lubricant film thickness compared to the microscale textures for a low load (W = 2.467) and small values of the flow factor (λ = 0.15, 0.075).

The HL simulations indicate that the lubricant film thickness increases with decreasing interstitial texture aspect ratio and interstitial texture density for every bearing operating condition simulated here. This is likely due to competing mechanisms in the multiscale textured bearing surface. On one hand, the interstitial texture features can accommodate a slightly increased volume of lubricant between the bearing surfaces and aid in generating additional hydrodynamic pressure; however, as these interstitial texture features increase in size, they can begin to have a detrimental impact on the hydrodynamic pressure by reducing the flat bearing surface area between microscale texture features. In a real bearing surface, additional factors not accounted for in the HL simulations (e.g., lubricant wettability, change in nominal contact area/pressure, or trapping wear debris) may have a larger impact, highlighting the need to perform physical experiments in addition to the simulations.

The results of the simulations indicate some general trends for the optimal texture design parameters in a multiscale textured bearing surface. The microscale texture features should be manufactured with a texture density around S_p = 0.20. The microscale texture aspect ratio can be tailored to the bearing operating conditions, with deeper texture features used for conditions that produce thicker lubricant films (i.e., smaller load-carrying capacity and larger flow factor). For the interstitial texture features, the interstitial texture feature density and aspect ratio should be minimized in every case, independent of the bearing operating conditions. The limits of the manufacturing technique should be considered when designing the smallest interstitial texture features that can be practically fabricated on the bearing surface.

3.4. Morphology of Laser-Textured Surfaces

Figure 6 shows optical images of workpiece surfaces after textures were created on a smooth Ti6Al4V surface using a solid-state laser. The untextured surface, designated by D000d00 in Figure 6a, serves as the primary reference. Images of textured surfaces show accurate round-shaped texture details. Figure 6b depicts a surface with just microscale textures, D100d00, which serves as a secondary reference measuring approximately 100 µm in diameter. Figure 6c–e exhibit the fabrication of interstitial texture features with diameters of 30, 45, and 60 µm, respectively, among the constant size micro texture feature.

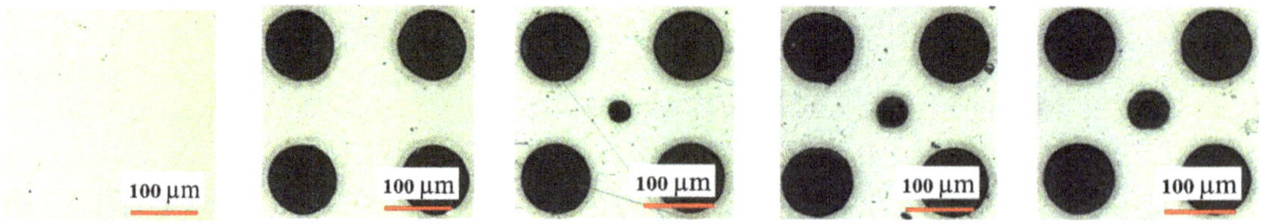

Figure 6. Optical microscopy images of surface designs: (**a**) D000d00; (**b**) D100d00; (**c**) D100d30; (**d**) D100d45; and (**e**) D100d60.

Figure 7a,b show the SEM images of microscale texture with 100 µm diameter and a sample for interstitial texture features with 30 µm diameter, respectively. The cross-sectional surface texture profile of microscale textures in Figure 7c and the interstitial textures in Figure 7d illustrate the surface roughness inside and around the surface texture features.

The average microscale texture feature depth is 2.2 µm, while the interstitial texture feature depth measures 0.73 µm. Negligible material buildup is observed around the texture edge, indicating that the ultra-fast laser pulse direct texturing method is a feasible approach for producing microscale and nano-scale texture features on metallic surfaces, and avoid the need for repolishing after laser texturing.

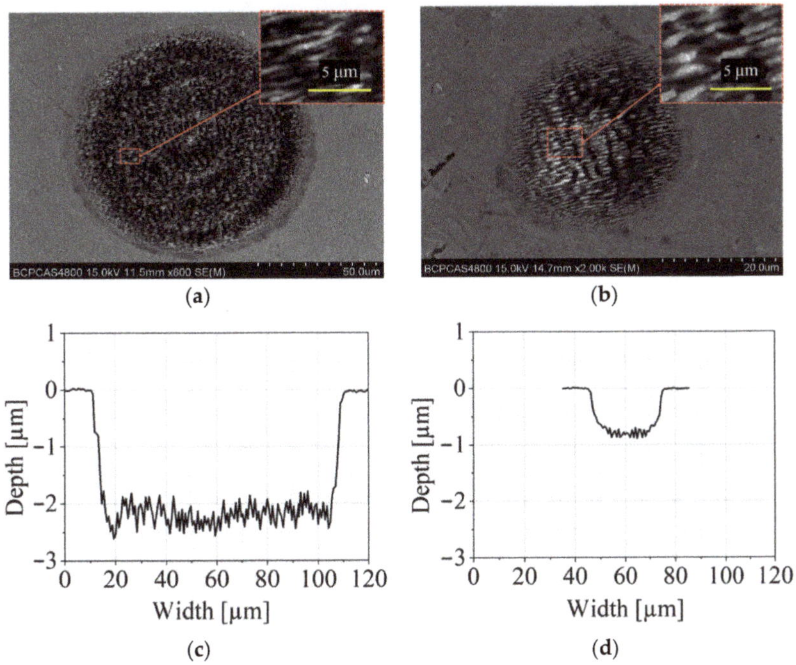

Figure 7. Illustration of laser-fabricated surface textures: SEM image of surface textures with a diameter of (**a**) 100 µm, and (**b**) 30 µm; cross-sectional profile of surface texture with a diameter of (**c**) 100 µm and (**d**) 30 µm.

Moreover, it is worth noting that with suitable laser parameters, laser-induced periodic surface structures (LIPSS) could be effectively generated inside the textured surface simultaneously during laser processing, as shown in the SEM images in Figure 7a,b inside of the surface texture features. Several studies reported that the formation of LIPSS significantly improves the wettability of textured surfaces compared to untextured surfaces [28,37].

3.5. Wettability Test Results

Wettability, indicating surface affinity for a lubricant, is determined by contact angle (CA) measurements, with angles below 90° denoting hydrophilic surfaces and those above 90° indicating hydrophobic surfaces. In general, a lower CA indicates enhanced wettability and good lubricity. Figure 8(a1–a3) display images of CA measurements following the initial dispense on surfaces of D000d00 (CA = 80.5°), D100d00 (CA = 62.7°), and D100d30 (CA = 53.1°). The contact angle decreases by 22.1% for just microscale texture, and by 34.0% for the multiscale textured surface compared to the untextured surface. These observations are in agreement with the Wenzel model [38] which states that in the case of a completely wetting liquid and a hydrophilic surface, increasing the roughness factor leads to a reduction in the contact angle. This presence of micro or nanoscale roughness, such as LIPSS depicted on Figure 7a,b, facilitates enhanced wetting and spreading of the liquid.

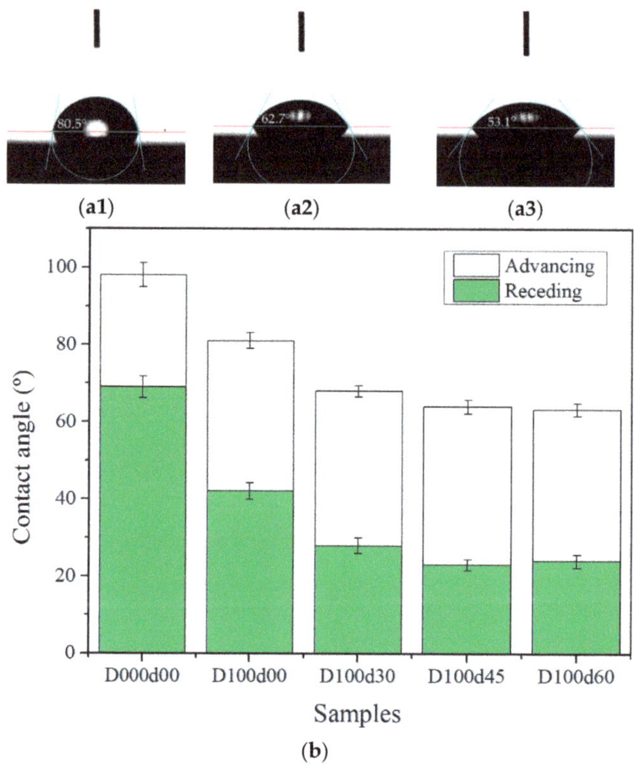

Figure 8. Contact angle measurements on the surfaces of untextured and textured discs (**a1**–**a3**) static measurements taken 15 s after initial drop on D000d00, D100d00 and D100d30, and (**b**) advancing and receding angles.

Figure 8b illustrates advancing and receding angles on both untextured and textured surfaces. It's noticeable that both advancing and receding angles decrease on textured surfaces compared to untextured ones. As the texture area density increases, both advancing and receding contact angles decrease, with D100d45 and D100d60 exhibiting similar angles. The variation between advancing and receding angles is notably large, peaking at 41° for D100d45. This underscores how multiscale texturing improves the wetting behavior of Ti6Al4V. In summary, the combination of a small contact angle and large dynamic hysteresis enhances liquid lubricant retention on textured surfaces, ultimately reducing friction [39,40].

3.6. Effect of Texture Parameters on Friction Coefficient

Figure 9 illustrates the temporal evolution of the friction coefficient during reciprocating sliding between Ti6Al4V discs with textured and untextured surfaces, and UHMWPE cylindrical pins under lubricated conditions. The test examines the run-in friction and wear conditions. The data were extracted from a stable region of a test with 1000 cycles and represent a single complete cycle, i.e., a forward and backward slide, to provide a vivid observation of the impact of various texture designs on friction and hydrodynamic lubrication.

From Figure 9, it is evident that the friction coefficient varied throughout the entire cycle. Higher friction coefficients were observed at the beginning of both forward and backward slides, reaching a minimum around the middle where constant linear sliding speed occurs [41]. Additionally, a similar friction evolution trend was noted between forward and backward slides, as depicted in Figure 9a,b. However, consistently higher friction was observed during the forward slide compared to the backward slide in all cases. This phenomenon could be attributed to surface finish and the run-in condition between the contacting pairs.

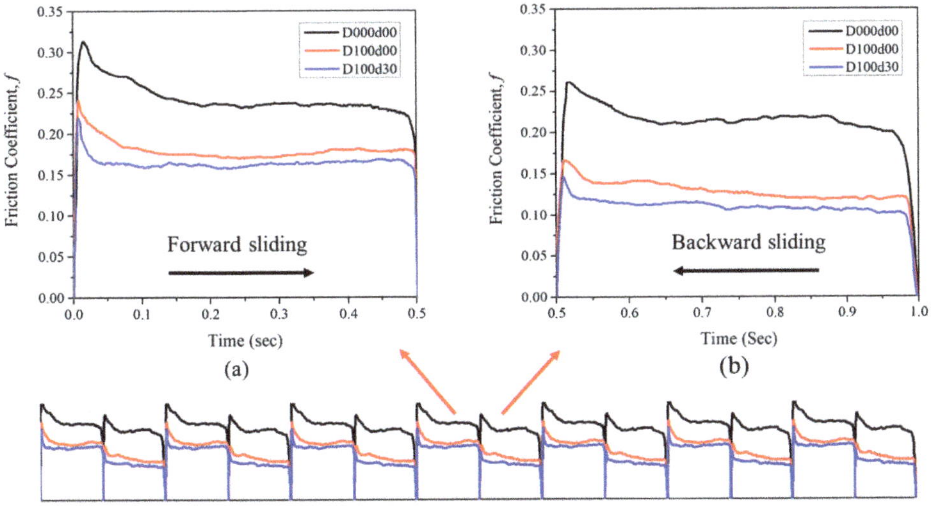

Figure 9. Friction coefficient of a single cycle between the Ti6Al4V and UHMWPE specimens as a function of time: (**a**) forward sliding; (**b**) backward sliding. The data is extracted from a long duration friction test.

In general, the surfaces with texture features exhibit lower friction coefficients compared to the untextured surfaces, attributed to intentionally designed and fabricated textures that boost hydrodynamic lubrication. The multiscale textures, in particular, achieved the lowest friction coefficient. Furthermore, on the textured surfaces, the friction coefficient reaches its lowest steady value within approximately 0.05 s after sliding begins, in contrast to the untextured surfaces, which reach the lowest steady friction coefficient value 0.20 s after sliding begins, as depicted in Figure 9a. This indicates that the sliding begins in the boundary/mixed lubrication regime and transitions towards the hydrodynamic lubrication regime as the sliding velocity increases. For the multiscale textures, the instant reduction in friction coefficient signifies a rapid surge in lubricant film thickness caused by the buildup of hydrodynamic pressure due to the surface textures. The instantaneous drop in the friction coefficient for the multiscale textured surfaces illustrates how, under identical operating conditions, the multiscale texture features can increase the time spent in the hydrodynamic lubrication regime and decrease the time in the boundary/mixed lubrication regime compared to the untextured and only microscale textured surfaces. This effect reveals the potential of multiscale texture features for significant friction and wear reduction.

Table 4 lists the average friction coefficient for forward and backward sliding and indicating one standard deviation. Minimum friction coefficient was observed for D100d30 during both forward and backward slide. Figure 10 illustrates the average friction coefficients of three types of multiscale textures compared to two reference samples—an untextured surface and just microscale textures. The untextured surface displays a higher friction coefficient in contrast to the textured surfaces with multiscale textures exhibiting the lowest fiction coefficient.

Forward sliding resulted in higher friction coefficient than return sliding for all tested conditions. In multiscale textures, the friction coefficient reduces as the size of the interstitial texture features decrease. In other words, decreasing the interstitial texture density reduces the friction coefficient. Accordingly, the experimental results align with the simulation outcomes, showing that an increase in lubricant film thickness increases the separation between articulating surfaces, leading to a reduction in the friction coefficient due to a decrease in solid-on-solid contact. However, when employing larger interstitial textures such as D100d60, multiscale textures may have a detrimental effect due to the absence of a sufficient load-carrying flat surface. Therefore, careful consideration of texture feature parameters is essential in the design of multiscale textures to achieve the desired results.

Table 4. Average friction coefficient results for forward and backward sliding.

Samples	Forward Slide	Backward Slide
D000d00	0.242 ± 0.023	0.211 ± 0.033
D100d00	0.179 ± 0.012	0.125 ± 0.021
D100d30	0.163 ± 0.009	0.109 ± 0.017
D100d45	0.166 ± 0.015	0.116 ± 0.018
D100d60	0.201± 0.016	0.172 ± 0.027

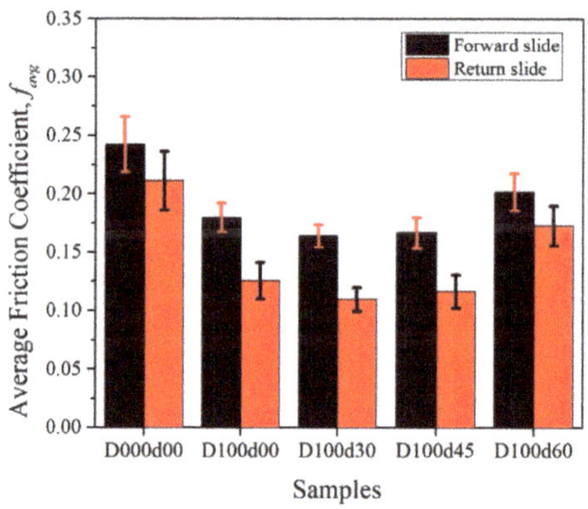

Figure 10. Average friction coefficient for textured and untextured samples in a single cycle.

3.7. Wear Analysis of UHMWPE Pin Surface

The major challenge in metal on polymer (MOP) tribo-pairs for prosthetic hip implant is the wear of softer materials like UHMWPE. Utilizing a textured surface is a promising solution to reduce both friction and wear. To analyze wear mechanisms and assess the impact of multiscale surface textures on UHMWPE wear, Figure 11 illustrates post-tribo-test images of the UHMWPE pins sliding against textured and untextured Ti6Al4V discs. Visible scratches appeared along the sliding direction of the UHMWPE surface. Various degrees of abrasive and adhesive wear were observed on the contacting surfaces of the UHMWPE pins, depending on the type of textured or untextured Ti6Al4V surface against which the pin slides. Three sample pins were selected because they represent the pins articulating with untextured, only microscale textured, and one from multiscale textured surface, which resulted in the lowest UHMWPE wear during the experiment.

The primary wear mechanism of the UHMWPE pin sliding on the smooth untextured surface, (D000d00) was abrasive wear and created deep scratches, as well as some adhesive wear as shown in Figure 11a. In contrast, the pin sliding on just the microscale textured surface (D100d00) exhibits fewer and shallower scratch marks, indicating less damage by abrasive wear, as depicted in Figure 11b. However, some noticeable adhesion marks are observed. The sample pin articulated with the multiscale textured surface (D100d30), shows faint wear marks compared to both untextured and just microscale textured surfaces (Figure 11c). The original machining and polishing marks (not parallel to the sliding direction) are still clearly apparent on the UHMWPE pin surface after sliding against the multiscale textured surface (Figure 11c) but have been largely worn off after sliding against the smooth surface (Figure 11a). The change in wear mechanism is caused by the multiscale surface textures creating thicker lubricant films and trapping wear debris compared to the untextured surface. Similar results were observed in [42,43] where UHMWPE pins articulating with untextured and microtextured CoCrMo discs.

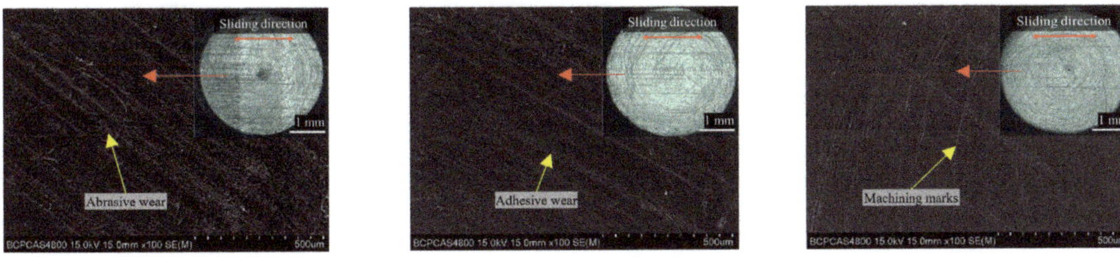

(**a**) (**b**) (**c**)

Figure 11. SEM and optical images of UHMWPE pin bottom surface after reciprocating sliding test against sample discs: (**a**) D000d00; (**b**) D100d00; (**c**) D100d30.

Figure 12 depicts the average surface roughness on the UHMWPE surface before and after the friction test. It is evident that the initial surface roughness (Ra 0.441 µm) of all pins increased after the pin-on-disc test due to the abrasive and adhesive wear damage. However, the change in surface roughness resulting from wear decreased by 58.7% for the multiscale textured surface D100d30 (Ra 0.859 µm), and by 36.74% for the only microscale textured surface D100d00 (Ra 1.317 µm), compared to the untextured surface D000d00 (Ra 2.082 µm). Within the multiscale textured surfaces, the change in surface roughness decreased as the texture density of interstitial textures reduced, aligning with both the friction coefficient and simulation results.

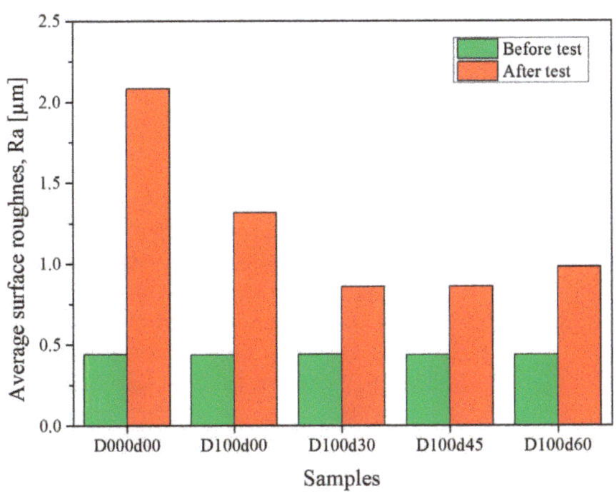

Figure 12. Average surface roughness, Ra, of the UHMWPE pins before and after the tribo-test.

In conclusion, surface texturing, especially multiscale textures, shows potential in maximizing lubricant film thickness by improving wettability and hydrodynamic lubrication. This, in turn, increases the separation of tribo-pairs, resulting in low and stable friction coefficient, reduced wear of UHMWPE, and ultimately, a noticeable improvement in the longevity of prosthetic hip implants.

4. Limitation and Accuracy

The current work utilizes an HL model and reciprocating pin-on-disc experiments to study the effect of multiscale texture features on the surface of biocompatible materials as a first step to indicate the suitability of the multiscale surface texturing technique to increase the durability of prosthetic hip implants by reducing the friction and wear. There are several simplifications of the current methods that differ from a real prosthetic hip implant. For example, our simulations assume parallel sliding surfaces to model a small section of a spherical prosthetic hip implant. The current HL model does not consider bearing surface deformation, which differs from hard-on-soft bearing surfaces such as metal on polymer or ceramic on polymer prosthetic hip implants, wherein the soft polymer surface deforms significantly more than the hard bearing surface. However, prior elasto-hydrodynamic lubrication (EHL) simulations that included the deformation of the polymer bearing surface showed similar trends as reported here for optimal texture design parameters [21,22].

Additionally, the Reynolds equation includes several assumptions such as fully flooded lubricant supply, steady-state, and Newtonian lubricant behavior, that may not always exist in a dynamically changing prosthetic hip implant with shear-thinning synovial fluid lubrication. However, the trends indicated in this study will help to increase the lubricant film thickness, which will increase the percentage of the time that full-film hydrodynamic lubrication will carry the applied load and decrease the percentage of time that the mixed and/or boundary lubrication regimes allow solid-on-solid contact between the bearing surfaces.

The physical friction experiments are carried out with pin-on-disc test set up which employed a simplified disc geometry, small pin contact area and fewer degrees of freedom compared to actual prosthetic hip implants. Moreover, constant loading and speed operating conditions used in this study do not replicate the in vivo test condition. However, the pin-on-disc experiment is useful for fast- and low-cost testing prior to conducting extensive tests. The results reveal the potential of multiscale texturing of hip implant bearing surface in reducing friction and wear of UHMWPE. Therefore, future research should consider conducting long-duration wear tests under dynamic loading and speed conditions on a spherical geometry (femoral head) with multiscale texture features on the bearing surface. Because the simulation results are independent of specific materials, this multiscale surface texture design could be investigated for other biomedical materials, such as CoCrMo or stainless steel.

5. Conclusions

We conclude that:

- A pattern of multiscale texture features shows promise, through simulation and experiment, in improving the lubricant film thickness in parallel sliding bearings, compared to smooth bearing surfaces and surfaces with just microscale texture features. The measured difference is most significant for the experimental results. Improvements are due to an increased lubricant wettability and hydrodynamic pressure.
- The optimal microscale texture design parameters change with the operating conditions, but the optimal interstitial texture design parameters are the same for every condition tested. Specifically, decreasing the aspect ratio and texture density of interstitial texture features leads to increased lubricant film thickness and reduced friction coefficient and wear.
- The multiscale textured surfaces with the smallest interstitial texture features (30 μm diameter, 0.73 μm depth) that could be reliably manufactured with picosecond pulsed laser ablation improved the static CA by 34%, reduced friction by 32.3%, and minimized running-in wear on UHMWPE surfaces by 58.7% compared to the untextured bearing surfaces, emphasizing the practical advantages of multiscale texturing in parallel sliding bearings and particularly for prosthetic hip implants.
- The surface texturing patterns change the relevant wear mechanisms on the UHMWPE pins from severe abrasive scratches and adhesive wear when sliding against the untextured surface, to some abrasive scratches and adhesive wear for the microscale textured surfaces, and only minor abrasive scratches for the multiscale texture features.

The change in wear mechanism is caused by the multiscale surface textures creating thicker lubricant films and trapping more wear debris than the other bearing surfaces.

Author Contributions: Conceptualization, F.B.T., Q.A. and T.Z.; methodology, F.B.T. and Q.A.; software, Q.A.; validation, F.B.T. and Q.A.; formal analysis, F.B.T. and Q.A.; investigation, F.B.T. and Q.A.; resources, T.Z.; data curation, F.B.T. and Q.A.; writing—original draft preparation, F.B.T. and Q.A.; writing—review and editing, F.B.T., Q.A. and T.Z.; visualization, F.B.T. and Q.A.; supervision, Q.A. and T.Z.; project administration, T.Z.; funding acquisition, T.Z. All authors have read and agreed to the published version of the manuscript.

Funding: This research was funded by the Science and Technology Major Project of Jiangxi Province, grant number 20213AAE02003, the Beijing Municipal Natural Science Foundation, grant number JQ20014, and the National Natural Science Foundation of China, grant numbers 51875043 and 52205440. The APC was funded by the Beijing Institute of Technology and Brigham Young University.

Data Availability Statement: The data presented in this study are available on request from the corresponding author.

Conflicts of Interest: The authors declare no conflicts of interest. The funders had no role in the design of the study; in the collection, analyses, or interpretation of data; in the writing of the manuscript; or in the decision to publish the results.

References

1. Learmonth, I.D.; Young, C.; Rorabeck, C. The operation of the century: Total hip replacement. *Lancet Lond. Engl.* **2007**, *370*, 1508–1519. [CrossRef]
2. Evans, J.T.; Evans, J.P.; Walker, R.W.; Sayers, A. How long does a hip replacement last? A systematic review and meta-analysis of case series and national registry reports with more than 15 years of follow-up. *Lancet Lond. Engl.* **2019**, *393*, 647–654. [CrossRef]
3. Allen, Q.; Raeymaekers, B. Surface Texturing of Prosthetic Hip Implant Bearing Surfaces: A Review. *J. Tribol.* **2021**, *143*, 040801. [CrossRef]
4. Rajaee, S.S.; Campbell, J.C.; Mirocha, J.; Paiement, G.D. Increasing Burden of Total Hip Arthroplasty Revisions in Patients between 45 and 64 Years of Age. *JBJS* **2018**, *100*, 449. [CrossRef] [PubMed]
5. Liu, A.; Richards, L.; Bladen, C.L.; Ingham, E.; Fisher, J.; Tipper, J.L. The biological response to nanometre-sized polymer particles. *Acta Biomater.* **2015**, *23*, 38–51. [CrossRef]
6. Capello, W.N.; D'Antonio, J.A.; Ramakrishnan, R.; Naughton, M. Continued improved wear with an annealed highly cross-linked polyethylene. *Clin. Orthop.* **2011**, *469*, 825–830. [CrossRef] [PubMed]
7. Oral, E.; Christensen, S.D.; Malhi, A.S.; Wannomae, K.K.; Muratoglu, O.K. Wear resistance and mechanical properties of highly crosslinked UHMWPE doped with Vitamin-E. *J. Arthroplasty* **2006**, *21*, 580–591. [CrossRef]
8. Cai, Y.; Yan, S. Development of ceramic-on-ceramic implants for total hip arthroplasty. *Orthop. Surg.* **2010**, *2*, 175–181. [CrossRef] [PubMed]
9. Skjöldebrand, C.; Tipper, J.L.; Hatto, P.; Bryant, M.; Hall, R.M.; Persson, C. Current status and future potential of wear-resistant coatings and articulating surfaces for hip and knee implants. *Mater. Today Bio.* **2022**, *15*, 100270. [CrossRef]
10. Rajak, D.K.; Kumar, A.; Behera, A.; Menezes, P.L. Diamond-Like Carbon (DLC) Coatings: Classification, Properties, and Applications. *Appl. Sci.* **2021**, *11*, 4445. [CrossRef]
11. Nsilani Kouediatouka, A.; Ma, Q.; Liu, Q.; Mawignon, F.J.; Rafique, F.; Dong, G. Design Methodology and Application of Surface Texture: A Review. *Coatings* **2022**, *12*, 1015. [CrossRef]
12. Tarabolsi, M.; Klassen, T.; Mantwill, F.; Schulz, A.P. Patterned CoCrMo and Al_2O_3 surfaces for reduced free wear debris in artificial joint arthroplasty. *J. Biomed. Mater. Res. A* **2013**, *101*. [CrossRef]
13. Tewelde, F.B.; Zhou, T.; Zhou, J.; Guo, W.; Zhao, B.; Ge, X.; Wang, W.; Wang, X.; Wang, X. Asymmetric surface texturing for directional friction control under dry sliding condition. *Tribol. Int.* **2023**, *181*, 108321. [CrossRef]
14. López-Cervantes, A.; Domínguez-López, I.; Barceinas-Sánchez, J.D.O.; García-García, A.L. Effects of surface texturing on the performance of biocompatible UHMWPE as a bearing material during in vitro lubricated sliding/rolling motion. *J. Mech. Behav. Biomed. Mater.* **2013**, *20*, 45–53. [CrossRef]
15. Shen, G.; Zhang, J.; Kang, C.; Fang, F. Study on surface texture patterns for improving tribological performance of bioimplants. *Surf. Coat. Technol.* **2021**, *422*, 127567. [CrossRef]
16. Kustandi, T.S.; Choo, J.H.; Low, H.Y.; Sinha, S.K. Texturing of UHMWPE surface via NIL for low friction and wear properties. *J. Phys. Appl. Phys.* **2009**, *43*, 015301. [CrossRef]
17. Liu, S.; Sai, Q.; Wang, S.; Williams, J. Effects of Laser Surface Texturing and Lubrication on the Vibrational and Tribological Performance of Sliding Contact. *Lubricants* **2022**, *10*, 10. [CrossRef]
18. Ito, H.; Kaneda, K.; Yuhta, T.; Nishimura, I.; Yasuda, K.; Matsuno, T. Reduction of polyethylene wear by concave dimples on the frictional surface in artificial hip joints. *J. Arthroplasty* **2000**, *15*, 332–338. [CrossRef]

19. Dong, Y.; Svoboda, P.; Vrbka, M.; Kostal, D.; Urban, F.; Cizek, J.; Roupcova, P.; Dong, H.; Krupka, I.; Hartl, M. Towards near-permanent CoCrMo prosthesis surface by combining micro-texturing and low temperature plasma carburising. *J. Mech. Behav. Biomed. Mater.* **2016**, *55*, 215–227. [CrossRef]
20. Borjali, A.; Monson, K.; Raeymaekers, B. Friction between a polyethylene pin and a microtextured CoCrMo disc, and its correlation to polyethylene wear, as a function of sliding velocity and contact pressure, in the context of metal-on-polyethylene prosthetic hip implants. *Tribol. Int.* **2018**, *127*, 568–574. [CrossRef]
21. Allen, Q.; Raeymaekers, B. The Effect of Texture Floor Profile on the Lubricant Film Thickness in a Textured Hard-on-Soft Bearing with Relevance to Prosthetic Hip Implants. *J. Tribol.* **2020**, *143*, 021801. [CrossRef] [PubMed]
22. Allen, Q.; Raeymaekers, B. Maximizing the Lubricant Film Thickness between a Rigid Microtextured and a Smooth Deformable Surface in Relative Motion, Using a Soft Elasto-Hydrodynamic Lubrication Model. *J. Tribol.* **2020**, *142*, 071802. [CrossRef] [PubMed]
23. Koch, K.; Bhushan, B.; Jung, Y.C.; Barthlott, W. Fabrication of artificial Lotus leaves and significance of hierarchical structure for superhydrophobicity and low adhesion. *Soft. Matter.* **2009**, *5*, 1386–1393. [CrossRef]
24. Bixler, G.D.; Bhushan, B. Fluid Drag Reduction with Shark-Skin Riblet Inspired Microstructured Surfaces. *Adv. Funct. Mater.* **2013**, *23*, 4507–4528. [CrossRef]
25. Wang, L.; Hui, Y.; Fu, C.; Wang, Z.; Zhang, M.; Zhang, T. Recent advances in *Gecko*—Inspired adhesive materials and application. *J. Adhes. Sci. Technol.* **2020**, *34*, 2275–2291. [CrossRef]
26. Longmore, R.B.; Gardner, D.L. The surface structure of ageing human articular cartilage: A study by reflected light interference microscopy (RLIM). *J. Anat.* **1978**, *126*, 353–365. [PubMed]
27. König, F.; Rosenkranz, A.; Grützmacher, P.G.; Mücklich, F.; Jacobs, G. Effect of single- and multi-scale surface patterns on the frictional performance of journal bearings—A numerical study. *Tribol. Int.* **2020**, *143*, 106041. [CrossRef]
28. Grützmacher, P.G.; Profito, F.J.; Rosenkranz, A. Multi-Scale Surface Texturing in Tribology—Current Knowledge and Future Perspectives. *Lubricants* **2019**, *7*, 95. [CrossRef]
29. Segu, D.Z.; Choi, S.G.; Choi, J.H.; Kim, S.S. The effect of multi-scale laser textured surface on lubrication regime. *Appl. Surf. Sci.* **2013**, *270*, 58–63. [CrossRef]
30. Resendiz, J.; Egberts, P.; Park, S.S. Tribological Behavior of Multi-scaled Patterned Surfaces Machined through Inclined End Milling and Micro Shot Blasting. *Tribol. Lett.* **2018**, *66*, 132. [CrossRef]
31. Zhang, K.; Li, H.; Zhang, C.; Han, Y.; Guo, X.; Liu, T. Effect of ion beam etching on the tribological performance of laser textured Co-Cr-Mo alloy. *Opt. Laser Technol.* **2023**, *160*, 109097. [CrossRef]
32. Hsu, S.M.; Jing, Y.; Hua, D.; Zhang, H. Friction reduction using discrete surface textures: Principle and design. *J. Phys. Appl. Phys.* **2014**, *47*, 335307. [CrossRef]
33. Hsu, S.M.; Jing, Y.; Zhao, F. Self-adaptive surface texture design for friction reduction across the lubrication regimes. *Surf. Topogr. Metrol. Prop.* **2015**, *4*, 014004. [CrossRef]
34. Allen, Q.; Raeymaekers, B. Soft EHL Simulations of Lubricant Film Thickness in Textured Hard-on-Soft Bearings Considering Different Cavitation Models, in the Context of Prosthetic Hip Implants. *Tribol. Lett.* **2021**, *69*, 118. [CrossRef]
35. Dobrica, M.B.; Fillon, M.; Pascovici, M.D.; Cicone, T. Optimizing surface texture for hydrodynamic lubricated contacts using a mass-conserving numerical approach. *Proc. Inst. Mech. Eng. Part J J. Eng. Tribol.* **2010**, *224*, 737–750. [CrossRef]
36. Gao, Y.; Wu, B.; Zhou, Y.; Tao, S. A two-step nanosecond laser surface texturing process with smooth surface finish. *Appl. Surf. Sci.* **2011**, *257*, 9960–9967. [CrossRef]
37. De Palo, R.; Mazzarone, A.E.; Volpe, A.; Gaudiuso, C.; Mezzapesa, F.P.; Spagnolo, V.; Ancona, A. Investigation of Laser-Induced Surface Structures (LIPSS) on quartz and evaluation of their influence on material wettability. *Opt. Laser Technol.* **2024**, *169*, 110097. [CrossRef]
38. Wenzel, R.N. Resistance of solid surfaces to wetting by water. *Ind. Eng. Chem.* **1936**, *28*, 988–994. [CrossRef]
39. Pratap, T.; Patra, K. Mechanical micro-texturing of Ti-6Al-4V surfaces for improved wettability and bio-tribological performances. *Surf. Coat. Technol.* **2018**, *349*, 71–81. [CrossRef]
40. Misyura, S.Y.; Kuznetsov, G.V.; Feoktistov, D.V.; Volkov, R.S.; Morozov, V.S.; Orlova, E.G. The influence of the surface microtexture on wettability properties and drop evaporation. *Surf. Coat. Technol.* **2019**, *375*, 458–467. [CrossRef]
41. Chyr, A.; Qiu, M.; Speltz, J.W.; Jacobsen, R.L.; Sanders, A.P.; Raeymaekers, B. A patterned microtexture to reduce friction and increase longevity of prosthetic hip joints. *Wear* **2014**, *315*, 51–57. [CrossRef] [PubMed]
42. Borjali, A.; Langhorn, J.; Monson, K.; Raeymaekers, B. Using a patterned microtexture to reduce polyethylene wear in metal-on-polyethylene prosthetic bearing couples. *Wear* **2017**, *392–393*, 77–83. [CrossRef] [PubMed]
43. Alvarez-Vera, M.; Ortega, J.A.; Ortega-Ramos, I.A.; Hdz-García, H.M.; Muñoz-Arroyo, R.; Díaz-Guillén, J.C.; Acevedo-Dávila, J.L.; Hernández-Rodriguez, M.A.L. Tribological and microstructural characterization of laser microtextured CoCr alloy tested against UHMWPE for biomedical applications. *Wear* **2021**, *477*, 203819. [CrossRef]

lubricants

MDPI

Article

A Non-Linear Optimization Model for Controlling the Real Area of Contact in Surface Texture Design

Sandra D. Ekşioğlu * and Min Zou

College of Engineering, University of Arkansas, 4183 Bell Eng. Bld., Fayetteville, AR 72701, USA
* Correspondence: sandra@uark.edu

Abstract: Motivated by the potential of surface texturing to enhance the tribological performance of micro- and nano-electromechanical systems (MEMS/NEMS), this study proposes a novel non-linear optimization approach for designing textured surfaces. This model minimizes the contact area between interacting surfaces and deformation during sliding under dry conditions by controlling key design parameters, such as the size and shape of the designed surface. We test the performance of the proposed model using the lotus leaf surface with dimensions of 248 $\times$ 136 micrometers. Due to the large size of the model, we propose a solution approach which consists of a data aggregation step, an optimization step, and a data disaggregation step. The optimization step decomposes the model into smaller models that are easier to solve. Via the sensitivity analysis, we highlight the trade-offs between data aggregation and model decomposition and their effect on the quality of the solutions found. In conclusion, our approach bridges the gap between fabrication capabilities and design requirements, paving the way for significant advances in tribological performance and surface engineering.

Keywords: surface texture design; integer linear program; non-linear optimization

1. Introduction

This research proposes a non-linear optimization model to design textured surfaces. Surface texturing is the process of creating micro- or nano-scale features on a surface. This technique modifies surface topography to improve tribological properties in micro- and nano-electromechanical systems (MEMS/NEMS) [1–3]. The objective of our proposed model is to determine the feature size, shape, and arrangement that minimize the real contact area between interacting surfaces and minimize deformation during sliding in dry conditions. Our motivation is the fact that, in small-scale systems, surface forces are highly sensitive to the contact area. Minimizing the contact area leads to minimizing adhesion and friction forces. Failure to do so amplifies surface forces that can lead to device failure [1,4]. There are numerous examples of small-scale mechanical devices in biomedical implants and precision manufacturing tools (such as pacemakers, wheelchairs, drills, etc.) that contain moving or sliding components where friction reduction via surface texturing could substantially improve performance and longevity [5,6]. The outcome of the proposed model is a surface texture that is specifically engineered to deliver improved performance in tribological applications by controlling the real contact area. Precise control of the design and fabrication of MEMS/NEMS can unlock the full potential of surface texturing for tribological applications.

Several techniques have been utilized to fabricate designed surfaces, including laser ablation, soft lithography, hot embossing, photolithography-based techniques, electron beam

https://doi.org/10.3390/lubricants13040163

lithography, nanoparticle spin coating, nanoparticle self-assembly, polymer jet printing, and "pulsed-mode" two-photon lithography [1,7]. These advanced fabrication technologies have enabled the creation of intricate surface textures with specific geometries. Despite notable advancements in fabrication technologies and theoretical modeling, a significant gap remains in techniques explicitly focused on the design of surface textures to reduce adhesion and friction. The design process—including determining the feature size, shape, and arrangement—is crucial for engineering surface textures that deliver superior performance in tribological applications. Research further highlights that the tribological performance of designed surfaces is significantly influenced by surface texturing, particularly the shape and specific parameters of the texture. However, existing studies have primarily examined surface texturing under lubricated and hydrodynamic conditions, with fewer investigations addressing dry sliding conditions.

Motivated by the potential of surface texturing to enhance tribological performance in miniaturized MEMS/NEMS, this study presents a novel approach to texture design under dry-sliding conditions. We develop a non-linear optimization model that minimizes both the contact area between interacting surfaces and texture deformation, while maintaining precise control over key design parameters such as feature size and shape. By leveraging advanced mathematical modeling and optimization, our work provides a robust framework for predicting and refining surface textures across a broad range of materials, including metals, ceramics, and polymers. In particular, stiff/hard ceramics commonly used in MEMS/NEMS stand to benefit greatly from minimal contact deformation, and metals similarly exhibit high stiffness that can accommodate the proposed design approach. By bridging the gap between manufacturing capabilities and performance requirements, this framework paves the way for major advancements in tribological performance and surface engineering.

The proposed model for surface texturing reduces friction and wear, thereby extending the lifespan of components in MEMS/NEMS. This leads to lower material consumption, reduced maintenance needs, and decreased energy usage associated with frictional losses. Furthermore, the model's focus on optimizing texture design ensures that fabrication resources are used efficiently, minimizing material waste and reducing the environmental footprint of manufacturing processes. By integrating sustainable design principles, the proposed approach supports the development of durable, high-performance surfaces that align with long-term ecological and economic objectives in advanced engineering applications.

2. Literature Review

In order to highlight the contribution of the proposed optimization model, we review the related literature focused on experimental, simulation, computational, and optimization methods for surface texturing design. We also provide a review of non-linear optimization methods and solution approaches. We summarize this literature in the next few paragraphs.

2.1. Methods for Surface Texturing Design

2.1.1. Experimental Methods

Afshar-Mohajer and Zou [1] use experimental methods to examine the effects of three distinct textures under dry sliding conditions on friction, contact area, and texture durability. They compared cones ($3D$ textures) with cylinders ($2.5D$-base textures) and rods ($2.5D$-tip textures), which were fabricated using two-photon lithography and replica molding with PDMS. Their findings show that $3D$ textures effectively reduced friction by minimizing the real contact area at the tip of the texture and showed higher load-bearing capacity and resistance to deformation due to their larger base. Evangelista et al. [6] present a comprehensive analysis of the impact that surface texturing has on the dry tribological performance of polymeric materials in medical devices. Their findings show that geometric shapes, such as grooves and pillars, are particularly effective in influencing the frictional behavior on polymer surfaces. Designed polymer surfaces, especially those with reduced effective contact areas, show a significant reduction in friction. Kang et al. [8] explored the frictional behavior of aluminum substrates textured with smooth, micro-dome, and nanoscale dimple designs. They found that nanotextured surfaces increase friction under dry conditions due to higher adhesion forces but significantly reduced friction under wet conditions.

2.1.2. Simulations and Computational Models

A study by Cheng et al. [9] investigates the friction behavior of cemented carbide surfaces textured through femtosecond laser processing. Finite element analysis of the friction process was conducted using Abaqus software. The research focuses on assessing the impact of different texture orientations and the angle between the sliding direction and texture pattern on the friction coefficient and wear resistance. Berardo et al. [10] investigate the dry sliding behavior of laser-textured SiC and carbon black surfaces. Their findings revealed that larger hole diameters and higher texture densities significantly reduced dynamic friction and heat generation, especially on the stiffer SiC surfaces. Finite element simulations further validated the critical role of hole diameter in optimizing tribological performance. Xing et al. [11] combined experimental testing and empirical analysis to study the effects of linear and wavy grooves on Al_2O_3/TiC ceramics, finding that wavy grooves with smaller spacing improved wear resistance by reducing contact area, increasing surface roughness, and trapping debris. He et al. [12] combined experimental and simulation methods to analyze micro-textures on cemented carbide surfaces, showing that circular pit geometries provided superior friction reduction and wear resistance by improving stress distribution and accommodating wear debris. Using response surface methodology, they optimized texture fabrication parameters, achieving a 20% reduction in friction coefficients. Yang et al. [13] use simulation to improve the milling performance of different V-groove micro-texture ball-end milling cutters for milling titanium alloys. By simulating thermal and force characteristics, they refined groove parameters to reduce cutting force, improve chip flow, and enhance wear resistance.

2.1.3. Optimization Technique

Ge et al. [14] evaluate five algorithms—linear regression, decision tree, gradient boosting, support vector machine, and neural network—for their ability to predict load-carrying capacity and friction force based on texture parameters including depth, side length, surface ratio, and shape.

2.2. Non-Linear Optimization

Non-linear optimization is a cornerstone of modern applied mathematics, with numerous applications spanning various engineering disciplines and industrial processes [15,16]. In the context of surface texturing, non-linear models excel at capturing and optimizing the complex interactions among geometry, material properties, and load conditions. These models often involve multiple, interdependent parameters—such as surface roughness, dimple geometry, coating characteristics, and elastic deformations—which together govern frictional and wear behavior [17].

For well-behaved problems where the objective function and constraints are differentiable, gradient-based methods are typically employed. However, when the design space is highly non-convex, lacks closed-form gradients, or includes mixed discrete-continuous variables—common traits in tribological surface optimization—heuristic and meta-heuristic methods (e.g., genetic algorithms, particle swarm optimization, simulated annealing) have gained widespread attention [18,19]. These population-based or stochastic techniques rely on iterative refinement and global exploration strategies to evade local minima and discover near-optimal solutions. In parallel, problem-specific heuristics may integrate domain knowledge (e.g., restricting texture parameters to physically feasible ranges) to guide searches more effectively.

For very large-scale or distributed formulations, decomposition methods present a valuable approach to reduce complexity [20,21]. These techniques split the original problem into smaller, more tractable subproblems, each focusing on a subset of design variables or constraints. Coordinating these subproblems through a master–subproblem workflow allows for the exploitation of specific structural features in tribological models, such as spatial partitioning of a textured surface or separate treatment of structural domains. When combined with heuristics or meta-heuristics, these decomposition approaches can yield near-global solutions in a fraction of the time required by direct, "all-in-one" optimization.

In tribology research, numerous theoretical studies seek to identify optimal texturing parameters—for instance, the aspect ratio of dimples, texture density, relative dimple depth, or texturing extent—to improve load-carrying capacity and reduce friction. Such designs often rely on non-linear optimization to balance the physical constraints imposed by governing equations (e.g., the Reynolds equation) and to manage the large number of variables characterizing surface geometry. By systematically refining these variables, researchers can effectively reconcile competing objectives—while minimizing surface contact—and thereby achieve more robust and efficient textured surfaces for real-world tribological applications [22].

2.3. Key Contributions to the Literature and Practical Relevance

Counterface-Aware Texture Optimization: Unlike existing approaches that focus solely on designing texture for one surface, our method systematically optimizes the counterface texture based on the topography of the given surface. This results in a more realistic and application-driven approach that directly impacts friction, wear resistance, and load distribution in real-world mechanical systems.

Multi-Factor Optimization for MEMS/NEMS and Large-Scale Applications: Our model is particularly relevant to MEMS/NEMS, as well as larger mechanical components where precise control over tribological properties is essential. The method takes a systems approach by incorporating multiple design parameters, making it a scalable and adaptable solution for high-precision industries such as aviation, manufacturing, and biomedical devices.

Beyond Empirical Design—A Data-Driven Predictive Tool: Many current texture designs rely on empirical methods, which are labor-intensive, time-consuming, and often suboptimal. Our computational optimization framework enables engineers to efficiently identify ideal texture parameters while significantly reducing the reliance on costly physical prototyping.

Compatibility with Modern Fabrication Techniques: The proposed method is compatible with widely used fabrication techniques such as photolithography, laser texturing, and other precision manufacturing methods. The ability to fine-tune texture designs computationally before fabrication makes this method highly relevant to practical industrial applications.

Statistical and Computational Validation: Our results demonstrate that enforcing symmetry in texture design increases contact area, which can negatively impact tribological performance. Through hypothesis testing and statistical analysis, we show that non-symmetric designs lead to significantly lower contact areas, thereby improving wear resistance and reducing friction.

The proposed model can be further refined by accounting for the mechanics of deformation during sliding, etc. However, we firmly believe that this work represents an important step forward in optimizing surface textures beyond conventional single-surface designs. By considering both the textured surface and the counterface, our approach provides a more comprehensive and practical solution for real-world applications. The integration of multi-factor optimization and computational modeling makes it a valuable predictive tool that enhances surface engineering in both research and industry.

The goal of the proposed research is to provide a robust framework that supports the effective design of textured surfaces for various engineering applications. This research will contribute to the ongoing efforts to improve the efficiency and reliability of MEMS/NEMS, paving the way for their broader adoption in industrial and commercial applications.

3. Modeling Approach

3.1. Problem Definition and Assumptions

Consider that we are given the width $\mathcal{W}$ and length $\mathcal{L}$ of a surface, and the corresponding height, h_{ij}, for each point $(i, j) \in \mathcal{W} \times \mathcal{L}$. We propose a mathematical model to design the texture of a new surface with dimensions $\omega \times \mathfrak{l}$ $(\omega < \mathcal{W}, \mathfrak{l} < \mathcal{L})$—that is, to determine the height Z_{ij} of each point $(i, j) \in \omega \times \mathfrak{l}$. We refer to this as the *designed surface*. The objective of the proposed model is to minimize the contact area between the two surfaces and deformation during a linear sliding of the designed surface along the X axis in dry conditions.

The model captures the unit contact between the two surfaces via G^x_{ijk}. Unit contact is measured at different points k, as the designed surface slides along the X axis. Let $d = \max_{(i,j)\in\mathcal{W}\times\mathcal{L}} h_{ij}$ represent the maximum distance between the lower and upper surfaces of solids in contact. Then,

$$G^x_{ijk} = \left[(h_{i,j+k-1} + Z_{ij}) - d\right]^+, \forall i \in [0, \omega], j \in [0, \mathfrak{l}], k = 1, \ldots, K. \tag{1}$$

In Equation (1), $[a]^+ = a$ if $a \geq 0$, $[a]^+ = 0$ if $a \leq 0$. We acknowledge that based on (1), the two surfaces are in contact when $G^x_{ijk} = 0$. For this reason, we adjust the definition of (1). Let π be a very small positive number such that if $G^x_{ijk} = 0$, then $G^x_{ijk} + \pi > 0$, and if $G^x_{ijk} < 0$, then $G^x_{ijk} + \pi < 0$. In this case,

$$G^x_{ijk} = \left[(h_{i,j+k-1} + Z_{ij}) - d + \pi\right]^+, \forall i = 1, \ldots, \omega, j = 1, \ldots, \mathfrak{l}, k = 1, \ldots, K. \tag{2}$$

Equations (2) and (3) enforce the unit contact to be a positive number smaller than π. This numerical manipulation is necessary to show mathematically that the two surfaces touch.

$$G^x_{ijk} \leq \pi, \quad \forall i = 1, \ldots, \omega, j = 1, \ldots, \mathfrak{l}, k = 1, \ldots, K. \tag{3}$$

Figure 1 illustrates our notation and the contact area as the designed surface slides along the X axis. Figure 1a illustrates the original surface (in the bottom) and the designed surface (in the top) at position $k = 1$, and before sliding begins. Figure 1b illustrates the contact area between two surfaces, notations we use, and our calculations of the contact area. Figure 1c,d illustrate the two surfaces in contact in positions $k = 2$ and $k = K$ during sliding of the designed surface along the X axis. This example illustrates a surface with dimples. Similar notation and modeling can be used for surfaces that have pillars.

In the next section, we present a mathematical model which we develop based on the following assumptions:

(i) There is minimum texture deformation during sliding in dry conditions;
(ii) The designed surface is smooth to minimize friction and wear;
(iii) The sliding distance of the designed surface is limited by the dimensions of the original surface;
(vi) The surfaces cannot commensurate to avoid interlocking of the surface features;
(v) The dimensions of the surface features are such that buckling does not occur under a given normal force.

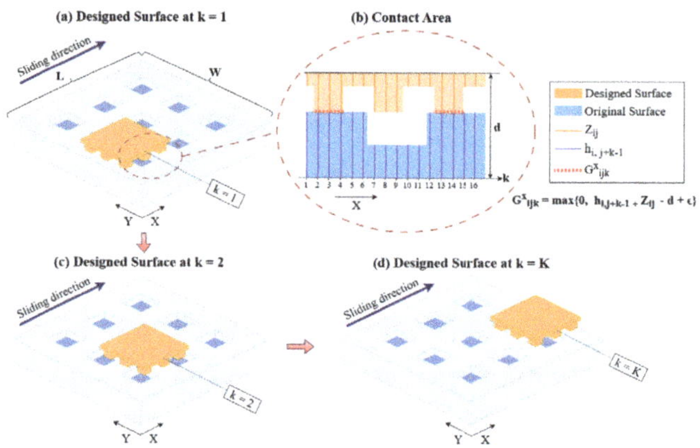

Figure 1. Illustration of the contact area between the sliding surfaces.

3.2. A Non-Linear Optimization Model

We propose a non-linear program (NLP) to model and solve the problem defined in Section 3.1. Table 1 summarizes the model parameters and the decision variables.

Table 1. List of parameters and decision variables used in the optimization model.

Parameters	Definition
$\mathcal{W}$	Width of the given surface. W is the corresponding number of discrete points.
ω	Width of the designed surface. w is the corresponding number of discrete points.
$\mathcal{L}$	Length of the given surface. L is the corresponding number of discrete points.
$\mathfrak{l}$	Length of the designed surface. l is the corresponding number of discrete points.
h_{ij}	Height of point $(i, j) \in \mathcal{W} \times \mathcal{L}$.
p	The normal load at contact.
y_m	The yield strength of material m.
d	The maximum the distance between the lower and upper surfaces of the two surfaces, $\max_{(i,j)\in\mathcal{W}\times\mathcal{L}} h_{ij}$.
M	A large number.
π	A small number.
ϵ	$= \pi p / y_m$.
Decision Variables	**Definition**
Z_{ij}	Height of point $(i, j) \in w \times l$.
G^x_{ijk}	Unit contact area.

The objective function (4) minimizes the total contact area at different points k, during sliding of the designed surface in dry conditions along the X axis.

$$\min : \sum_{k=1}^{K} \int_0^{\omega} \int_0^{\mathfrak{l}} (1/\pi) G_{ijk}^{x} d_j d_i \tag{4}$$

To minimize texture deformation, the contact pressure should remain below the yield strength of the surface material. Let y_m represent the yield strength of material m, and let p represent the normal load of contact area.

$$\text{If } \int_0^{\omega} \int_0^{\mathfrak{l}} G_{ijk}^{x} d_j d_i > 0, \text{ then, } y_m \geq p / \left(\int_0^{\omega} \int_0^{\mathfrak{l}} (1/\pi) G_{ijk}^{x} d_j d_i \right), \quad \forall k = 1, \ldots, K. \tag{5}$$

This is equivalent to

$$\int_0^{\omega} \int_0^{\mathfrak{l}} G_{ijk}^{x} d_j d_i \geq \pi p / y_m, \quad \forall k = 1, \ldots, K. \tag{6}$$

Symmetrical shapes ensure uniform distribution of stresses and loads across the surface, reducing uneven wear, and simplify the manufacturing process, reducing production costs. We use constraints (7)–(9) to impose symmetry in the designed surface. Constraints (7) and (8) impose symmetry along the edges of the surface. Constraints (9) impose symmetry along the diagonal of the designed surface and are therefore used to enforce the symmetry of square shapes. Collectively, these constraints impose the full symmetry of the designed surface.

$$Z_{i1} = Z_{il}, \qquad \forall i = 1, \ldots, \omega, \tag{7}$$

$$Z_{1j} = Z_{wj}, \qquad \forall j = 1, \ldots, \mathfrak{l}, \tag{8}$$

$$Z_{ij} = Z_{ji}, \quad \forall i = 1, \ldots, \omega, j = 1, \ldots, \mathfrak{l}, i < j. \tag{9}$$

Figure 2 illustrates these symmetry patterns for a surface with dimensions $\omega, \mathfrak{l} = 5$. In this figure, squares with the same number have the same value of Z_{ij}.

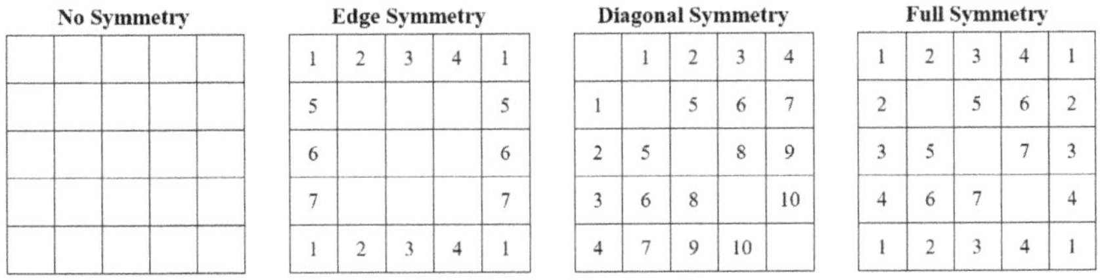

Figure 2. Patterns of symmetry.

In surface texturing, asperity (that is, sharp edges) can create points of high stress concentration that could lead to fractures and failure of the material under load and during operations [1,7]. Hierarchical micro/nano structures that feature longer nanohairs or nanopillars on top of the micropillars are particularly susceptible to deformation and breakage during sliding friction tests [7]. In addition, manufacturing designed surfaces with sharp edges can be more challenging and costly. To ensure a smoother surface, we modify the objective function by including two additional terms:

$$\min : \sum_{k=1}^{K} \int_0^{\omega} \int_0^{\mathfrak{l}} (1/\pi) G_{ijk}^{x} d_j d_i + \int_0^{\omega} \int_0^{\mathfrak{l}} \left(\alpha F_{ij}^{x} + \beta F_{ij}^{y} \right) d_j d_i \tag{10}$$

where $\alpha, \beta \geq 0$, and

$$F^x_{ij} = \|Z_{i,j} - Z_{i,j+1}\|, \quad \forall i = 1, \dots, \omega, j = 1, \dots, \mathfrak{l}, \tag{11}$$

$$F^y_{ij} = \|Z_{i,j} - Z_{i+1,j}\|, \quad \forall i = 1, \dots, \omega, j = 1, \dots, \mathfrak{l}, \tag{12}$$

$$Z_{ij} \geq 0, \quad \forall i = 1, \dots, \omega, j = 1, \dots, \mathfrak{l}. \tag{13}$$

F^x_{ij} and F^y_{ij} represent the absolute difference between neighboring points Z_{ij} with $Z_{i,j+1}$ and with $Z_{i+1,j}$, respectively. The parameters α and β are penalty terms. The values of α and β determine the relative importance of the different terms in the objective. For example, if α is a very large number, the model will find solutions for which the value of F^x_{ij} is minimized. In fact, $F^x_{ij} = 0$ when $Z_{ij} = Z_{i,j+1}$, resulting in a flat surface.

Conversely, when $\alpha = 0$, the magnitude of F^x_{jk} does not influence the objective function, allowing for the solution with large differences between neighboring points and resulting in non-smooth surfaces. The parameter β has a similar effect. Via our numerical analysis, we determine α, β to ensure that the designed surface is smooth but not flat.

In summary, our proposed NLP model minimizes (10) subject to (2), (3), (6)–(9) and (11)–(13). The full problem formulation can be found in the Appendix A. Notice that the focus of this model is on minimizing the contact area as the designed surface slides in dry conditions along the X-axis. This model can be adopted to minimize contact area as the designed surface moves along the Y-axis. In this case, the objective function minimizes G^y_{ijk}.

$$G^y_{ijk} = \left[(h_{i+k-1,j} + Z_{ij}) - d + \pi\right]^+, \forall i = 1, \dots, \omega, j = 1, \dots, \mathfrak{l}, k = 1, \dots, K. \tag{14}$$

3.3. A Linear Optimization Model

The non-linearity of the NLP model is due to the objective function (10) and constraints (2), (6), (11) and (12).

We approximate the continuous domains of the integrations in (10) by discretizing them into a grid of points. For example, we discretize the integration region $[0, \omega] \times [0, \mathfrak{l}]$ of $\int_0^\omega \int_0^\mathfrak{l} G^x_{ijk} d_j d_i$ by dividing the interval $[0, \omega]$ into w sub-intervals of width Δi; and by dividing $[0, \mathfrak{l}]$ into l sub-intervals of width Δj. The corresponding sub-intervals have dimensions $\Delta i \times \Delta j$. Thus, $\int_0^w \int_0^l G^x_{ijk} d_j d_i \approx \sum_{i=1}^{w} \sum_{j=1}^{l} G^x_{ijk} \Delta i \Delta j$. We follow a similar approach to discretize $\int_0^w \int_0^l F^x_{ij}$ and $\int_0^w \int_0^l F^y_{ij}$. The following is the resulting linear objective function:

$$\min : \sum_{k=1}^{K} \sum_{i=1}^{w} \sum_{j=1}^{l} (1/\pi) G^x_{ijk} \Delta i \Delta j + \alpha \sum_{i=1}^{w} \sum_{j=1}^{l-1} F^x_{ij} \Delta i \Delta j + \beta \sum_{i=1}^{w-1} \sum_{j=1}^{l} F^y_{ij} \Delta i \Delta j \tag{15}$$

Constraint (6) becomes

$$\sum_{i=1}^{w} \sum_{j=1}^{l} G^x_{ijk} \Delta_i \Delta_j \geq \pi p / y_m, \quad \forall k = 1, \dots, K. \tag{16}$$

Constraints (2), (11) and (12) can be linearized by introducing binary variables and additional constraints [23]. We linearize constraints (2) by substituting them with the following constraints (for all $i = 1, \dots, w, j = 1, \dots, l, k = 1, \dots, K$).

$$G_{ijk}^x \geq h_{i,j+k-1} + Z_{ij} - d + \pi, \quad (17)$$

$$G_{ijk}^x \leq M * Y_{ijk}, \quad (18)$$

$$G_{ijk}^x \leq h_{i,j+k-1} + Z_{ij} - d + \pi + M * (1 - Y_{ijk}), \quad (19)$$

$$G_{ijk}^x \geq 0, \quad (20)$$

$$Y_{ijk} \in \{0,1\}. \quad (21)$$

We linearize constraints (11) and (12) by substituting them with the following constraints (for all $i = 1, \ldots, w, j = 1, \ldots, l$).

$$F_{ij}^x \geq Z_{ij} - Z_{i,j+1}, \quad (22)$$

$$F_{ij}^x \geq -Z_{ij} + Z_{i,j+1}, \quad (23)$$

$$F_{ij}^y \geq Z_{ij} - Z_{i+1,j}, \quad (24)$$

$$F_{ij}^y \geq -Z_{ij} + Z_{i+1,j}. \quad (25)$$

The proposed integer linear program (ILP) minimizes (15) subject to (3), (7)–(9), (13) and (16)–(25). This is an exact approximation of the NLP optimization model; therefore, solutions to ILP are also solutions to the NLP model. A full formulation of the ILP can be found in the Appendix B.

3.4. An Illustrative Example

Figure 3 presents three dimpled surfaces, each with dimensions $W = 7$ mm, $L = 21$ mm. The shape and size of the dimples are different for each surface. We use the following values for model parameters, $\alpha = 0$, $\beta = 0$, $w = 7$ mm, $l = 7$ mm, $\pi = 1$ mm. The material used in these surfaces is assumed to have a yield strength of 2 MPa. We impose full symmetry of the designed surface via constraints (7)–(9).

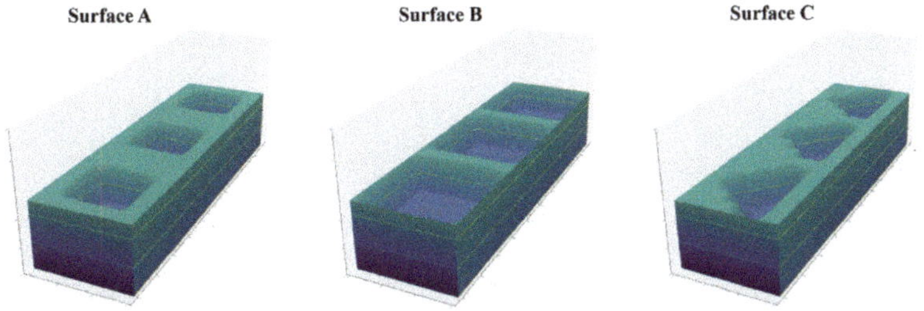

Figure 3. Samples of dimpled surfaces.

Figure 4a illustrates the contact area between the two surfaces at different positions during sliding of the designed surface along the X axes. Figure 4b presents views of the designed surface from different angles and Figure 4c presents views of the dimpled surface from different angles.

We conduct a sensitivity analysis to determine the impact of the normal unit load on contact (p) on the average contact area. In these experiments, the value of p varies from 0 to 1.6 Newton (N). The total load depends on the dimensions of the designed surface. The results of these experiments are summarized in Figure 5. These results indicate that the average contact area increases with p. Additionally, the maximum unit load for Surfaces A and C is 1.6 N and the average contact area is 0.8 mm. For Surface B, the corresponding values are 0.9 N and 0.5 mm. Larger loads would lead to deformation of the texture.

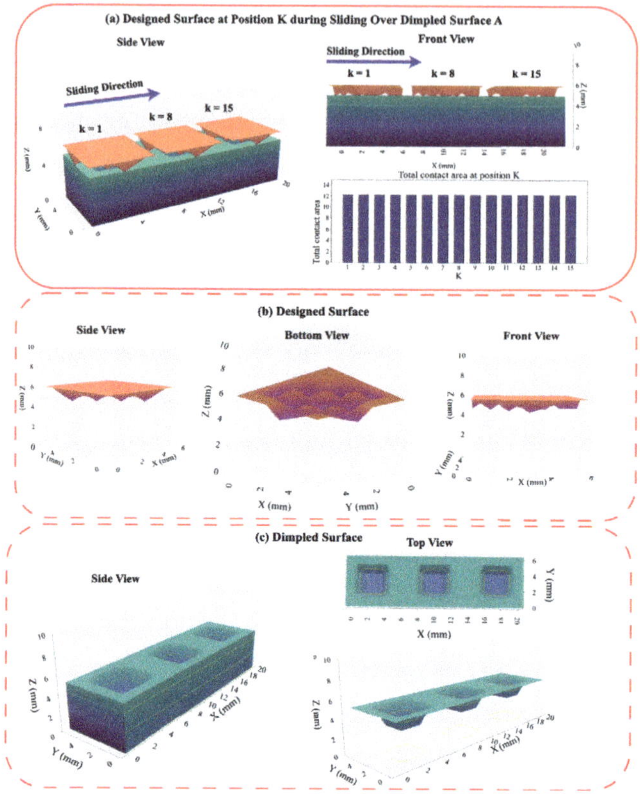

Figure 4. (**a**) Contact area at different positions during sliding along X axes. (**b**) Views of the designed surface from different angles. (**c**) Views of the dimpled surface from different angles.

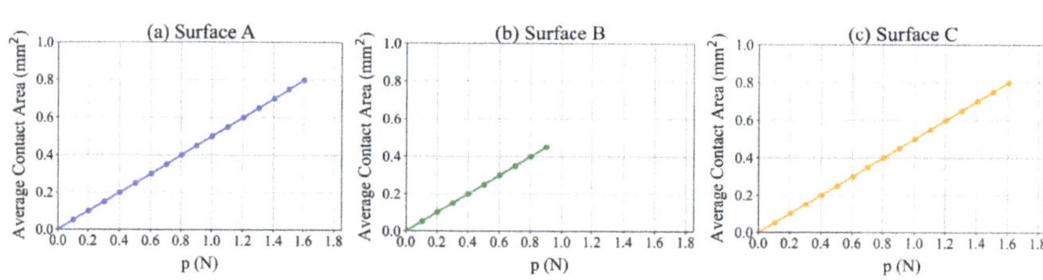

Figure 5. The average contact area for p between 0 and 1.6 N.

Figure 6 gives 3D representations of the designed textured surfaces for $p = 0.1$ N and $p = 0.5$ N. Notice the deeper indentations and more pronounced height variations of the designed surface for higher loads. When the normal load increases beyond $p = 0.5$ N, the designed surface becomes flat to maximize the contact area.

Table 2 presents the distribution of the contact area for different surfaces and different values of p. Based on these results, the contact area increases with load to minimize texture deformation. Furthermore, the shape of the dimples impacts the distribution of the contact area.

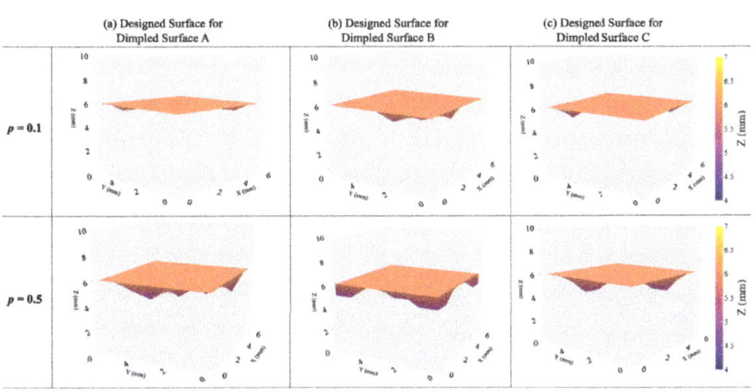

Figure 6. Three-dimensional representations of the designed textured surfaces for p = 0.1 N and $p = 0.5$ N.

Table 2. The distribution of the contact area for $p = 0.1$ N and $p = 0.5$ N.

Contact Area (mm^2)	Frequency					
	$p = 0.1$ N			$p = 0.5$ N		
	Surface A	Surface B	Surface C	Surface A	Surface B	Surface C
$[0, 0.25)$	615	675	690	495	491	466
$[0.25, 0.50)$	0	0	0	0	64	132
$[0.50, 0.75)$	120	0	15	0	0	105
$[0.75, 1)$	0	60	0	120	0	20
$[1, 1.1)$	0	0	30	120	180	12

In order to demonstrate the advantage of using the proposed optimization model, we conduct the following experiment. We calculate the contact area between Surface A and a flat surface at different positions during its horizontal sliding. We assume that the flat surface has the same dimensions as the designed surface, $w = 7$ mm, $l = 7$ mm. Figure 7 presents the distribution of the total contact area between Surface A and the designed surface (Figure 7a), and between Surface A and the flat surface (Figure 7b). The overall total contact area is larger for the flat surface as compared to the designed surface. Additionally, the contact area for the designed surface is always the same—12 mm^2—which presents a sturdier design as compared to the flat surface for which, at 4 (out of 12) positions, the contact area is 10 mm^2.

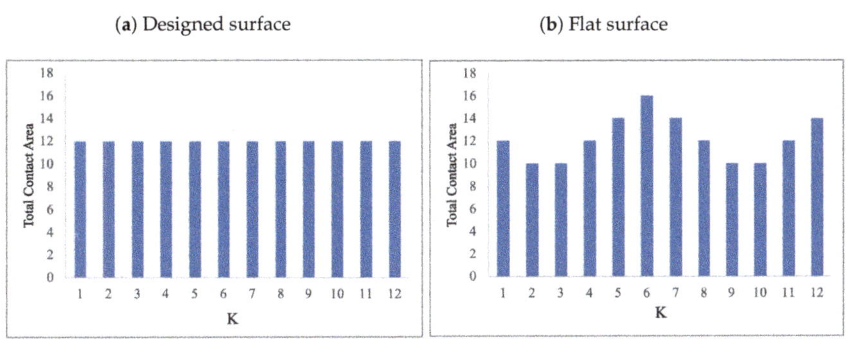

Figure 7. Distribution of the total contact area between Surface A and the designed surface and the flat surface.

4. Solution Approach for Large-Scale Applications

The proposed ILP model poses significant computational challenges for large-scale applications. To address these challenges and improve the computational efficiency of the optimization model, we propose the following solution approach. This approach consists of a data aggregation step, an optimization step, and a data disaggregation step. The proposed approach is summarized in Figure 8.

I. Data Aggregation

In this step, we aggregate the data to reduce its dimensionality. Let S denote the set of heights h_{ij} of all discrete points on the surface. Thus, $S = \{h_{ij} | i = 1, 2, \ldots W; j = 1, 2, \ldots, L\}$. We divide this set into $\overline{W} \times \overline{L}$ groups, and each group contains $c * e$ data points. Thus, $\overline{W} = \lceil W/c \rceil$ and $\overline{L} = \lceil L/e \rceil$. Each group of data is represented by $\overline{h}_{ij}$. To determine the value of $\overline{h}_{ij}$, we consider the group maximum value, minimum value, and average value, as shown below.

$$\begin{aligned} \overline{h}_{ij} = & \max h_{i'j'}, \quad i' = (i-1) * e + 1, (i-1) * e + 2, \ldots, i * e, \\ & j' = (j-1) * c + 1, (j-1) * c + 2, \ldots, j * c, \end{aligned} \tag{26}$$

$$\begin{aligned} \overline{h}_{ij} = & \min h_{i'j'}, i' = (i-1) * e + 1, (i-1) * e + 2, \ldots, i * e, \\ & j' = (j-1) * c + 1, (j-1) * c + 2, \ldots, j * c, \end{aligned} \tag{27}$$

$$\overline{h}_{ij} = \left(\sum_{i'=(i-1)*e+1}^{i*e} \sum_{j'=(j-1)*c+1}^{j*c} h_{i'j'} \right) / (c * e). \tag{28}$$

After aggregation, our data set reduces to $\overline{S} = \{\overline{h}_{ij} | i = 1, 2, \ldots \overline{W}; j = 1, 2, \ldots, \overline{L}\}$. By our experimental analysis, we evaluated the impact of different representations of $\overline{h}_{ij}$ on the quality of the solutions obtained.

II. Optimization

To reduce computational effort, we decompose the ILP model into $\mathcal{N}$ models of smaller size. The overall solution of ILP is an aggregation of the solutions found from solving these smaller sized models. Each model uses the aggregated data from Step I.

The motivation for decomposing the model is that high levels of data aggregation (as $c \rightarrow W$ and $e \rightarrow L$) lead to solutions of poor quality. We explore via our experiments the trade-offs between data aggregation and model decomposition on the quality of the solutions found.

We solve a total of N optimization sub-models and, for each sub-model, the designed surface has dimensions $\overline{w} \times \overline{l}$. We refer to this as *unit designed surface*. We set $\mathcal{N} = \lceil \overline{W}/\overline{w} \rceil$ and $K = \overline{L} - \overline{l} + 1$. For problem $n \in \mathcal{N}$, the corresponding data set includes $\mathcal{S}_n = \{\overline{h}_{ij} | i = (n-1) * \overline{w} + 1, (n-1) * \overline{w} + 2, \ldots, n * \overline{w}; j = 1, 2, \ldots, \overline{L}\}$, such that $\overline{S} = \bigcup_{n=1}^{\mathcal{N}} \mathcal{S}_n$ and $\bigcap_{n=1}^{\mathcal{N}} \mathcal{S}_n = \emptyset$. The solution of the optimization model determines the values of Z_{ij} for $(i, j) \in \overline{w} \times \overline{l}$.

Each optimization model determines the values of Z_{ij} that minimize the total contact area between the unit designed surface and the surface with dimensions $\overline{w} \times L$ as the unit designed surface slides along the X axis. We assume that this design will repeat itself several times. We aggregate the solutions of the N optimization models to determine the final design of the designed surface. The designed surface with dimensions $\overline{W} \times \overline{L}$ consists of $\mathcal{N} \times M$ unit surfaces with dimensions $\overline{w} \times \overline{l}$, for $M < \lceil \overline{L}/\overline{l} \rceil$. Figure 8 summarizes our solution approach.

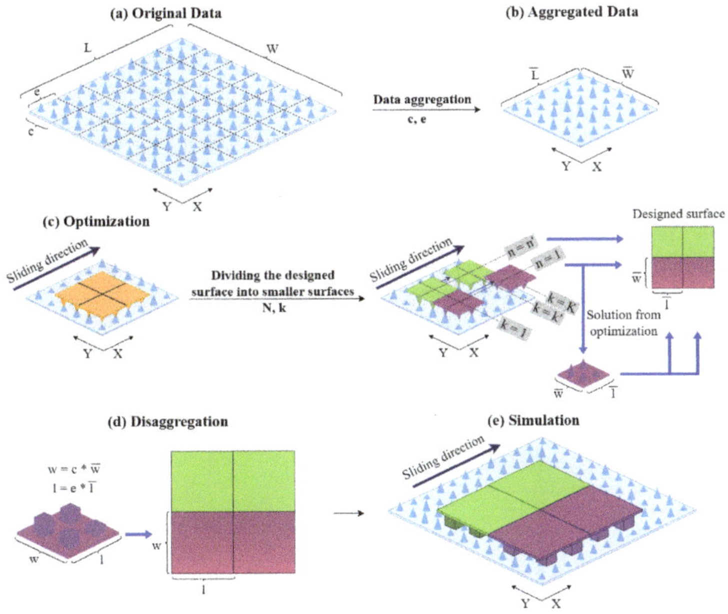

Figure 8. Schematic representation of the solution approach for $\mathcal{N} = 2, M = 2$.

III. Data Disaggregation

In this step, we use the optimization results to determine the design of the surface to be fabricated. Since the solutions Z_{ij} are derived from the aggregated data, we assign the value Z_{ij} to a group $c \times e$ of adjacent points.

5. Numerical Analysis

5.1. Lotus Leaf Case Study

To evaluate the performance of the proposed model and solution approach, we use the lotus leaf surface shown in Figure 9. This surface is divided into a grid consisting of $W = 2670$ rows and $L = 1461$ columns. The dimensions of each point in this grid are ($\Delta_i =$) 92.808 by ($\Delta_j =$) 92.808 nanometers (nm). The data provide the height (h_{ij}) of each point on the grid and are stored in a comma-separated value (CSV) file.

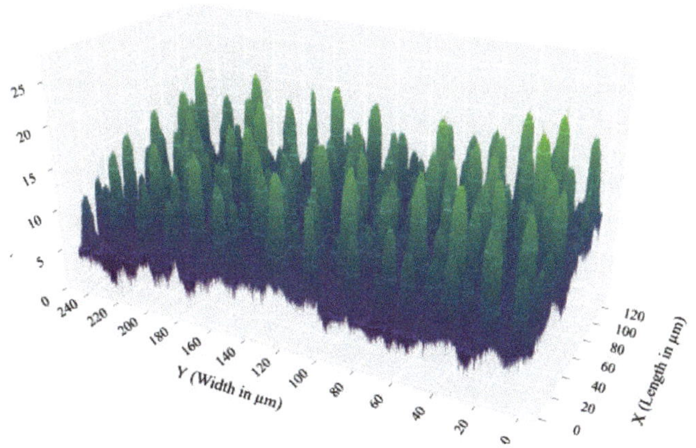

Figure 9. 3D visualization of the lotus leaf surface.

5.2. Experimental Setup

The experiments presented in this section are performed using Python 3.11.5. The ILP optimization model is solved using Gurobi version 11.0.0. We use the High-Performance Computing (HPC) system at the University of Arkansas, which is equipped with Dual Intel Xeon 6130 processors (16 cores per CPU, 32 cores per node) and 192 GB of RAM. We utilized eight threads to solve each model, and we solved our models in parallel.

Figure 10 illustrates our solution approach for sample data of size $W = 400$ and $L = 200$ from the surface of the lotus leaf. This corresponds to 37.12 μm × 18.56 μm (micrometers). Figure 10a presents the original data, while Figure 10b presents the aggregated data ($c = 10$ and $e = 10$). As a result, $\overline{W} = 40$ and $\overline{L} = 20$. Figure 10c presents the results of the optimization model for $n = 1$, along with the unit designed surface with dimensions $\overline{w} = 5$ and $\overline{l} = 5$. Figure 10d presents the results of disaggregation step.

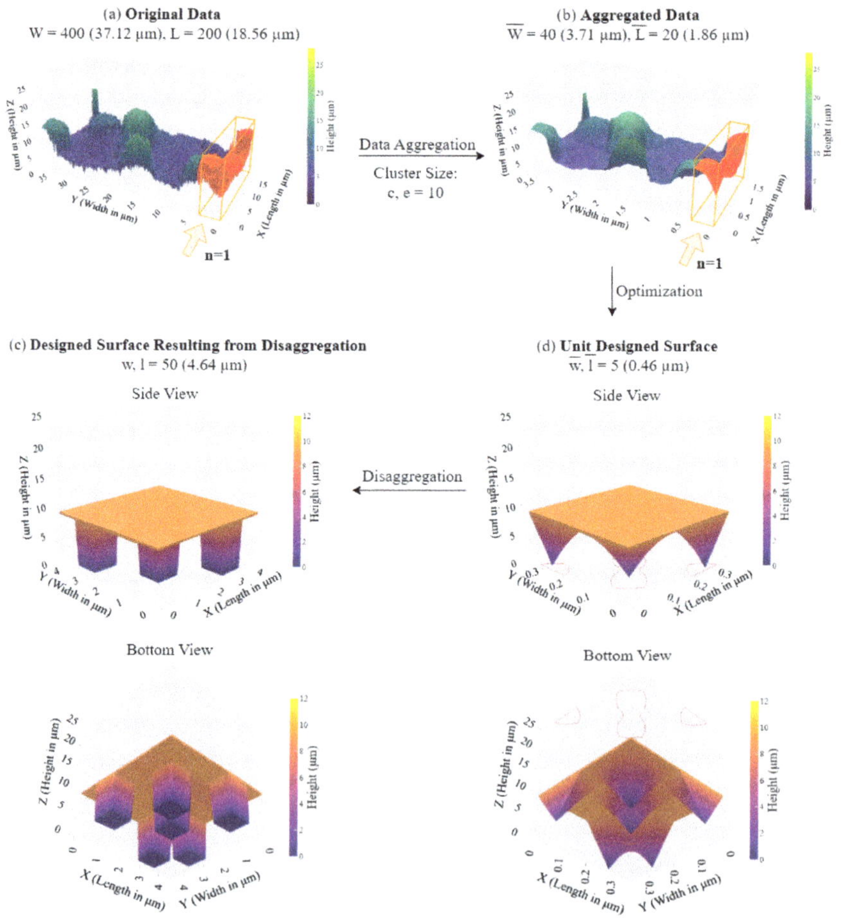

Figure 10. A demonstration of the proposed solution approach using lotus leaf surface.

Table 3 presents the distribution of the contact area and Figure 11 presents the total contact area at position $k \in K$ during the sliding of the designed surface along the X axis. The results of Figure 11a reveal that the contact area varies from 0 to 8 μm and 72% of the observations have contact area of zero.

Table 3. Distribution of the contact area

Contact Area (mm^2)	Frequency
$[0, 1)$	339
$[1, 2)$	20
$[2, 3)$	8
$[3, 4)$	3
$[4, 5)$	2
$[5, 6)$	2
$[6, 7)$	8
$[7, 8)$	14

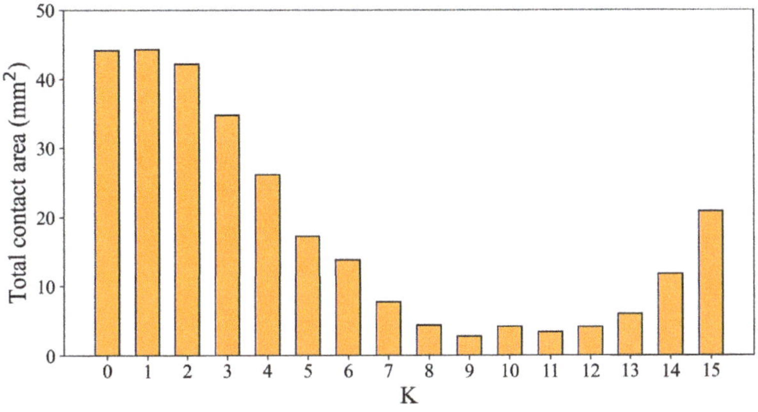

Figure 11. Evaluating solution quality for the example of Figure 10.

5.3. Sensitivity Analysis

The focus of this analysis is to evaluate the impact of the modeling and solution approaches we propose on the designed surface. We report the total CPU time to solve the ILP model using Gurobi. A terminating criterion for this solver is a solution error gap of 5%.

To validate the quality of the solutions obtained via the proposed approach, we conduct simulations. The designed surface used in the simulation model is created using the results of the disaggregation step (described in Section 4). We simulate the sliding of the designed surface and calculate the corresponding contact area.

We use the following parameter values for our base case model: $\pi = 0.02$, $\alpha = 1$, $\beta = 1$, $c = 10$, $e = 10$. The values of $\overline{h}_{ij}$ are calculated using (26). Of the symmetry-imposing constraints, the ILP model solved includes only (7) and (8). In the discussion below, we will only mention the parameters whose values are changed during the sensitivity analysis.

5.3.1. Value of Unit Load in Contact p

In constraint (16), changes in p impact $\epsilon = \pi p / y_m$. We tested the model for different values of ϵ between 0.02 μm^2 and 2.0 μm^2. In these experiments, we tested units designed surfaces with different dimensions: $\overline{w}, \overline{l} = 50$ (4.64 μm), $\overline{w}, \overline{l} = 60$ (5.57 μm), $\overline{w} = 50$ (4.64 μm) and $\overline{l} = 70$ (6.50 μm), and $\overline{w} = 70$ (6.50 μm) and $\overline{l} = 50$ (4.64 μm). Figure 12 presents the average contact area during simulation and the CPU time to solve the corresponding ILP models. Based on these results, increasing the value of p increases the average contact area for every surface. This increase is necessary to reduce surface deformation as unit load p increases. Figure 12b shows an increase in the run time of ILP with values of p, w, and l.

The lines in Figure 12a show small fluctuations since the solutions presented are within 5% of optimality. This is mainly because the problems are difficult to solve, as demonstrated via the results of Figure 12b. There are several factors that affect the run time of the Gurobi solver, which explain the fluctuations observed in Figure 12b. For example, (i) using parameters such as optimality gap tolerance—in our case—affects when the solver stops, leading to variations in run time; (ii) the number of nodes explored by the Gurobi solver and the efficiency of pruning can vary significantly depending on the problem instance; (iii) the Gurobi solver uses various heuristics and cutting planes that may lead to non-uniform performance across problem instances; (iv) variations in processor load, memory usage, and parallel thread execution can impact runt imes.

In order to maintain a low run time, we set ϵ to 0.2 µm^2 on the remaining experiments we present.

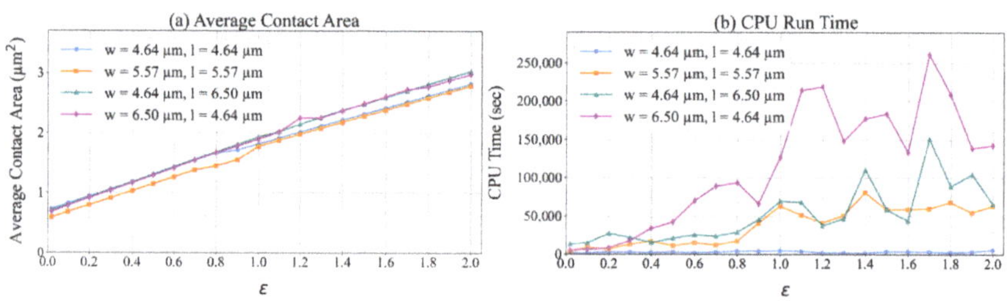

Figure 12. Evaluating the impact of $\epsilon(\pi p/y_m)$ on the average contact area and the run time of ILP.

5.3.2. Values of α and β

In the objective function (10), the values of α and β affect the texture (smoothness) of the surface. We tested our solution approach for different values of α and β between 0 and 10 for designed surfaces with dimensions $\overline{w}, \overline{l} = 50$ (4.64 µm). Figure 13 presents the average contact during simulation and the run time of the ILP model for different combinations of α and β. Based on the results of these experiments, the average contact area increases with α and β. No discernible trend in CPU time is evident.

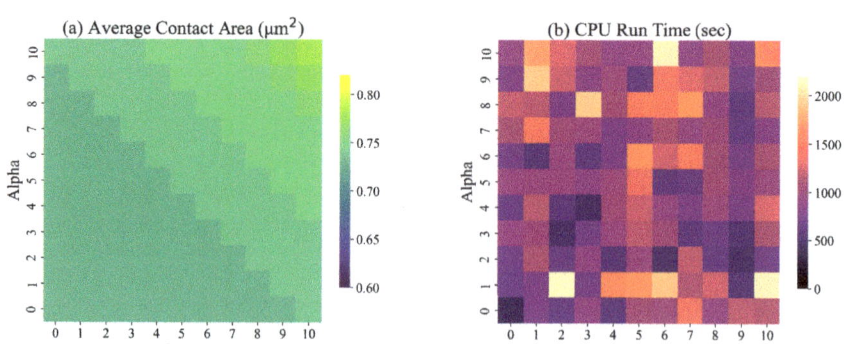

Figure 13. The average contact area and the run time of ILP for different values of α and β: partially symmetric designed surface.

We repeated the experiments by including constraints (9) for a fully symmetric designed surface. The results are presented in Figure 14.

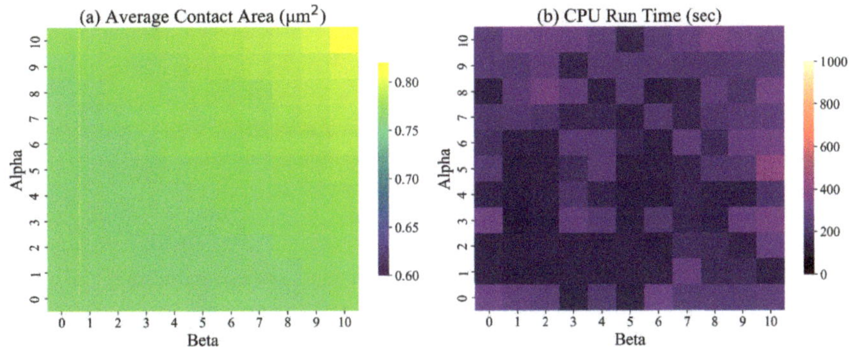

Figure 14. The average contact area and the run time of ILP for different values of α and β: fully symmetric designed surfaces.

We repeated the experiments by excluding all symmetry imposing constraints, (7)–(9). The results of these experiments are presented in Figure 15.

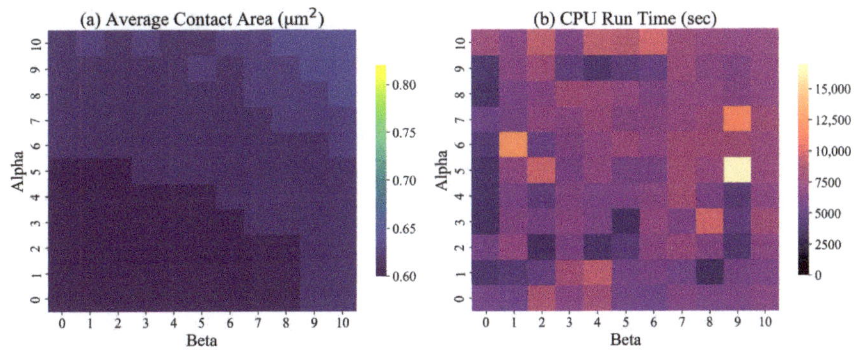

Figure 15. The average contact area and the run time of ILP for different values of α and β: no symmetry imposing constraints.

The results of Figures 13–15 show that the average contact area is smallest if the model does not impose shape symmetry. However, the corresponding run time of the model is the largest. Based on these findings, to minimize the average contact area and maintain low CPU times, we set $\alpha = \beta = 1$.

We conducted hypothesis testing to determine whether the difference in the average contact area for fully symmetric and partially symmetric and fully symmetric and non-symmetric designed surfaces are statistically significant. We conducted paired t-tests. The hypothesis testing of partially and fully symmetric surfaces yielded a t-statistic of 0.21 and a p-value of 0.83, indicating no statistically significant difference in the contact area for the two surfaces. In contrast, the hypothesis testing of fully and non-symmetric surfaces yielded a t-statistic of 367.0 and a p-value of 6.19×10^{-185}. This result indicates a statistically significant difference in the contact area for the two surfaces.

The results of these experiments highlight the trade-offs between solution quality, run time of the ILP model, and practicality and cost of manufacturing the designed surfaces. The ILP model that imposes full symmetry of the designed surface is easier to solve. Fully symmetrical shapes ensure an even distribution of stresses and loads across the surface, lowering the risk of surface failure. Fully symmetrical shapes are easier to manufacture, reducing production time and costs. However, the corresponding contact area is the largest.

5.3.3. Size and Shape of the Unit Designed Surface

We test square and rectangular shapes of different dimensions by changing the values of $\overline{w}$ and $\overline{l}$. Due to the high computational demands of solving the ILP model for $\overline{w}$ = 100 (9.28 μm), $\overline{l}$ = 90 (8.35 μm), we use an error gap of 6% as a stopping criterion for Gurobi. For the remaining experiments, we use an error gap of 5%. Figure 16 summarizes the results of these experiments.

Based on the results of these experiments, the average contact area decreases, and the ILP run time quickly increases with the size of the designed surface. We also observe that square-shaped surfaces exhibit a smaller average contact area compared to rectangular-shaped surfaces with similar surface areas. To strike a balance between the computational time and solution quality, we use $\overline{w}$ = 80 (7.43 μm), $\overline{l}$ = 80 (7.43 μm) in our experiments.

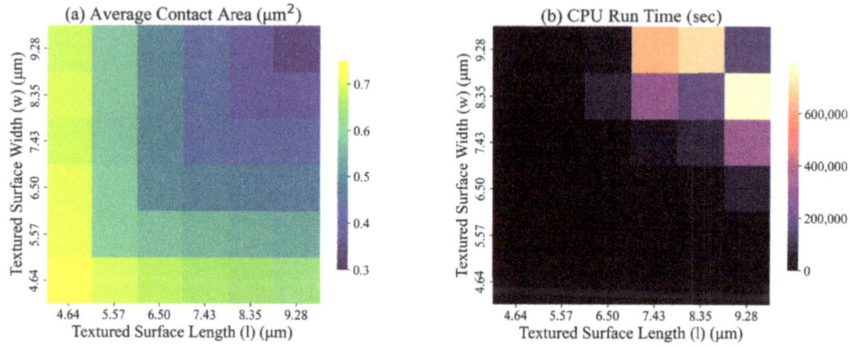

Figure 16. The average contact area and the run time of ILP for different values of w and l.

5.3.4. Symmetry of the Designed Surface

We compare different symmetry patterns of the unit designed surface of size $\overline{w}, \overline{l}$ = 50 (4.64 μm). The symmetry patterns considered are full symmetry, partial symmetry, and no symmetry. Based on the results presented in Figure 17, the average contact area is smallest in models without symmetry-imposing constraints. However, the CPU run time for these models is the largest. The average contact area for models that impose symmetry along the diagonal is close to that of structures without symmetry-imposing constraints. The corresponding CPU run time for these models is low. Models that impose full symmetry lead to solutions with the highest contact area and lowest CPU run time. These results are consistent with the results presented in Figures 13–15.

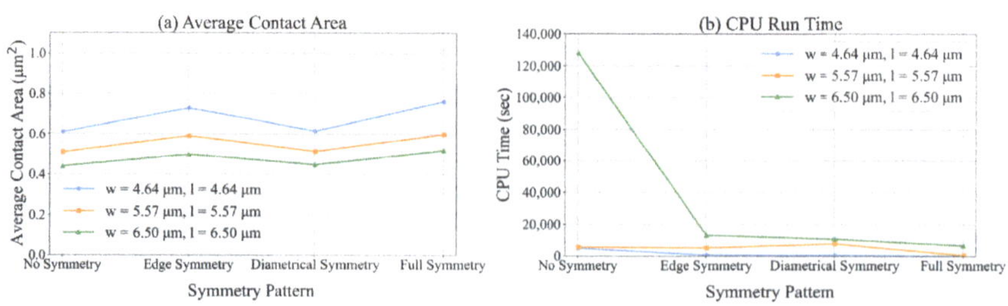

Figure 17. The average contact area and the run time of ILP for unit texture surfaces of different size and symmetry patterns.

5.3.5. Clustering Method

We employed three methods for representing the aggregated data. We use Equations (26)–(28) to calculate the maximum, minimum, and average values within each data cluster. We test the performance of each method on the average contact area. We also test the impact of the dimensions of unit designed surface on the average contact area. In these experiments, the unit designed surface is fully symmetric.

In Figures 18 and 19 we present the average contact area and computational time for different values of n. We observe that the average contact area decreases and the CPU run time increases with the dimensions of the unit designed surface. Using the average of the data to represent the cluster (e.g., using (28) to determine $\overline{h}_{ij}$) results in the lowest average contact area. Using the average and minimum of the data to represent the cluster results in the lowest CPU run time.

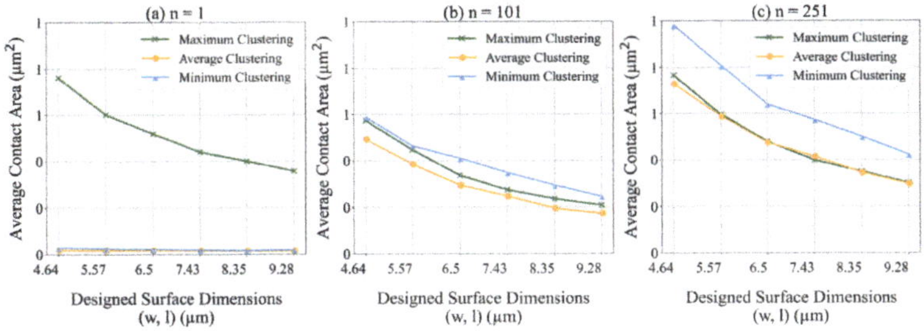

Figure 18. The average contact area for different clustering methods and different values of n.

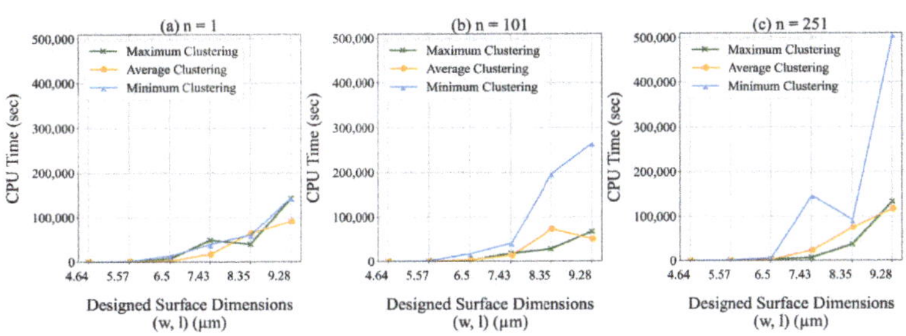

Figure 19. The total CPU time for different clustering methods and different values of n.

5.3.6. Cluster Size

We test the impact of the size of data aggregation on the average contact area and CPU run time for fully symmetric unit designed surface of different sizes. We change the values of c and e from 10 to 20. The results of these experiments are presented in Figure 20. These results indicate that a high level of data aggregation deteriorates the quality of the solutions found. However, aggregating the data reduces the computational time required to solve the ILP model.

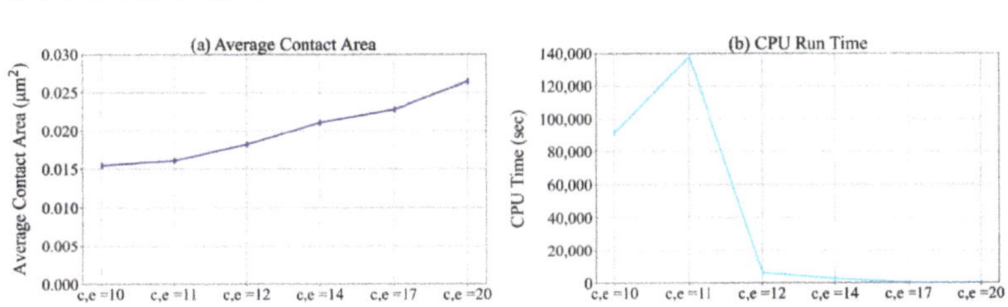

Figure 20. The average contact area and the run time of ILP for different clustering sizes (c and e) and dimensions of the designed surface

6. Conclusions

We propose a novel non-linear program (NLP) to design textured surfaces to enhance the tribological performance of micro- and nano-electromechanical systems (MEMS/NEMS). The objective of our proposed model is to determine feature size, shape, and arrangement that minimize the real contact area between interacting surfaces and minimize deformation during sliding in dry conditions. Model constraints capture (i) the relationships among load at contact, the size of the contact area, and the yield strength of the material, (ii) the relationship between the symmetry of shapes in the textured surface and the size of the contact area, and (iii) the relationship between the smoothness of the textured surface and the size of the contact area. Since solving the proposed NLP is challenging, we develop a linear approximation that, for small-size problems, can be solved using commercial optimization software. We also propose a solution approach to solve large-size problems.

We conducted several experiments to test the performance of the proposed model. The results of these experiments highlight the relationships that exist among the load in contact, shape symmetry, surface smoothness, and the average contact area, the size and shape of the unit-designed surface and the average contact area, and the problem size and run time of the proposed solution approach.

The results of our experiments show that this modeling approach provides a robust framework that supports the effective design of textured surfaces for various engineering applications.

A limitation of the proposed model is that it does not account for the mechanics of deformation during sliding. While the study provides valuable insights into tribological behavior, the model does not explicitly incorporate how the surface textures deform under applied loads during sliding contact. This omission means that potential stress concentrations, material fatigue, and texture wear mechanisms are not fully captured. To improve upon this, future work could integrate deformation mechanics into the model.

Author Contributions: Conceptualization, M.Z.; Methodology, S.D.E.; Validation, S.D.E. and M.Z.; Formal analysis, S.D.E.; Resources, S.D.E.; Writing—original draft, S.D.E.; Writing—review & editing, M.Z.; Project administration, S.D.E. All authors have read and agreed to the published version of the manuscript.

Funding: This research received no external funding.

Data Availability Statement: The authors will provide the data used upon request.

Acknowledgments: This research is supported by the Arkansas High Performance Computing Center, which is funded through multiple National Science Foundation grants and the Arkansas Economic Development Commission.

Conflicts of Interest: The authors declare no conflicts of interest.

Abbreviations

The following abbreviations are used in this manuscript:

ILP	integer linear program
MEMS/NEMS	micro- and nano-electromechanical systems
μm	micrometer
mm	milimeter
nm	nanometer
MPa	megapascal
NLP	non-linear program
Pa	pascal

Appendix A. Non-Linear Programming Model

$$\min : \sum_{k=1}^{K} \int_0^{\omega} \int_0^{\mathfrak{l}} 1/\pi G_{ijk}^x d_j d_i + \int_0^{\omega} \int_0^{\mathfrak{l}} (\alpha F_{ij}^x + \beta F_{ij}^y) d_j d_i \tag{A1}$$

$$G_{ijk}^x = \left[(h_{i,j+k-1} + Z_{ij}) - d + \pi\right]^+, \forall i \in [0,\omega], j \in [0,\mathfrak{l}], k = 1,\ldots,K, \tag{A2}$$

$$\int_0^{\omega} \int_0^{\mathfrak{l}} G_{ijk}^x d_j d_i \geq (\pi p)/y_m, \forall k = 1,\ldots,K, \tag{A3}$$

$$Z_{i1} = Z_{i\mathfrak{l}}, \forall i = 1,\ldots,\omega, \tag{A4}$$

$$Z_{1j} = Z_{\omega j}, \forall j = 1,\ldots,\mathfrak{l}, \tag{A5}$$

$$Z_{ij} = Z_{ji}, \forall i = 1,\ldots,\omega, j = 1,\ldots,\mathfrak{l}, i < j, \tag{A6}$$

$$F_{ij}^x = \|Z_{i,j} - Z_{i,j+1}\|, \forall i = 1,\ldots,\omega, j = 1,\ldots,\mathfrak{l}, \tag{A7}$$

$$F_{ij}^y = \|Z_{i,j} - Z_{i+1,j}\|, \forall i = 1,\ldots,\omega, j = 1,\ldots,\mathfrak{l}, \tag{A8}$$

$$Z_{ij} \geq 0, \forall i = 1,\ldots,\omega, j = 1,\ldots,\mathfrak{l}, \tag{A9}$$

$$0 \leq G_{ijk}^x \leq \pi, \quad \forall i = 1,\ldots,\omega, j = 1,\ldots,\mathfrak{l}, k = 1,\ldots,K. \tag{A10}$$

Appendix B. Integer Linear Programming Model

$$\min : \sum_{k=1}^{K} \sum_{i=1}^{w} \sum_{j=1}^{l} 1/\pi G_{ijk}^x + \alpha \sum_{i=1}^{w} \sum_{j=1}^{l-1} F_{ij}^x + \beta \sum_{i=1}^{w-1} \sum_{j=1}^{l} F_{ij}^y \tag{A11}$$

$$G_{ijk}^x \leq M * Y_{ijk}, \forall i = 1,\ldots,w, j = 1,\ldots,k = 1,\ldots,K, \tag{A12}$$

$$G_{ijk}^x \leq h_{i,j+k-1} + Z_{ij} - d + \pi + M * (1 - Y_{ijk}), \forall i = 1,\ldots,w, j = 1,\ldots,k = 1,\ldots,K, \tag{A13}$$

$$G_{ijk}^x \geq h_{i,j+k-1} + Z_{ij} - d + \pi, \forall i = 1,\ldots,w, j = 1,\ldots,k = 1,\ldots,K, \tag{A14}$$

$$\sum_{i=1}^{w} \sum_{j=1}^{l} G_{ijk} \geq (\pi p)/y_m, \forall k = 1,\ldots,K, \tag{A15}$$

$$Z_{i1} = Z_{il}, \forall i = 1,\ldots,w, \tag{A16}$$

$$Z_{1j} = Z_{wj}, \forall j = 1,\ldots,l, \tag{A17}$$

$$Z_{ij} = Z_{ji}, \forall i = 1,\ldots,w, j = 1,\ldots,l, i < j, \tag{A18}$$

$$F_{ij}^x \geq Z_{ij} - Z_{i,j+1}, \forall i = 1,\ldots,w, j = 1,\ldots,l, \tag{A19}$$

$$F_{ij}^x \geq -Z_{ij} + Z_{i,j+1}, \forall i = 1,\ldots,w, j = 1,\ldots,l, \tag{A20}$$

$$F_{ij}^y \geq Z_{ij} - Z_{i+1,j}, \forall i = 1,\ldots,w, j = 1,\ldots,l, \tag{A21}$$

$$F_{ij}^y \geq -Z_{ij} + Z_{i+1,j}, \forall i = 1,\ldots,w, j = 1,\ldots,l, \tag{A22}$$

$$Z_{ij} \geq 0, \forall i = 1,\ldots,w, j = 1,\ldots,l, \tag{A23}$$

$$G_{ijk}^x \leq \pi, \quad \forall i = 1,\ldots,w, j = 1,\ldots,l, k = 1,\ldots,K, \tag{A24}$$

$$G_{ijk}^x \geq 0, \forall i = 1,\ldots,w, j = 1,\ldots,k = 1,\ldots,K, \tag{A25}$$

$$Y_{ijk} \in \{0,1\}, \forall i = 1,\ldots,w, j = 1,\ldots,k = 1,\ldots,K. \tag{A26}$$

References

1. Afshar-Mohajer, M.; Zou, M. Multi-Scale In Situ Tribological Studies of Surfaces with 3D Textures Fabricated via Two-Photon Lithography and Replica Molding. *Adv. Mater. Interfaces* **2020**, *7*, 2000299.
2. Kim, S.H.; Asay, D.B.; Dugger, M.T. Nanotribology and MEMS. *Nano Today* **2007**, *2*, 22–29. [CrossRef]
3. Mo, Y.; Turner, K.; Szlufarska, I. Friction Laws at the Nanoscale. *Nature* **2009**, *457*, 1116–1119. [CrossRef] [PubMed]
4. Zou, M.; Cai, L.; Wang, H. Adhesion and friction studies of a nano-textured surface produced by spin coating of colloidal silica nanoparticle solution. *Tribol. Lett.* **2006**, *21*, 25–30. [CrossRef]

5. Neves, D.; Diniz, A.E.; de Lima, M.S.F. Efficiency of the laser texturing on the adhesion of the coated twist drills. *J. Mater. Process. Technol.* **2006**, *179*, 139–145.
6. Evangelista, I.; Wencel, D.; Beguin, S.; Zhang, N.; Gilchrist, M.D. Influence of surface texturing on the dry tribological properties of polymers in medical devices. *Polymers* **2023**, *15*, 2858. [CrossRef] [PubMed]
7. Afshar-Mohajer, M.; Yang, X.; Long, R.; Zou, M. Understanding the friction and deformation behavior of micro/nano-hierarchical textures through in-situ SEM observation and mechanics modeling. *Tribol. Int.* **2022**, *165*, 107271. [CrossRef]
8. Kang, M.; Park, Y.M.; Kim, B.H.; Seo, Y.H. Micro- and nanoscale surface texturing effects on surface friction. *Appl. Surf. Sci.* **2015**, *345*, 344–348. [CrossRef]
9. Cheng, H.; Zhou, F.; Fei, Z. Dry Friction Properties of Friction Subsets and Angle Related to Surface Texture of Cemented Carbide by Femtosecond Laser Surface Texturing. *Coatings* **2023**, *13*, 741. [CrossRef]
10. Berardo, A.; Borasso, M.; Gallus, E.; Pugno, N.M. Finite element simulations and statistical analysis for the tribological design of mutually textured surfaces and related experimental validation. *J. Tribol.* **2025**, *147*, 031701. [CrossRef]
11. Xing, Y.; Deng, J.; Wu, Z.; Wu, F. High friction and low wear properties of laser-textured ceramic surface under dry friction. *Opt. Laser Technol.* **2017**, *93*, 24–32.
12. He, C.; Yang, S.; Zheng, M. Analysis of synergistic friction reduction effect on micro-textured cemented carbide surface by laser processing. *Opt. Laser Technol.* **2022**, *155*, 108343. [CrossRef]
13. Yang, S.; Guo, C.; Ren, W. Research on optimization of milling performance of V-groove micro-texture ball-end milling cutter. *J. Mech. Sci. Technol.* **2022**, *36*, 2849–2860. [CrossRef]
14. Ge, Z.; Hu, Q.; Wang, R.; Fei, H.; Zhu, Y.; Wang, Z. Machine Learning-Driven Optimization of Micro-Textured Surfaces for Enhanced Tribological Performance: A Comparative Analysis of Predictive Models. *Coatings* **2024**, *14*, 1539. [CrossRef]
15. Bartholomew-Biggs, M. *Nonlinear Optimization with Engineering Applications*; Springer Science & Business Media: Berlin/Heidelberg, Germany, 2008; Volume 19.
16. Ruszczynski, A. *Nonlinear Optimization*; Princeton University Press: Princeton, NJ, USA, 2011.
17. Gachot, C.; Rosenkranz, A.; Hsu, S.; Costa, H. A critical assessment of surface texturing for friction and wear improvement. *Wear* **2017**, *372*, 21–41.
18. Sonmez, F.O. Shape optimization of 2D structures using simulated annealing. *Comput. Methods Appl. Mech. Eng.* **2007**, *196*, 3279–3299.
19. Shen, Z.; Li, S.; Wu, Z.; Dong, B.; Luo, W.; Zeng, L. Optimization design of irregular grooved texture on the surface of sliding pair based on adaptive genetic algorithm. *Ind. Lubr. Tribol.* **2023**, *75*, 1208–1218.
20. Grothey, A. Decomposition Methods for Nonlinear Nonconvex Optimization Problems. Ph.D. Thesis, The University of Edinburgh, Edinburgh, UK, 2001.
21. Nsilani Kouediatouka, A.; Ma, Q.; Liu, Q.; Mawignon, F.J.; Rafique, F.; Dong, G. Design methodology and application of surface texture: A review. *Coatings* **2022**, *12*, 1015. [CrossRef]
22. Gropper, D.; Wang, L.; Harvey, T.J. Hydrodynamic lubrication of textured surfaces: A review of modeling techniques and key findings. *Tribol. Int.* **2016**, *94*, 509–529. [CrossRef]
23. Chan, M.; Yin, Y.; Amado, B.; Williams, P.; Xiao, D. *Optimization with Absolute Values*; Cornell University Computational Optimization Open Textbook: Ithaca, NY, USA, 2020.

lubricants

Article

Analysis of Lubrication Characteristics and Friction Test of Texture Topography of Angular Contact Ball Bearing Based on Computational Fluid Dynamics

Zhi Li [1,*], Shijie Yin [1], Qisheng Zhang [2], Xiqing Zhang [1] and Hong Zhang [3]

[1] School of Vehicle and Transportation Engineering, Taiyuan University of Science and Technology, Taiyuan 030024, China; yinshijie@stu.tyust.edu.cn (S.Y.); xiqingz@tyust.edu.cn (X.Z.)
[2] Engineering Research Center Heavy Machinery Ministry of Education, Taiyuan University of Science and Technology, Taiyuan 030024, China; 2022078@tyust.edu.cn
[3] School of Mechanical Engineering, Taiyuan University of Science and Technology, Taiyuan 030024, China; hexie007@tyust.edu.cn
* Correspondence: lizhi@tyust.edu.cn

Abstract: A textured surface topography can be used to improve the lubrication performance of bearings. These improvements are closely related to the design of the textured topography. Therefore, studying the effect of the textured topography of rolling bearings on lubrication performance is significant. This study used computational fluid dynamics (CFD) technology to simulate and analyze the lubrication of an angular contact ball bearing under different working conditions. We studied the influence of a textured topography with different area occupancy rates on the oil-phase volume fraction, as well as the lubrication effect of the textured surface on the bearing's inner ring and chamber at different rotational speeds and oil inlet speeds. We conducted friction characteristic experiments on point–contact friction pairs using a friction and wear tester. The effects of different loads and rotational speeds on the friction characteristics and surface wear of textured and smooth surfaces were analyzed. The results indicate that the oil-phase volume fraction is always higher than that of the conventional bearing in the inner ring and chamber of a textured bearing. The textured bearing exhibited better lubrication and friction performance. Different textured topographies have different positive effects on lubrication performance, and the influence of the working conditions should be fully considered to achieve these improvements.

Keywords: angular contact ball bearing; surface texture; elastohydrodynamic lubrication; friction characteristics; computational fluid dynamics

1. Introduction

Bearings are widely used in mechanical systems. Their performance directly affects the reliability and life of mechanical equipment [1,2]. With the increasing demand for mechanical systems that can achieve high speeds of over 1000 r/min and bear heavy loads of over 10 kN, the lubrication performance of rolling bearings has attracted extensive attention [3]. Good lubrication can improve the performance and reliability of rolling bearings during operation [4,5] and increase their service life [6]. In order to explore the lubrication mechanism of rolling bearings, researchers have conducted modeling and analyses of the factors that may affect the lubrication of rolling bearings. Considering the complex situations in which bearings actually operate, some researchers have introduced models that are closer to the actual working conditions of rolling bearings to analyze the effects of different parameters on the formation of lubricating films, changes in thickness, and the distribution of pressure [7–10]. In addition to simulation modeling and analysis, experimental research is also an important means of exploring the lubrication mechanism of rolling bearings. By using equipment to observe and verify the lubrication characteristics of bearings, experimental research can complement simulation models and perfect the system used to determine the characteristic lubrication mechanisms of rolling bearings [11–13].

The advent of bionics prompted researchers to direct their attention toward the impact of surface texture on the lubrication, friction, and wear of mechanical components. In the realm of surface texture research, the fractal complexity of the texture is intricately associated with lubricity. It mirrors the complexity of the surface's micro-geometric shape [14,15]. Studies have found that this micro-geometric shape exerts an influence on the flow and distribution of the lubricating oil within the frictional contact zone [16,17]. The field of research pertaining to surface texture technology is extensive, encompassing numerous disciplines. This technology has improved the lubrication performance and load capacity of mechanical components and has reduced friction and wear. Therefore, it has extended the service life of equipment and reduced the energy consumed and emitted during energy use [18,19]. With regard to bearings, it is more common to fabricate surface textures for sliding bearings [20]. This can accommodate debris, store lubricating oil for secondary replenishment, generate additional hydrodynamic pressure, and increase the thickness of the oil film, thus further improving the lubrication performance of bearings [21–24]. In recent years, the implementation of surface texture technology in the context of rolling bearings has significantly increased. Vidyasagar et al. [25] used a nanosecond fiber laser machine to produce a microstructure similar to the skin of the black mamba snake on FAG deep-groove ball bearings. The results showed that the friction characteristics of the bearings with a snakeskin texture were better than those of untextured bearings. Bhardwaj et al. [26] analyzed the use of circumferential micro-grooves on thrust ball bearings. The results showed that the micro-groove bearings significantly reduced the friction torque, increase in temperature in the ring, and vibration compared with traditional bearings. Wu et al. [27] assessed the use of three textures, namely, dimple, groove, and gradient groove, on thrust ball bearings. It was found that the gradient groove texture produced a one-way guiding effect during oil droplet diffusion and had the best damping effect. The texture effectively reduced the friction torque during operation, thereby enhancing the lubricating performance of the bearings. Gouda et al. [28] manufactured a circumferential micro-groove on the static outer ring of radial ball bearings for experiments. The experiments showed that the friction torque and vibration were reduced compared with traditional bearings and that the thickness of the film was increased. Han et al. [29] changed the texture parameters to study the oil film load capacity of rolling bearings. The results showed that the size of the texture was positively correlated with the oil film load capacity. However, the texture depth first exhibited a positive correlation with the oil film load capacity and then a negative correlation. Among the textures, the texture with a diameter of 80 μm and a depth of 3~5 μm maximized the oil film load capacity.

The above research mostly focused on the design and optimization of texture and its impact on the surface texture technology of rolling bearings. In these studies, researchers changed the shape and size of bearings to analyze the different textures. Thus, further research on the surface texture in bearing applications should fully consider the effect of lubrication under different working conditions. Therefore, based on the theory of multiphase flow and using fluid dynamics technology, the oil film distribution of untextured rolling bearings under different rotational speeds and oil inlet speeds was analyzed in this study. It was found that the uneven distribution of lubricating oil and a thin oil film can affect the lubrication effect of rolling bearings. Then, textured bearings were introduced to explore the influence of textured bearings with different area occupancy ratios on the oil-phase volume fraction. The alterations in the distribution and volume fraction of the oil phase in various components of the bearing after the introduction of textures were analyzed. Finally, textured surfaces with varying area occupancy ratios were fabricated by utilizing a laser-marking machine. Subsequent to the execution of friction performance tests, the ability of different surface textures to enhance the lubrication performance of angular contact ball bearings was substantiated. The purpose of this study was to explore the lubrication mechanism implicated in the effect of surface textures on rolling bearings under different working conditions, thereby providing a reference for the practical application of surface textures on rolling bearings.

2. Fluid Domain Model of Angular Contact Ball Bearing

2.1. Geometric Model and Meshing

In this study, a 7008C angular contact ball bearing was selected to establish a bearing fluid domain model and analyze the distribution of fluid in the bearing chamber accurately. The structural parameters are shown in Table 1.

Table 1. Main structural parameters of angular contact ball bearing.

Parameter	Value
Inner diameter	40 mm
Outer diameter	68 mm
Width	15 mm
Ball diameter	3.5 mm
Number of balls	18
Contact angle	15°
Ring materials	GCr15
Ball materials	GCr15

A fluid domain model of a bearing with an oil inlet and pressure inlet was established. The whole model was divided into the nozzle part and the bearing chamber part. The bearing fluid domain structure is shown in Figure 1. The periodic mesh in ICEM CFD was used to divide the fluid domain of 1/18, and then the complete model was obtained by connecting the interface. The meshing model is shown in Figure 2.

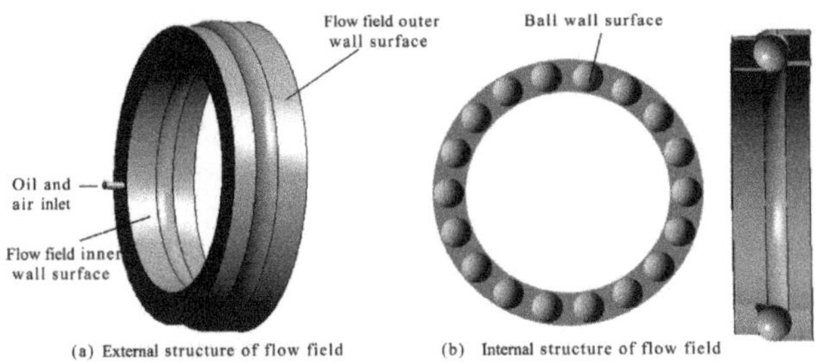

Figure 1. Flow field model of bearing.

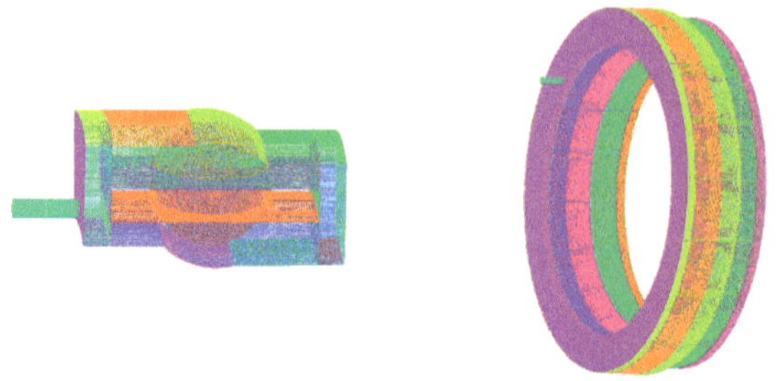

(a) 1/18 fluid domain meshing model (b) Complete fluid domain mesh model

Figure 2. Fluid domain meshing model.

2.2. Boundary Conditions and Solution Method Settings

According to the running state of the bearing, the outer ring was set as a non-slip wall. The inner ring and the ball were set as moving walls. For the characteristics of the elastohydrodynamic oil film, the elastohydrodynamic oil film was regarded as a medium that could effectively transfer the force between the rolling element and the inner and outer rings, with the motion of the components in the rolling bearing being simplified to pure rolling. The parameters of each component were determined according to Formulas (1) and (2) [30], as shown in Table 2.

$$n_c = \frac{1}{2}[(n_e + n_i) + (n_e - n_i)\frac{D_b}{D_m}\cos\alpha] \tag{1}$$

$$n_b = \frac{1}{2}[\frac{D_m}{D_b} - \frac{D_b}{D_m}\cos^2\alpha](n_e - n_i) \tag{2}$$

where n_c and n_b are the revolution and rotation speed of the rolling element; n_i and n_e are the rotational speed of the inner and outer rings of the bearing; α is the contact angle; D_m is the diameter of the bearing pitch circle; and D_b is the diameter of the rolling element.

Table 2. Parameters of each component of the bearing.

Inner Ring Speed (r/min)	Pitch Diameter (mm)	Rolling Element Diameter (mm)	Contact Angle (°)	Rolling Element Revolution Speed (r/min)	Rolling Element Rotation Speed (r/min)
800	54	3.5	15	424	−3221
1000	54	3.5	15	518	−3774
1200	54	3.5	15	623	−4529

The simulation environment was based on Fluent 2022R1. The multiphase flow was simulated using the VOF (volume-of-fluid) model. Due to the factors that rolling bearings are subjected to during actual operation, such as pressure differences, bubbles are produced inside the fluid. The oil–air two-phase flow of the bearings can better reflect the actual working conditions of the bearings. The VOF model is based on the concept of volume fraction, which divides the fluid domain into regions with different phases and simulates the behavior of multiphase flow by tracking the movement and deformation of the phase interface. During the high-speed operation of the bearing, the two-phase flow of the internal oil and air can be regarded as two incompatible fluid media. Therefore, the simulation of the fluid inside the bearing can be completed by the VOF model. Due to the disturbance caused by the rolling elements, the fluid of each phase is in a turbulent flow state. Considering the rotation and swirling flow, the fluid domain of the bearing chamber adopted the RNG k-ε model. Since the solution of the flow field model was the final stable result, the volume discretization scheme was set as the implicit volume force. The air phase was set as the compressible main phase. The oil phase was set as the incompressible secondary phase. VG68 lubricating oil, comprising mainly polyalkylene glycol (PAG), was chosen. The corresponding fluid parameters are shown in Table 3. The initialized bearing chamber oil was set to zero, and the pressure phase was set using the PRESTO format. The parameters of each fluid domain change with time; thus, the calculation model was solved by PISO, and the time was set as transient.

Table 3. Oil and air material parameters.

Parameters	Air	Oil
Density ρ (kg/m^3)	1.225	876
Specific heat C_p (J/(kg·K)	1013	1995
Dynamic viscosity η (Pa·s)	1.79×10^{-5}	0.0525
Temperature T (K)	298.15	298.15

3. Simulation Results and Analysis

3.1. Oil Diffusion Process

Taking a rotational speed of 800 r/min as an example, Figure 3 shows the oil-phase distribution of the lubricating oil on the surface of the rolling element when bearing rolling commenced. The simulation analysis shows that the calculation time was 2.6 T (T represents the time when the bearing rotated in a circle; T is 0.075 s). The bearing was initially filled with air, and the proportion of the oil phase was zero. At t = 0.065 T, a small amount of oil was attached to the surface of the rolling element, and the distribution of the lubricating oil was not uniform. When t = 1.04 T, the lubricating oil adhered to a thin film on the surface of the rolling element as the bearing continued to run. When t = 1.3 T and t = 2.6 T, the oil diffused with the help of compressed air. The surface of each rolling element had the adhesion of oil. The lubricant was not uniformly distributed in the chamber when the bearing was started. The part near the nozzle could be directly sprayed and covered, enhancing the lubrication. Lubrication away from the nozzle may be insufficient.

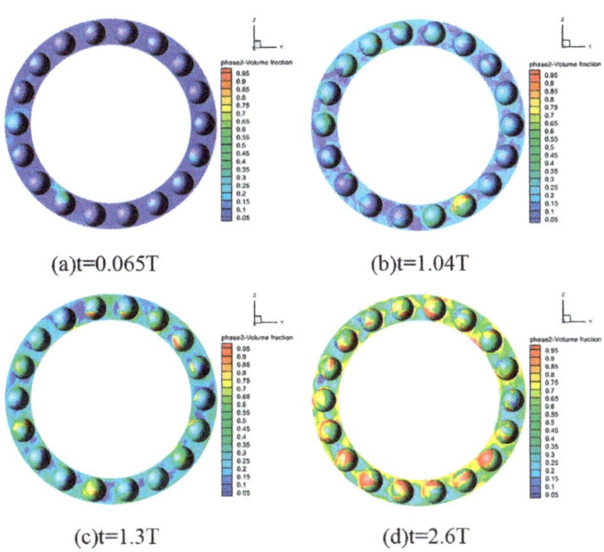

(a)t=0.065T (b)t=1.04T (c)t=1.3T (d)t=2.6T

Figure 3. Variation process of oil-phase distribution.

3.2. Phase Interface Diagram Analysis

Diagrams of the oil–air two-phase interface in the bearing chamber at different oil inlet speeds are shown in Figure 4.

Figure 4. Oil–air two-phase interface diagram in 800 r/min bearing chamber (oil inlet speeds: 2 m/s, 4 m/s, and 6 m/s).

This figure demonstrates that an increase in the oil inlet speed resulted in an increase in the oil supply. Meanwhile, a more uniform trend was exhibited by the distribution of the oil film in the bearing chamber, and the film thickness also increased. Due to the fact that the oil supply was insufficient when the oil inlet speed was slow, the oil film was only concentrated in some areas of the bearing, resulting in insufficient lubrication in other areas. Then, more lubricating oil could be distributed evenly throughout the bearing chamber with an increase in the oil inlet speed.

3.3. Effects of Rotational Speed and Oil Inlet Speed on Oil Film State

The various distributions of the oil phase within the bearing chamber at different rotational speeds when the oil inlet speed was maintained at 6 m/s are illustrated in Figure 5a. The rotational speed was 800 r/min, and the oil-phase distribution of the bearing chamber at different oil inlet speeds is shown in Figure 5b. Figure 5a shows that when the rotational speed was low, the oil film near the inner and outer ring channels and the rolling element was thin and nonuniformly distributed, and the lubrication of the inner ring was poor. As the speed increased, the thickness of the oil film gradually increased. The maximum oil film thickness was attained at a rotational speed of 1000 r/min. This is attributed to the fact that as the rotational speed of the inner ring increases, the increase in the velocity of the fluid flow generates sufficient pressure to support the load. When relative motion occurs in the inner ring, the outer ring, and the rolling elements of the rolling bearing, the lubricating oil generates dynamic pressure due to its viscosity and this relative motion. With the enhancement of the hydrodynamic effect, the thickness of the oil film also increases. The airflow vortex near the rolling element is too high when the rotational speed continues to increase. Airflow vortices can lead to a reduction in the local pressure and velocity disorder. This increases the instability of the flow and then interferes with the formation and maintenance of the oil film as the flow of the lubricating oil is impeded. At the same time, the viscosity of the lubricating oil is closely related to temperature. The high temperature caused by the increase in the rotational speed also reduces the viscosity of the lubricating oil. A viscosity that is too low has an influence on the thickness of the oil film and the lubrication performance. As Figure 5b shows, there was an insufficient amount of lubricating oil in the bearing chamber at a low oil inlet speed. The oil film produced during the two-phase flow was thin and incompletely distributed in an oil-starved state. The lubrication was insufficient and, therefore, not conducive to the normal operation of the bearing. With an increase in the oil inlet speed, the oil-phase volume fraction in the bearing chamber increased, the thickness of the lubricating oil film increased, and the distribution became more uniform.

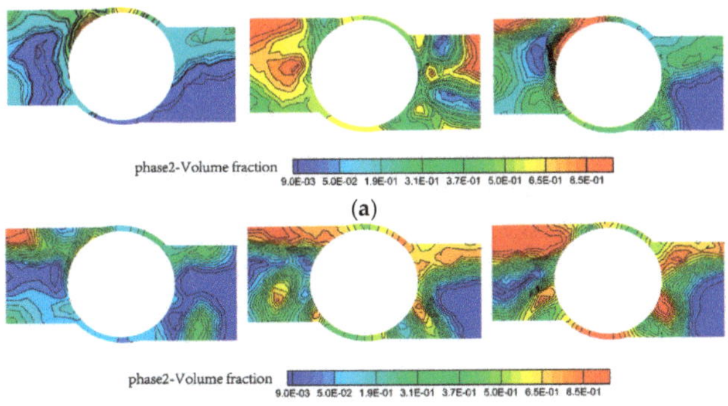

Figure 5. Oil-phase distribution in bearing chamber at different rotational speeds and oil inlet speeds. (**a**) Oil-phase distribution at different rotational speeds (800 r/min, 1000 r/min, and 1200 r/min); (**b**) oil-phase distribution at different oil inlet speeds (2 m/s, 4 m/s, and 6 m/s).

4. Study on Lubrication Characteristics of Textured Rolling Bearing

To improve the uneven distribution of lubricating oil in rolling bearings, increase the thickness of the oil film, and enhance maintenance, surface texture technology was introduced to establish a bearing model with a textured inner ring. This was in order to analyze the lubrication performance of the textured bearings.

4.1. Simplified Model and Meshing

First, the surface texture on the inner ring of the bearing was prepared. The distance between the two pits was 750 μm, the depth of the pit was 30 μm, the diameter of the circular pit was 120 μm, and the angle between the two pits was 10°. To facilitate the analysis and calculation, the influence of the cage and bearing fillets on the lubrication performance was ignored. The 1/18 model of the bearing was meshed and calculated. Due to the bearing model being only 1/18, the mesh division provided by the finite element software was used directly. The solid model and fluid domain model of the textured bearing are shown in Figures 6 and 7, respectively.

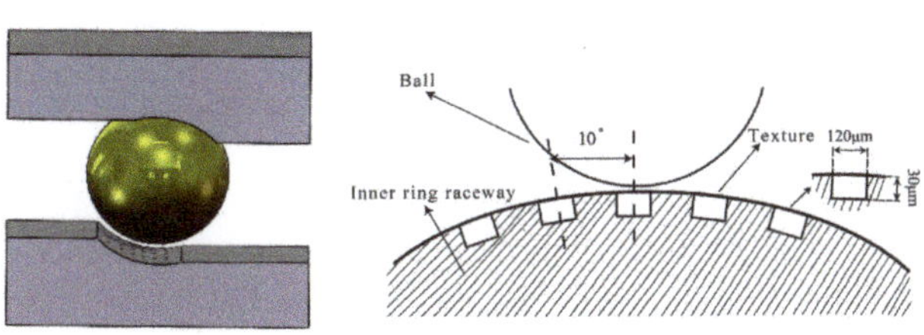

Figure 6. Solid model of textured bearing.

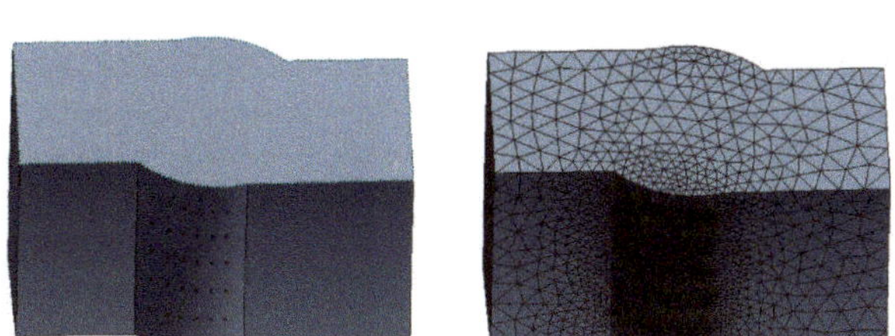

Figure 7. Fluid domain model of textured bearing.

4.2. Effect of Surface Texture on Bearing Oil-Phase Distribution

As illustrated in Figure 8, at a rotational speed of 800 r/min, there was a difference in the oil-phase distribution between the conventional bearing and the bearing with texture on the inner raceway. Due to the friction between the air and the surface of the rotating inner and outer rings, the air curtain effect and centrifugal force were produced in the untextured bearing. These factors made it difficult for the lubricating oil to reach the surface of the rolling element directly. Consequently, the outer ring contained a greater quantity of lubricating oil, whilst the inner ring contained a comparatively smaller amount. However, by texturing the inner raceway of rolling bearings, the lubricating oil could be dispersed more evenly throughout all parts of the bearing chamber. Compared with the untextured bearings, the content of lubricating oil near the contact area of the textured inner ring was significantly increased. This may be due to the fact that the presence of textures leads to variations in the fluid pressure and flow velocity. The resulting differences in the fluid pressure and velocity reinforce the hydrodynamic effect. Meanwhile, the arrangement of textures can guide the flow of the lubricating oil, thereby enabling a more uniform distribution of the lubricating oil. This indicates that texturing treatment can optimize the distribution of the lubricating oil and enhance the lubrication effect on the inner ring.

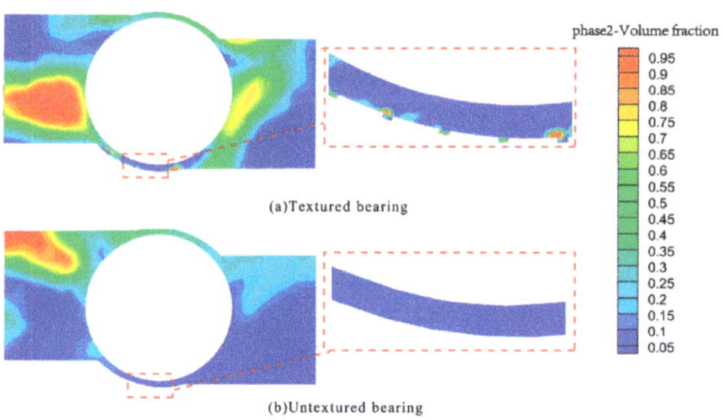

Figure 8. Comparison of oil-phase distribution of fluid domain section.

4.3. Effect of Texture Area Occupancy Ratio on Lubrication Performance

Figure 9 shows the oil-phase volume fraction of the bearing chamber under different texture area occupancy ratios. The results show that the oil-phase volume fraction of the bearing chamber on the textured surface was greater than that on the smooth surface. This was caused by the lubricating oil passing through the micro-pits under the drive of the rolling elements and the micro-pits changing the distance inside the bearing. Due to the viscosity of the fluid, under specific geometric shapes, the speed distribution of the fluid changed, resulting in the generation of a pressure difference. The change in distance enhanced the hydrodynamic pressure effect. Therefore, the oil-phase volume fraction increased under the influence of the texture. However, as the area occupancy ratio of the texture increased, the oil-phase volume fraction decreased. Due to a reduction in the spacing between the micro-pits, the fluid flow was restricted, resulting in a decrease in the dynamic pressure effects and oil-phase volume fraction. When designing the surface texture, it is essential to ensure that increasing the texture area occupancy ratio does not excessively reduce the fluid dynamic pressure effect; meanwhile, it is important to consider its impact on the storage capacity of the lubricating oil, thereby maintaining the good lubrication performance of the bearing.

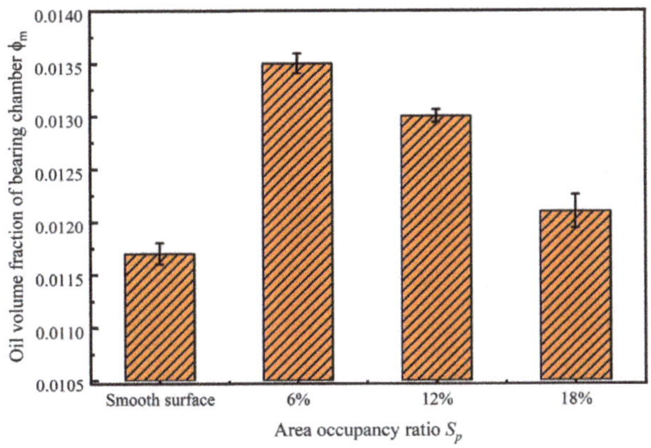

Figure 9. Effect of texture area occupancy ratio on bearing lubrication performance.

4.4. Effect of Oil Inlet Speed and Rotational Speed on Lubrication Performance

Figure 10a illustrates the influence of different oil inlet speeds on the volume fraction of the bearing oil phase at a fixed rotational speed of 800 r/min. It is shown that an increase in the oil inlet speed increased the oil-phase volume fraction in the inner ring and bearing chamber. The effect of texture on the oil-phase volume fraction in the inner ring was more obvious than that in the bearing chamber. This is because the textured micro-pits arranged in the inner ring can store more lubricating oil and, thus, provide continuous secondary lubrication. This situation was more obvious when the oil inlet speed was higher. Compared with textured bearings, untextured bearings have a poorer oil storage performance in the inner ring; as the supply of oil increases, the difference in the content of the oil increases. Figure 10b shows the effect of the oil-phase volume fraction when the oil inlet speed was fixed at 2 m/s and the rotational speed was changed. The oil content first increased and then decreased. However, the oil content in the textured bearing areas was always higher than that in the untextured bearing areas, whether it was in the inner ring of the bearing or the bearing chamber. This was because as the rotational speed gradually increased, the lubricating oil originally stored in the texture began to flow out from the inside of the texture due to the synergy between the hydrodynamic effect, centrifugal force, and other multiple factors. This caused the oil-phase volume fraction to increase. However, when the rotational speed was too high, due to the decrease in the viscosity of the lubricating oil, the lubricating oil decreased. This led to a reduction in the oil-phase volume fraction. Compared with untextured bearings, textured bearings can provide local support and secondary lubrication to the oil film, ensuring that the lubricating oil is stably distributed during the operation of the bearing and preventing the premature rupture of the oil film. In addition, textured bearings can also effectively guide the flow of the lubricating oil. Therefore, the oil-phase volume fraction of the textured inner ring and bearing chamber is always higher than that of the untextured bearings. This indicates that texture significantly improves the friction characteristics of bearings.

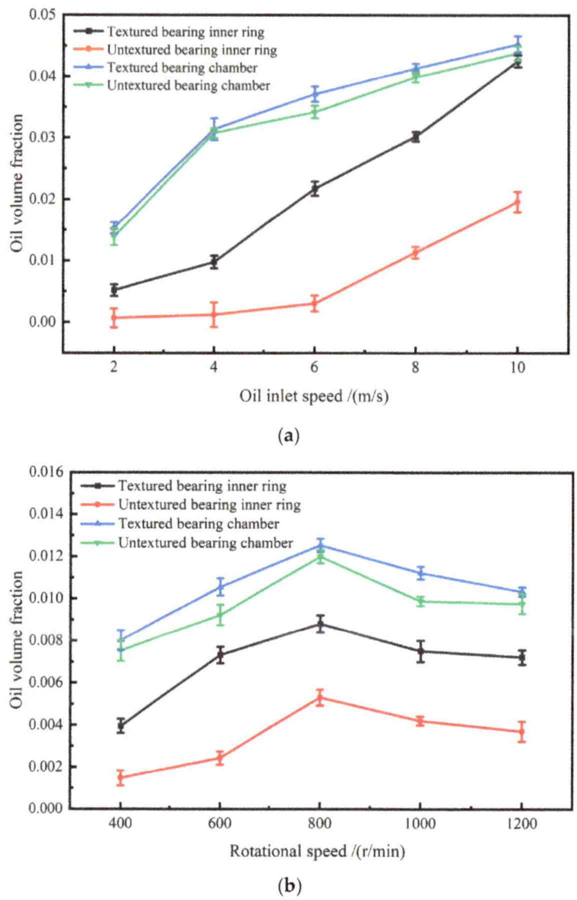

Figure 10. Effect of bearing oil inlet speed and rotational speed on oil-phase volume fraction. (**a**) Effect of oil inlet speed on oil-phase volume fraction; (**b**) effect of rotational speed on oil-phase volume fraction.

5. Frictional Experiment

From the above research, it is evident that surface textures can optimize the mechanisms and characteristics of lubrication and improve the friction and wear conditions in the contact area. In order to further explore the mechanism of this effect, textures were designed and processed on a simplified raceway contact surface. Moreover, the friction performance of point contact friction pairs was assessed.

5.1. Design and Processing of Textured Specimens

Ignoring the curvature of the inner ring raceway in rolling bearings, the raceway contact surface was simplified to a rough plane. The 40 Cr steel ring with good mechanical properties and wear resistance was selected as the specimen material. According to the simulation model presented in the previous section, micro-pit textures were designed. The pit diameter was 120 μm, and the area occupancy ratios were 6%, 12%, and 18%, respectively. Table 4 shows the detailed parameters and numbering of the textured specimens.

Table 4. Specimen parameters and numbering.

Texture Shape	Texture Diameter (d)/μm	Area Occupancy Ratio (S_p)/%	Numbers (n)	Depth (h)/μm
Circular pit	120	6	4774	4
Circular pit	120	12	9549	4
Circular pit	120	18	14,323	4

Textured specimens with different area occupancy ratios were manufactured using a laser-processing machine, as shown in Figure 11. The laser-processing machine consisted of a laser lens, a laser beam, and a working platform and was connected to a computer. The device applied a continuous laser beam with a high power density to the processed material via an optical element. Once the energy on the surface of the material in the irradiation area had been absorbed, it was focused onto a small spot using a higher temperature. Then, textured micro-pits were formed.

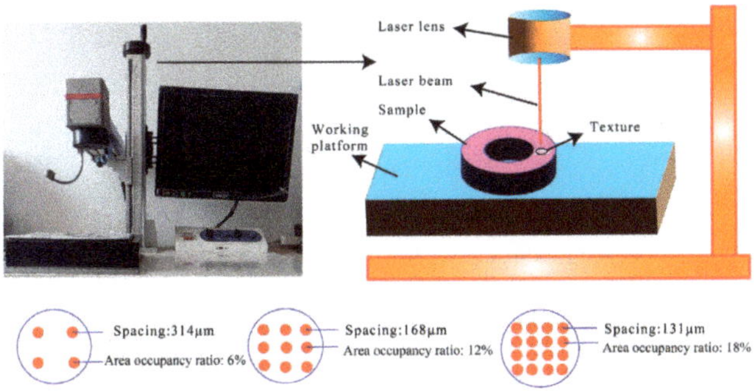

Figure 11. Laser-processing schematic diagram.

While ensuring that the other parameters remained unchanged, four laser-processing parameters, namely, the laser power, scanning speed, pulse frequency, and scanning time, were set to control the spacing and depth-to-diameter ratio of the texture. The optimal laser-marking parameters are shown in Table 5.

Table 5. Laser-marking parameters.

Equipment	Laser Power /W	Pulse Frequency /kHz	Scanning Speed /(mm/s)	Scanning Time
Nanosecond laser	30	20	200	1

Once the texture had been processed, the specimens were polished with metallographic sandpaper to remove the debris left on the surface of the texture after laser marking. Then, the effect of polishing was observed using a metallographic microscope and a two-dimensional profilometer. Finally, the specimens were placed in an acetone solution for ultrasonic cleaning. The textured specimens with different area occupancy ratios were then tested, as shown in Figure 12.

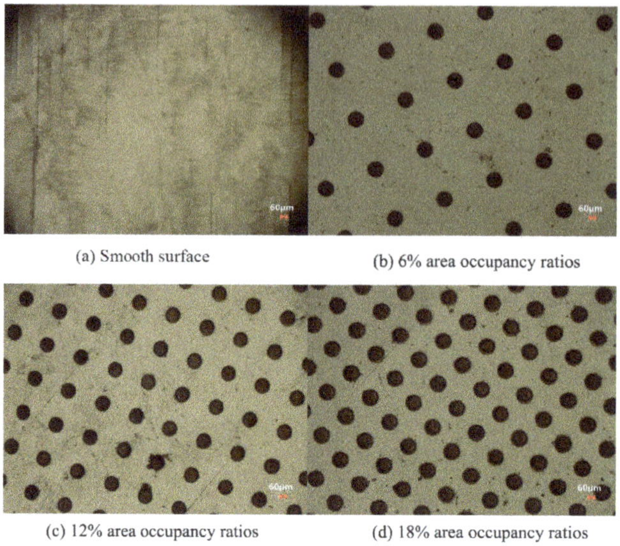

Figure 12. Textured specimens with different area occupancy ratios.

5.2. Experimental Method

The MMW-1000 friction and wear-testing machine was used for the experimental study of point contact friction pairs. A ball-on-disc friction pair composed of upper and lower specimens was employed, as shown in Figure 13. The upper specimen was a rotating GCr15 (d = 12.5 mm) steel ball, and the lower specimen was a fixed 40Cr ring specimen. The outer diameter of the ring specimen was 28.36 mm, the inner diameter was 13.84 mm, the environmental temperature was 22 °C, and the humidity was 36%. The experiment used standard Mobil 10W-40 lubricating oil with a density of 840 kg/m^3 and a viscosity of 6.95×10^{-3} mm^2/s at 40 °C.

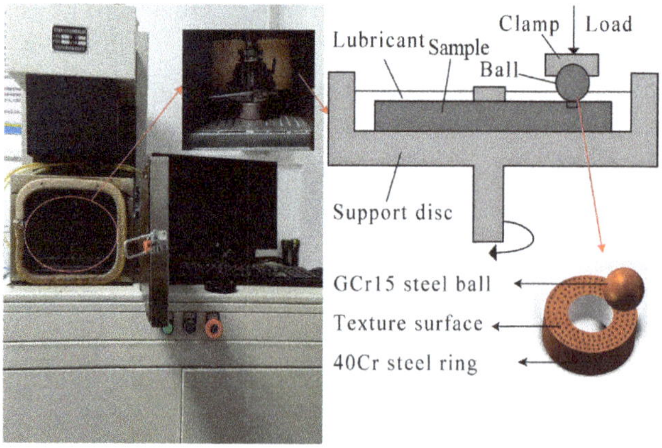

Figure 13. Experimental bench and friction pair.

5.3. Effect of Load on Friction Characteristics of Textured Surface

When multiple experiments with different durations were conducted, it was found that the friction coefficients of some specimens changed significantly after 400 s. A shorter experimental duration failed to effectively show the friction and wear properties of the specimens in the stable stage. If the experimental duration was extended and the influence of various factors, such as equipment limitations, on the test results were excluded, the curve was expected to maintain the same trend for a long period of time. Considering the accuracy and efficiency of the experiment, 720 s was chosen as the experimental duration. The friction coefficient curves obtained at a rotational speed of 540 r/min and loads of 30 N, 90 N, and 150 N, respectively, are shown in Figure 14. This figure shows that the friction coefficient of each surface increased with the change in load. However, the friction coefficients of the textured surfaces were generally smaller than those of the smooth surface. As demonstrated in Section 4.2, the simulation analysis results indicate that the presence of texture leads to an increase in the oil-phase volume fraction of textured surfaces. It appears that sufficient lubricating oil can ensure that the lubrication performance of the textured surfaces remains good, thus reducing the friction coefficient. From the figure, it is clear that the average friction coefficient of the smooth surface was not its lowest under a low load. This may be due to the fact that the friction coefficient was greatly affected by the roughness under a low load. The friction coefficient of the textured surface progressively became lower than that of the smooth surface as the load applied increased. This is consistent with the oil storage performance of micro-pits and indicates that the surface texture can significantly improve the performance of friction pairs under high load.

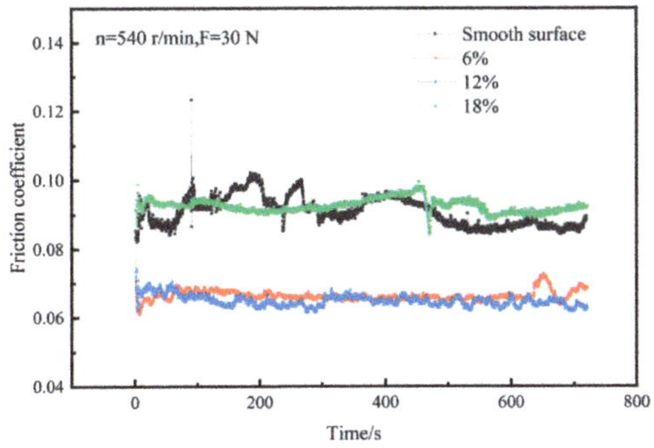

(**a**) n = 540 r/min; F = 30 N.

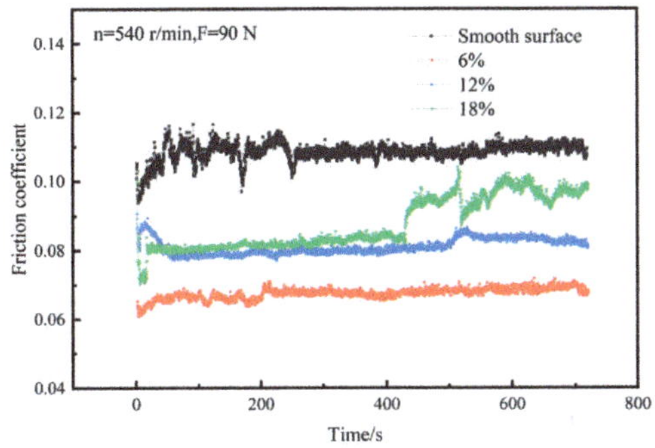

(**b**) n = 540 r/min; F = 90 N.

Figure 14. *Cont.*

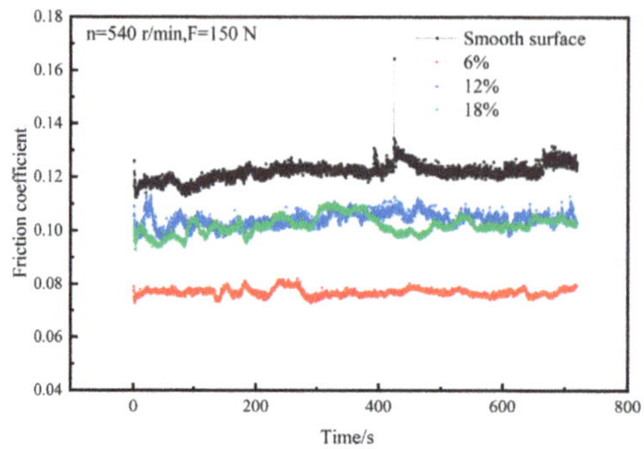

(**c**) n = 540 r/min; F = 150 N.

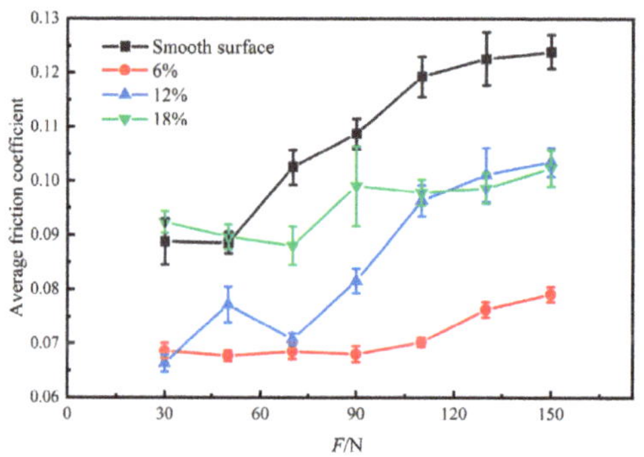

(**d**) Average friction coefficient.

Figure 14. Friction coefficients under different loads.

5.4. Effect of Rotational Speed on Friction Characteristics of Textured Surface

The friction coefficient curves at a load of 90 N and rotational speeds of 180 r/min, 660 r/min, and 900 r/min, respectively, are shown in Figure 15. It can be observed that the larger the area occupancy ratio at a low rotational speed, the worse the anti-friction effect on the textured surface. This is because at low rotational speeds, when the friction pair enters the elastohydrodynamic contact zone, the surface roughness with a larger area occupancy ratio increases; meanwhile, the contact area decreases, resulting in a decrease in the oil film′s bearing capacity and the inability to form a stable lubricating film. The friction coefficient of the textured surface was negatively correlated with the rotational speed as the rotational speed increased, while the smooth surface showed the opposite trend. In accordance with the simulation results in Section 4.4, as the rotational speed increased, the surface texture underwent secondary lubrication. This increased the oil-phase volume fraction in the contact area, which enhanced the dynamic pressure effect on the contact surface, making the distribution of the oil phase more uniform. Meanwhile, as the experiment progressed, the lubricant on the contact pair of the smooth surface gradually evaporated, and dry friction may have occurred. By analyzing the test results for the four surfaces, it is clear that the surface texture can effectively improve the lubrication performance of bearings and reduce their friction coefficient at a high rotational speed.

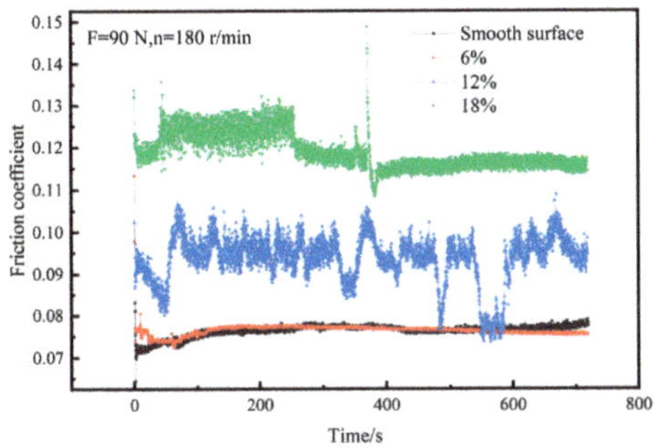

(**a**) F = 90 N; n = 180 r/min.

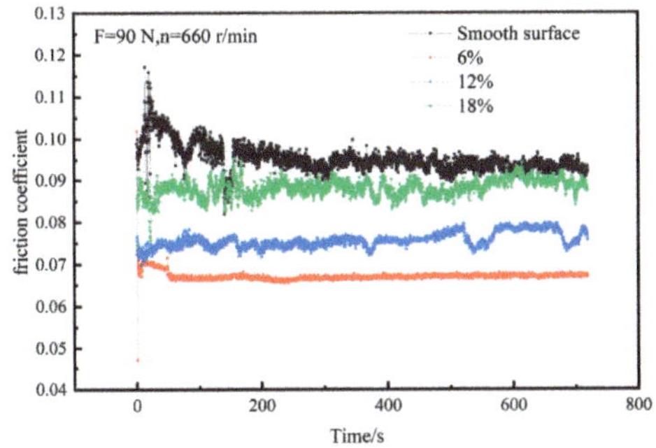

(**b**) F = 90 N; n = 660 r/min.

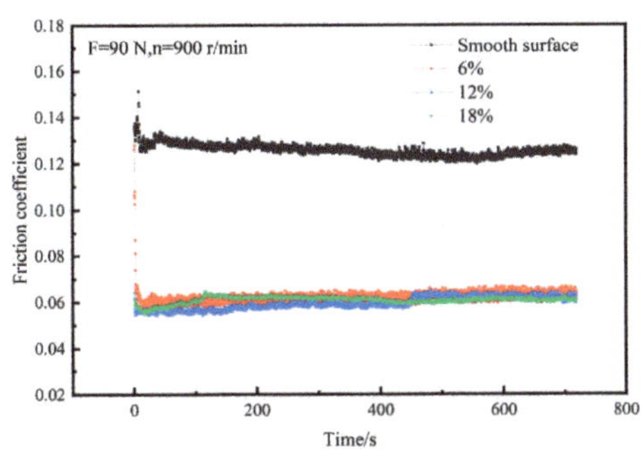

(**c**) F = 90 N; n = 900 r/min.

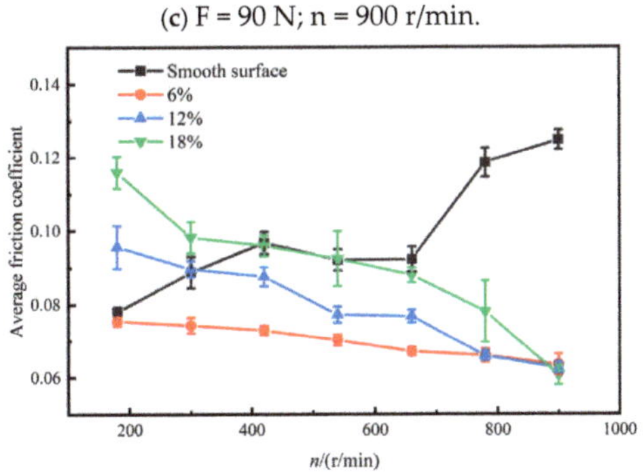

(**d**) Average friction coefficient.

Figure 15. Friction coefficients at different rotational speeds.

5.5. Analysis of Surface Wear and Wear Traces

Figure 16 shows the wear on each specimen after the completion of the experimental process. It can be seen that the textured surface performed outstandingly in terms of wear. In accordance with the simulation results, the textured surface with a 6% area occupancy ratio had the best lubrication effect. The function of secondary lubrication led to an increase in the oil-phase volume fraction, which improved the lubrication performance and reduced the material wear. The smooth surface quality was reduced by 0.0116 g, the textured surfaces with occupancy ratios of 12% and 18% were reduced by 0.0049 g and 0.0048 g, and the textured surface with an occupancy ratio of 6% was reduced by 0.0011 g.

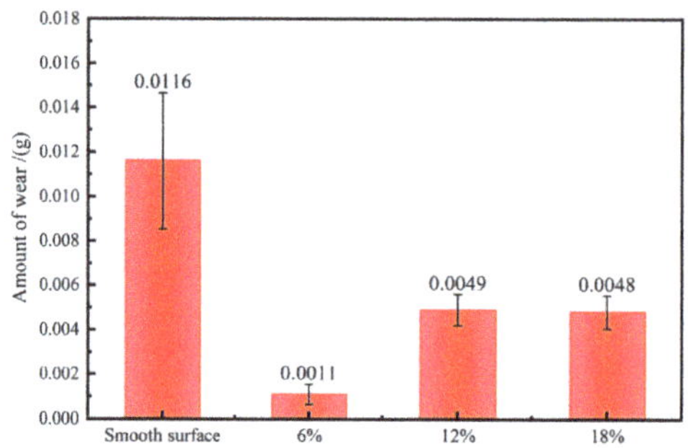

Figure 16. Wear amounts on different surfaces.

As shown in Figure 17, a comparison of the wear on the smooth surface and the textured surface is presented. On the smooth surface, there are relatively obvious abrasion marks, and the wear traces are distributed in a rather scattered manner. In contrast, the wear traces on the textured surface are more concentrated. This may be because on the smooth surface, wear debris was generated after long-term friction, and the oil film was damaged, thus forming abrasion marks. However, on the textured surface, the pits contributed to the collection of wear debris and enhanced lubrication. As a result, the wear traces were more concentrated, and there were no obvious abrasion marks.

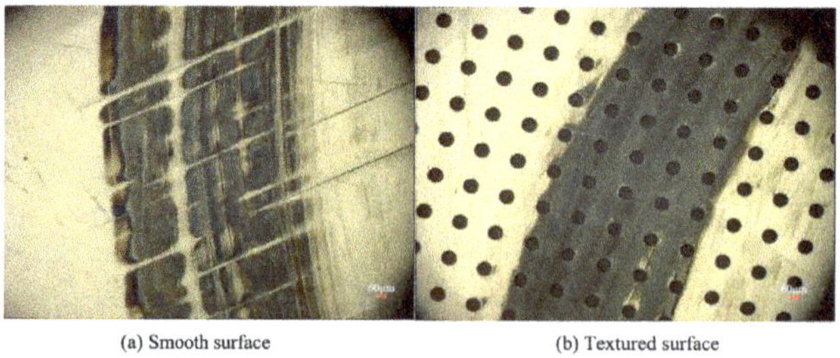

Figure 17. Comparison of surface wear traces.

6. Conclusions

In this study, the changes in the oil–air two-phase flow of an angular contact ball bearing were analyzed. The positive effect of the textured surface topography on the lubrication performance of the bearing under different working conditions was studied. The frictional experiment was carried out on a simplified raceway. It was verified that a textured surface can effectively reduce the friction coefficient and wear under a high load and high rotational speed. The following conclusions are drawn:

1. From the two-phase flow model of the untextured bearing, it can be seen that the oil film of the angular contact ball bearing was nonuniformly distributed during operation and that the lubrication conditions were poor in some areas. Under appropriate rotational speeds and oil inlet speeds, the oil-phase distribution in the bearing chamber became gradually uniform, and the lubrication state of the bearing became relatively good. However, some areas continued to experience poor lubrication.
2. From the CFD model of the textured bearing, it is clear that compared with the untextured bearing, the textured bearing exhibited a relatively good lubricating performance under the same working conditions and could effectively overcome the poor lubrication of the bearings. The influence of texture on the oil-phase volume in the inner ring was greater than that on the bearing chamber. Among the three different texture area occupancy ratios, the oil-phase volume fraction in the bearing chamber was the highest when the texture area occupancy ratio was 6%. However, the oil-phase volume fraction in the bearing chamber of all textured bearings was higher than that in the untextured bearing.
3. As demonstrated by the experimental findings, the textured surface was effective in reducing friction and enhancing wear resistance. Among the surfaces, the textured surface with a 6% area occupancy ratio showed the best anti-friction and anti-wear performance, and the data are in line with the simulation results. Due to the increase in the oil-phase volume fraction on the textured surface, the thickness of the oil film increased. The lubrication effect of the specimens was improved, and the friction coefficient was, consequently, reduced. The test results for three textured surfaces show that a textured surface is more conducive to the improvement of tribological properties under a high load and high rotational speed.

Author Contributions: Conceptualization, Z.L.; methodology, Q.Z. and X.Z.; software, Z.L. and S.Y.; validation, S.Y.; formal analysis, S.Y.; investigation, Q.Z. and X.Z.; resources, H.Z.; writing—original draft preparation, S.Y.; writing—review and editing, Z.L.; visualization, Q.Z. and X.Z.; supervision, Z.L.; project administration, H.Z.; funding acquisition, H.Z. All authors have read and agreed to the published version of the manuscript.

Funding: This research was funded by the National Natural Science Foundation of China, grant number 52475117.

Data Availability Statement: The original contributions presented in this study are included in this article.

Conflicts of Interest: The authors declare no conflicts of interest.

References

1. Liu, Y.; Chen, Z.; Zhai, W.; Lei, Y. Investigation on skidding behavior of a lubricated rolling bearing with fluid-solid-heat coupling effect. *Mech. Syst. Signal Process.* **2024**, *206*, 110922. [CrossRef]
2. Qin, Y.; Li, C.; Wu, X.; Wang, Y.; Chen, H. Multiple-degree-of-freedom dynamic model of rolling bearing with a localized surface defect. *Mech. Mach. Theory* **2020**, *154*, 104047. [CrossRef]
3. Meng, F.; Zheng, Y.; Liu, Y.; Gong, J.; Wang, B. Multi-ellipsoid contact elastohydrodynamic lubrication performance for deep groove ball bearing. *Tribol. Int.* **2020**, *150*, 106367. [CrossRef]
4. Yan, K.; Zhang, J.; Hong, J.; Wang, Y.; Zhu, Y. Structural optimization of lubrication device for high speed angular contact ball bearing based on internal fluid flow analysis. *Int. J. Heat Mass Transfer.* **2016**, *95*, 540–550. [CrossRef]
5. Yao, Q.; Dai, L.; Tang, J.; Wu, H.; Liu, T. High-speed rolling bearing lubrication reliability analysis based on probability box model. *Probab. Eng. Mech.* **2024**, *76*, 103612. [CrossRef]
6. Kim, S.; Seo, Y.H.; Park, J. Transformer-based novel framework for remaining useful life prediction of lubricant in operational rolling bearings. *Reliab. Eng. Syst. Saf.* **2024**, *251*, 110377. [CrossRef]
7. Wang, Y.; Wang, W.; Li, Y.; Zhao, Z. Lubrication and Thermal Failure Mechanism Analysis in High-Speed Angular Contact Ball Bearing. *ASME J. Tribol.* **2018**, *140*, 031503.

8. Liu, J.; Ni, H.; Xu, Z.; Pan, G. A simulation analysis for lubricating characteristics of an oil-jet lubricated ball bearing. *Simul. Model. Pract. Theory* **2021**, *113*, 102371. [CrossRef]
9. Liu, C.; Li, W.; Guo, F.; Wong, P.; Li, X. Effect of Oil Dispersion on Lubricating Film Thickness Generation under Oil Droplet Supply Conditions. *Lubricants* **2023**, *11*, 512. [CrossRef]
10. Yu, X.; Sun, Y.; Wu, S. A complete deterministic model of point contact mixed lubrication coupling dual-mode effects of surface fractal characteristics. *Int. J. Non. Linear Mech.* **2024**, *161*, 104669. [CrossRef]
11. Zhang, X.; Glovnea, R. Grease film thickness measurement in rolling bearing contacts. *Proc. Inst. Mech. Eng. Part J J. Eng. Tribol.* **2020**, *235*, 1430–1439. [CrossRef]
12. Wen, B.; Li, Y.; Wang, M.; Yang, Y. Measurement for Lubricant Distribution in an Angular Contact Ball Bearing and Its Influence Investigation. *Lubricants* **2023**, *11*, 63. [CrossRef]
13. Liang, H.; Zhang, Y.; Wang, W. Influence of the cage on the migration and distribution of lubricating oil inside a ball bearing. *Friction* **2022**, *10*, 1035–1045.
14. Su, B.; Zou, X.; Wang, Z.; Huang, L. Numerical analysis of lubrication properties on a bio-inspired parabolic textured surface. *Meccanica* **2024**, *59*, 987–1000. [CrossRef]
15. Peta, K. Multiscale Wettability of Microtextured Irregular Surfaces. *Materials* **2024**, *17*, 5716. [CrossRef] [PubMed]
16. Fang, J.; Liu, X.; Wang, T.; Song, Z. Micro Lubrication and Heat Transfer in Wedge-Shaped Channel Slider with Convex Surface Texture Based on Lattice Boltzmann Method. *Nanomaterials* **2024**, *14*, 295. [CrossRef]
17. Wang, W.; He, Y.; Li, Y.; Wei, B.; Hu, Y.; Luo, J. Investigation on inner flow field characteristics of groove textures in fully lubricated thrust bearings. *Ind. Lubr. Tribol.* **2018**, *70*, 754–763.
18. Adeyemi, K.; Sun, B.; Xue, W.; Liu, W.; Cao, Y. Friction and Wear Characteristics Modification via Laser Surface Textured Grooves. *Surf. Eng.* **2020**, *37*, 658–668. [CrossRef]
19. Bonse, J.; Kirner, S.V.; Griepentrog, M.; Spaltmann, D.; Krüger, J. Femtosecond Laser Texturing of Surfaces for Tribological Applications. *Materials* **2018**, *11*, 801. [CrossRef]
20. Zhao, X.; Zhang, Y. Analysis of the tribological and dynamic performance of textured bearings under contaminated conditions. *Tribol. Int.* **2023**, *187*, 108732.
21. Kango, S.; Sharma, R.K.; Pandey, R.K. Thermal analysis of microtextured journal bearing using non-Newtonian rheology of lubricant and JFO boundary conditions. *Tribol. Int.* **2014**, *69*, 19–29. [CrossRef]
22. Zhao, X.; Zhang, Y. Tribological and dynamic performance analysis of rolling bearings with varied surface textures operating under lubricant contamination. *Wear* **2023**, *532–533*, 205109. [CrossRef]
23. Sep, J.; Tomczewski, L.; Galda, L.; Dzierwa, A. The study on abrasive wear of grooved journal bearings. *Wear* **2017**, *376–377*, 54–62. [CrossRef]
24. Liu, W.; Ni, H.; Wang, P.; Chen, H. Investigation on the tribological performance of micro-dimples textured surface combined with longitudinal or transverse vibration under hydrodynamic lubrication. *Int. J. Mech. Sci.* **2020**, *174*, 105474. [CrossRef]
25. Vidyasagar, K.E.C.; Pandey, R.K.; Kalyanasundaram, D. Improvement of Deep Groove Ball Bearing's Performance Using a Bionic Textured Inner Race. *J. Bionic. Eng.* **2021**, *18*, 974–990. [CrossRef]
26. Bhardwaj, V.; Pandey, R.K.; Agarwal, V.K. Experimental exploration for the performance improvement of a thrust ball bearing using circumferential micro-grooved races. *Surf. Topogr. Metrol. Prop.* **2021**, *9*, 035017. [CrossRef]
27. Wu, C.; Yang, K.; Ni, J.; Lu, S.; Yao, L.; Li, X. Investigations for vibration and friction torque behaviors of thrust ball bearing with self-driven textured guiding surface. *Friction* **2023**, *11*, 894–910. [CrossRef]
28. Gouda, B.; Tandon, N.; Pandey, R.K.; Babu, C.K. Performance improvement of a radial ball bearing using a micro-groove on stationary outer race. *Proc. Inst. Mech. Eng. Part C J. Mech. Eng. Sci.* **2022**, *236*, 11521–11536. [CrossRef]
29. Han, K.; Li, J.; Wang, Q.; Chen, W.; Xue, J. Study on oil film pressure distribution and load capacity of textured rolling bearings. *Ind. Lubr. Tribol.* **2020**, *72*, 961–967. [CrossRef]
30. Wen, S.; Huang, P.; Tian, Y.; Ma, L. *Principles of Tribology*, 5th ed.; Tsinghua University Press: Beijing, China, 2018; p. 96.

lubricants

MDPI

Article

Numerical Simulation of the Lubrication Performance of the Stator and Rotor Friction Pair Surface Rhombus-like Texture in Screw Pumps

Xiangzhi Shi [1,†], Xinfu Liu [1,*,†], Chunhua Liu [2], Zhongxian Hao [3], Shouzhi Huang [3], Yi Sun [1] and Xinglong Niu [1]

[1] Key Lab of Industrial Fluid Energy Conservation and Pollution Control (Ministry of Education), Qingdao University of Technology, Qingdao 266520, China; xzshi1999@163.com (X.S.)
[2] College of Mechanical and Electronic Engineering, China University of Petroleum (East China), Qingdao 266580, China
[3] Research Institute of Exploration & Development, PetroChina, Beijing 100083, China
* Correspondence: upcdoctor@126.com
† These authors contributed equally to this work.

Abstract: To address wear failure in screw pump stator and rotor friction pairs, this study constructed a numerical model of a rhombus-like micro-dimple texture on friction pair surfaces based on the scale structure of rhombus rattlesnakes. The model was based on the fluid dynamic pressure lubrication mechanism. The CFD method was used to calculate the bearing capacity, friction coefficient, flow field pressure distribution, and flow trace distribution of an oil film carrying surface. The effects of the area rate, depth, shape, and angle of the rhombus-like dimple texture and the actual well fluid viscosity of shale oil on the surface lubrication performance of screw pump stator and rotor friction pairs were analyzed. The results demonstrated that increasing the texture area rate and the angle of the long sides and decreasing the texture angle resulted in a decrease in the oil film surface friction coefficient and an increase in the average pressure and net bearing capacity as well as the hydrodynamic lubrication performance. The average pressure increased and then decreased as the texture depth increased, while the friction coefficient of the oil film surface initially decreased and then increased. At a texture depth of 20 μm, the friction coefficient reached its lowest value while the average pressure and net bearing capacity of the oil film reached their highest value, which resulted in optimal hydrodynamic lubrication performance. When the texture depth became greater than 20 μm, vortices were gradually formed within the texture, which decreased the hydrodynamic lubrication performance. When the area rate of the rhombus-like dimple texture, depth, angle between long sides, and angle were, respectively, equal to 27%, 20 μm, 74°, and 0°, the net bearing capacity of the oil film was maximized, the friction coefficient was minimized, and the hydrodynamic lubrication performance and anti-wear effect reached their highest values. The increase in the viscosity of the actual well fluid could enhance the dynamic pressure lubrication performance and improve the bearing capacity.

Keywords: screw pumps; rhombus-like texture; dynamic pressure lubrication; oil film carrying capacity; texture parameters

1. Introduction

With the deepening of oilfield development, the proportion of thick oil wells, sandy oil wells, medium and low production oil wells, and high gas content wells has increased.

Screw pumps are widely used in major oilfields due to their unique advantages [1–6]. Their core components are the metal rotor and rubber stator, which are tightly meshed to form a number of closed chambers for extracting crude oil [7–9]. In a high-temperature and high-pressure downhole environment, the rubber stator undergoes prolonged periodic extrusion and friction from the metal rotor, resulting in wear failure, which negatively affects the service life of the screw pump and decreases the efficiency of oil recovery [10–14]. Therefore, to address the problem of wear failure in the stator and rotor friction pairs of screw pumps, new technology is needed to improve friction performance. This could establish a foundation for improving the oil recovery efficiency of screw pumps.

Many studies have shown that a reasonable design of surface texture could increase the friction performance of material surfaces [15–19]. Therefore, the design of a suitable micro-texture on the surface of screw pump stator and rotor friction pairs would be able to avoid the wear failure between the stator and rotor. In addition, it has been demonstrated that the wear-reducing performance of surface textures is affected by their size, shape, area rate, depth, and arrangement [20–23]. Jia et al. [24] employed a laser machining system to process circular micro-pit textures on plunger pumps and demonstrated that these micro-textures effectively reduced wear through a multifunctional friction and wear tester. Carl et al. [25] fabricated linear and square textures on the surface of Ti3SiC2–TiC-type MAX phase composites using the laser surface texturing technique, and their results showed that the textures exhibited excellent wear reduction performance. Wang et al. [26] demonstrated that when the friction reciprocating frequency was low, symmetrical textures exhibited superior friction reduction performance, and the friction reduction effect of hexagonal textures surpassed that of triangular textures. However, when the friction reciprocating frequency was high, triangular textures demonstrated better performance. Peng et al. [27] examined the influence of a microgroove texture on the friction performance of the friction surface of the rotor in an oil recovery screw motor and discovered that the friction coefficient increased with the increase in depth of a certain texture angle. Wang et al. [28] studied the influence of surface texture on the friction behavior of 42CrMo metal and rubber friction pairs and discovered that the presence of texturing significantly improved the wear performance of the mating pair. Joshua et al. [29] demonstrated that the effectiveness of micro-textures was closely related to their depth characteristics. Cui et al. [30] analyzed the tribological properties of diamond-shaped textures at different depths under water lubrication conditions, and the coefficient of friction could be seen to be 1 mm pit > no texture > 0.5 mm pit. Otero et al. [31] identified that the surface texture served as an effective method to reduce the friction coefficient of the point contact, particularly under low-speed and high-pressure conditions, thereby promoting boundary/mixed lubrication.

With the development of numerical simulation technology, researchers have used numerical optimization methods to obtain micro-texture parameters with better tribological properties. Figure 1 illustrates the research progress in numerical simulation methods for surface textures. Guo et al. [32] conducted a systematic investigation into the influence of geometrical parameters and sliding speeds on tribological properties by establishing a numerical computational model of a continuous groove texture and found that the texture had better lubrication properties at lower area densities. Shen et al. [33] investigated the effects of geometrical parameters as well as the distribution and arrangement of textures on lubrication properties by numerical simulation. The results showed that geometric parameters such as depth, length, width, and angle had a significant effect on lubrication performance. Among them, the effect of area density was the most obvious, and the effect of length was the least obvious. Liu et al. [34] performed a computational investigation into micro-depressed surface features, with findings indicating that the load-bearing capacity improved as the area density increased. Hua et al. [35] created a computational model using MATLAB and utilized an equation derived from the finite element method to evaluate the pressure distribution. Their analysis revealed that pressure distribution depended largely on the texture parameters. Wei et al. [36] conducted a numerical simulation to assess how fan-shaped surface textures influence the dynamic pressure lubrication performance of fluids. Their numerical simulation found that the optimal fan-shaped texture significantly outperforms traditional textures (circular, square, rhombus, v-shaped, triangular, anti-triangular) and anti-fan textures with the same radius, depth, and operating conditions. Li et al. [37] employed mass-conserving cavitation boundary conditions and a non-dominated sorting genetic algorithm to optimize the herringbone texture. Wang et al. [38] investigated the tribological properties of smooth surfaces, square-textured surfaces, triangular-textured surfaces, and circular-textured surfaces. Their research revealed that surface texture could substantially enhance friction performance under consistent operational parameters. In particular, the square texture design exhibited superior friction performance to triangular and circular patterns. Additionally, the optimal friction behavior was observed when the surface texture depth ranged from 20 μm to 50 μm. Zhong et al. [39] employed MATLAB software to study computational models of hexagonal textures with two directional configurations and examined how geometric attributes and operational parameters influence their tribological behavior. Mao et al. [40] designed a herringbone surface texture and verified its anti-wear effect by simulation.

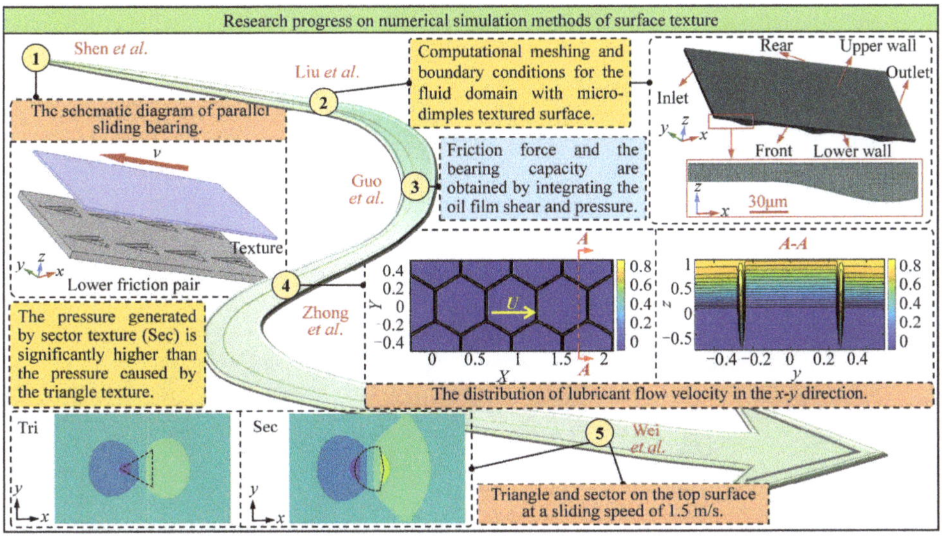

Figure 1. The research progress of numerical simulation methods for surface texture. Adapted from [32–34,36,39].

In summary, numerous studies have investigated circular, rectangular, triangular, and asymmetric textures, primarily applied in bearings, cutting tools, and other fields. However, existing studies on the reduction of wear on the stator–rotor friction surface of screw pumps have mainly focused on improvements in stator rubber properties [41–43], while those on the texture of stator and rotor friction surfaces are limited. Therefore, the introduction of surface texturing technology to increase the tribological performance of the stator and rotor friction pairs of screw pumps is of practical significance.

Based on the above analysis, to investigate the effects of micro-texture on the surface lubrication performance of screw pump stator and rotor friction pairs, inspired by bionics principles, this study proposed designing a rhombus-like texture that mimicked rattlesnake scales on the surface of screw pump stator and rotor friction pairs. A finite element model for the stator and rotor friction pairs of rhombus-like dimple-textured screw pumps was developed based on the principle of fluid lubrication. The single factor method was used to analyze the bearing capacity, friction coefficient, pressure distribution, and flow trace of different area rates, depths, shapes, and angle textures. The parameter combination of the texture exhibiting the best wear reduction effect was then selected. It is expected to decrease wear in screw pump stator and rotor friction pairs, enhance the wear performance of their surfaces, and extend the service life of screw pumps.

2. Model Analysis

2.1. Geometric Model of the Screw Pump Stator Surface Texture

The scale structure on the surface of snakes exhibits strong wear resistance, effectively reducing the friction caused by external objects on the skin. In this study, the surface texture pattern of snakeskin was selected to design the surface texture. Figure 2a shows the rhombus rattlesnake skin texture and bionic texture profiles. Figure 2b shows a schematic diagram of the friction pair of a screw pump metal rotor and rubber stator, Figure 2c presents the axial cross section of an electric submersible screw pump, and Figure 2d provides a schematic diagram of a rhombus-like micro-texture unit and its cross section on the surface of a rubber stator. The rhombus-like texture is centered within the texture unit, which has a side length (L) of 1 mm. The rhombus-like dimple-textured unit has long axes a and c and short axes b, h and α, which represent the depth and the long-side angle of the texture unit, respectively. β, h_0, and u are, respectively, the angle between the center axis of the texture and the x-direction axis of its unit, the oil film gap (equal to 20 μm), and the flow velocity of the extracted fluid in the screw pump. The area rate (S_p) of the rhombus-like texture is expressed as follows:

$$S_p = \frac{S_t}{S} \times 100\% \tag{1}$$

where S_t is the area of the rhombus-like texture (m^2), and S is the area of the texture unit (m^2).

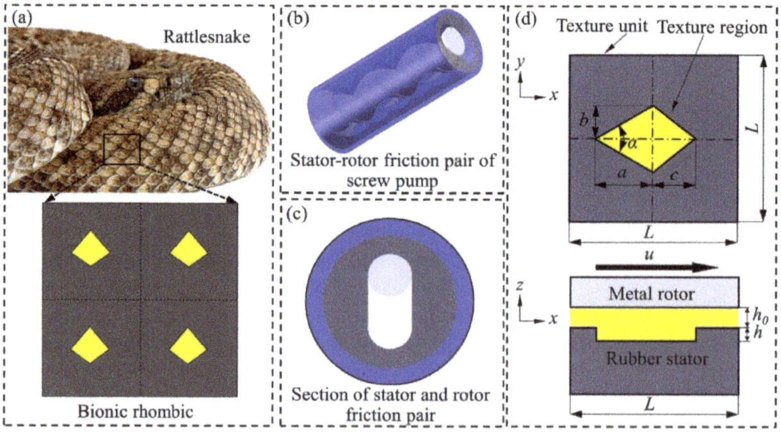

Figure 2. Geometric model of the rhombus-like texture (**a**) rhombus rattlesnake skin texture and bionic texture, (**b**) stator and rotor friction pair, (**c**) section of stator and rotor friction pari, (**d**) cross-sectional profile of dimple.

2.2. Theoretical Model for the Numerical Simulation

The actual well fluid from a shale oil well delivered by an oil recovery screw pump is considered an incompressible Newtonian fluid. Its density and viscosity were kept constant in the simulation. The calculation process only considered the impact of inertial forces while disregarding the temperature effects. The fluid dynamics calculation method based on N–S equations was used. The momentum and continuity equations were as follows [44]:

$$\frac{\partial(\rho u)}{\partial t} + \mathrm{div}(\rho\eta u) = \frac{\partial\tau_{xx}}{\partial x} + \frac{\partial\tau_{yx}}{\partial y} + \frac{\partial\tau_{zx}}{\partial z} - \frac{\partial p}{\partial x} + F_x \tag{2}$$

$$\frac{\partial(\rho v)}{\partial t} + \mathrm{div}(\rho\eta v) = \frac{\partial\tau_{xy}}{\partial x} + \frac{\partial\tau_{yy}}{\partial y} + \frac{\partial\tau_{zy}}{\partial z} - \frac{\partial p}{\partial y} + F_y \tag{3}$$

$$\frac{\partial(\rho w)}{\partial t} + \mathrm{div}(\rho\eta w) = \frac{\partial\tau_{xz}}{\partial x} + \frac{\partial\tau_{yz}}{\partial y} + \frac{\partial\tau_{zz}}{\partial z} - \frac{\partial p}{\partial z} + F_z \tag{4}$$

$$\frac{\partial u}{\partial x} + \frac{\partial v}{\partial y} + \frac{\partial w}{\partial z} = 0 \tag{5}$$

where ρ is the density of the actual well fluid of a shale oil well (kg/m^3); t is time (s); u, v, and w are the components of fluid velocity in the three directions of x, y, and z, respectively (m/s); η is the kinetic viscosity of the crude oil medium (Pa·s); τ_{xx}, τ_{xy}, and τ_{xz} are the viscous action τ in the respective directions of the components; p is the pressure of the oil film (Pa); and F_x, F_y, and F_z are the components of the volume force of the crude oil medium in the x, y, and z directions (N).

The dynamic pressure lubrication performance was described by the oil film carrying capacity and the friction force between the stator and rotor friction pairs of the screw pump. The carrying capacity of the oil film can be determined by integrating the oil film pressure, which is expressed as follows [34]:

$$F_z = \iint p(x,y)ds \tag{6}$$

The friction force of the oil film between the stator and rotor friction pairs of the screw pump can be determined by integrating the viscous shear force, which is expressed as follows:

$$F_x = \iint \tau(x,y)ds \tag{7}$$

where $p(x,y)$ is the static pressure distribution function, and $\tau(x,y)$ is the shear stress distribution function.

The friction coefficient between the stator and rotor friction pair of the screw pump is the ratio of the friction force to the bearing capacity:

$$f = F_z/F_x \tag{8}$$

Figure 3 shows the single screw pump rotor movement trajectory diagram based on Equations (9)–(11), which can be calculated using the screw pump stator–rotor friction pair between the point of engagement point speed (v_{x1} and v_{y1}), the line engagement point speed v' (v'_{x1} and v'_{y1}), and the maximum speed of the engagement point $v_{\max}$ and minimum speed $v_{\min}$ [45]:

$$\begin{cases} v_{x1} = R\omega - 2e\omega\sin\omega t_1 \\ v_{y1} = -R\omega - 2e\omega\sin\omega t_1 \end{cases} \tag{9}$$

$$\begin{cases} v'_{x1} = R\omega \\ v'_{y1} = -R\omega \end{cases} \quad (10)$$

$$\begin{cases} v_{\max} = R\omega + 2e\omega \\ v_{\min} = R\omega - 2e\omega \end{cases} \quad (11)$$

where R is the radius of the semicircle (m), e is the eccentricity distance (m), ω is the angular velocity of the rotor (rad/s), and t_1 is the rotation time of the rotor (s).

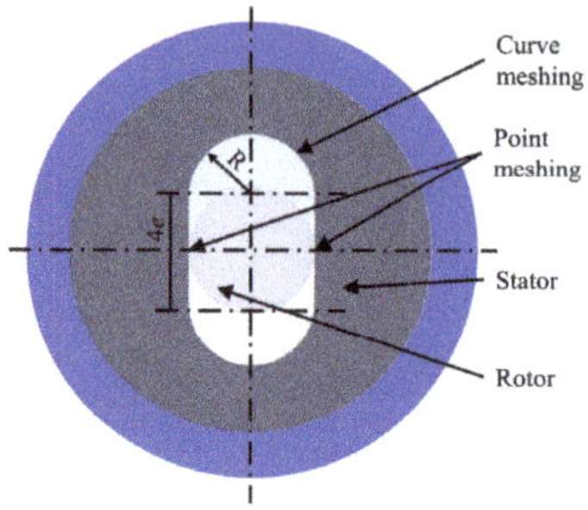

Figure 3. Single screw pump rotor movement trajectory.

3. Numerical Simulation Method

3.1. Mesh Division and CFD Solution Settings

Fluent Meshing 2020 was employed to generate a polyhedral mesh for the computational fluid domain. The number of meshes can significantly affect the numerical results, and therefore, a mesh independence test was conducted. Based on the geometric complexity of the computational domain, an initial coarse mesh of 224,162 elements was first generated. To ensure mesh convergence, the mesh was systematically refined by a factor of 1.5 to 2 times the number of the previous mesh elements, resulting in subsequent mesh numbers of 327,679, 496,174, 810,470, 1,534,590, and 2,279,097. To verify mesh independence, the pressure distribution on the symmetrical axis of the texture-bearing surface was analyzed. As shown in Figure 4, it was evident that when the number of meshes exceeded 800,000, the numerical solution demonstrated high stability and convergence. Therefore, this study used an 800,000-element mesh for the calculations to ensure high accuracy and stability.

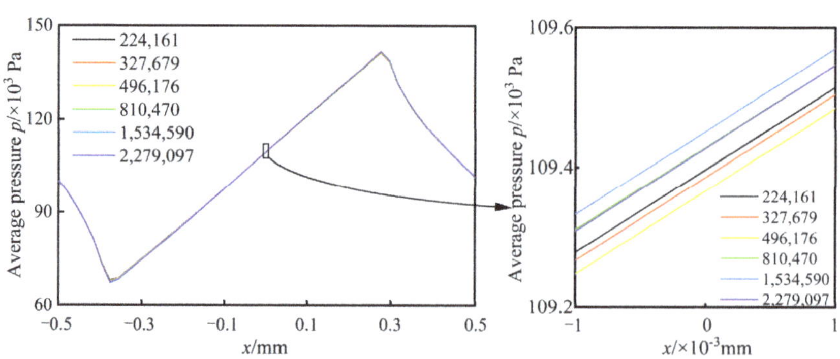

Figure 4. Verification of mesh independence based on pressure distribution on the axis of an oil film carrying surface.

The 3D model was analyzed using the CFD software Fluent 2020. The pressure-based steady-state double precision solver was used. The turbulence model selected for this study was the k-epsilon model, as used in the study by Liu et al. [46,47]. The discrete equations were in second-order windward difference format, and the coupled algorithm was adopted for pressure–velocity coupling. In the fluid domain, the upper and lower walls were in the no-slip condition, the upper wall moved in the x-direction with a velocity (u) of 1.8 m/s, the lower wall was stationary, and inlet and outlet pressure boundary conditions were adopted. In the numerical calculation, the density of the actual well fluid in the shale oil well and the viscosity were equal to 870 kg/m^3 and 0.08 Pa·s, respectively.

3.2. Validation of the Numerical Simulation Method

To verify the effectiveness of the numerical simulation method, the square dimple texture documented in reference [48] was numerically calculated by using the calculation method in this study. Figure 5 illustrates the fluid domain from reference [48]. The side length (l) of the fluid domain unit, side length (a) of the square texture, depth (h_1), oil film thickness (h_2), moving speed (u_0), lubricating oil density, and viscosity were equal to 1 mm, 400 μm, 10 μm, 50 μm, 6 m/s, 891 kg/m^3, and 0.098 Pa·s, respectively. The pressure data of the center line of the wall on the fluid domain were extracted, and the results obtained by the proposed approach were compared with those from reference [48]. This comparison is illustrated in Figure 6. It can be seen that the two approaches have consistent results [48], which verifies that the proposed numerical calculation method has high accuracy.

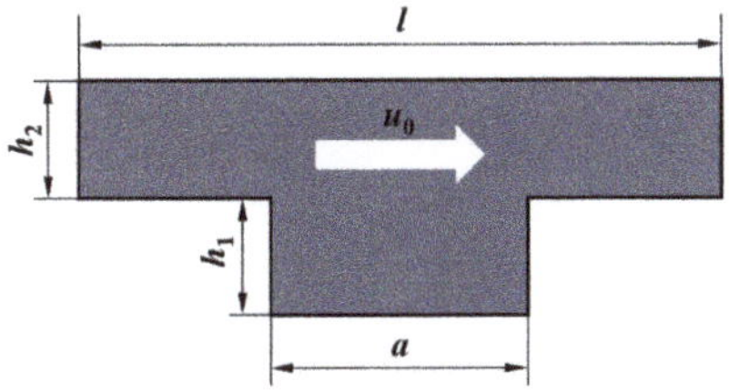

Figure 5. Schematic diagram of the computational domain model in reference [48].

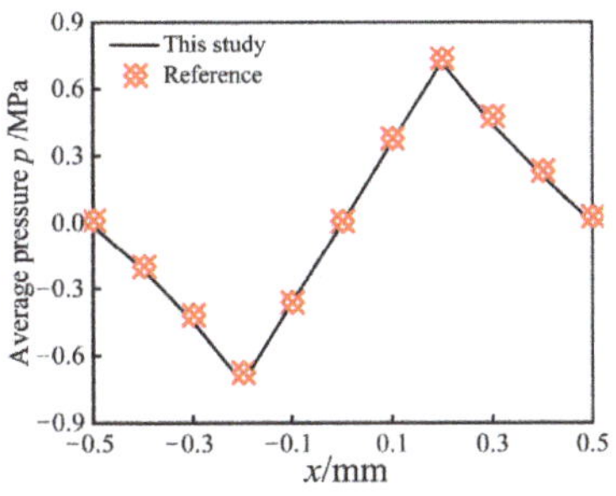

Figure 6. Comparison of pressures obtained by the computational methods in this study and reference [48].

3.3. Determination of the Flow Direction of the Medium

The fixed texture long axes a and c were 400 μm and 300 μm, respectively. The short axis b and depth h were 300 μm and 50 μm, respectively. The angle β between the central axis of the texture and the x-direction axis of the texture unit was set to 0. When the flow direction of the medium was along the x, $-x$, and y axes, the pressure of the texture-carrying surface was analyzed. The obtained results are shown in Figure 7. The results demonstrated that the inflow direction of the medium significantly affected the pressure distribution on the carrying surface. When the flow direction of the medium was x and y, the area of the negative pressure zone of the bearing surface was smaller than that of the positive pressure zone. On the contrary, when the flow direction was $-x$, the area of the negative pressure zone was larger than that of the positive pressure zone, which indicated variations in net pressure and dynamic pressure lubrication performance generated by the texture. Therefore, determining the flow direction of the medium was essential.

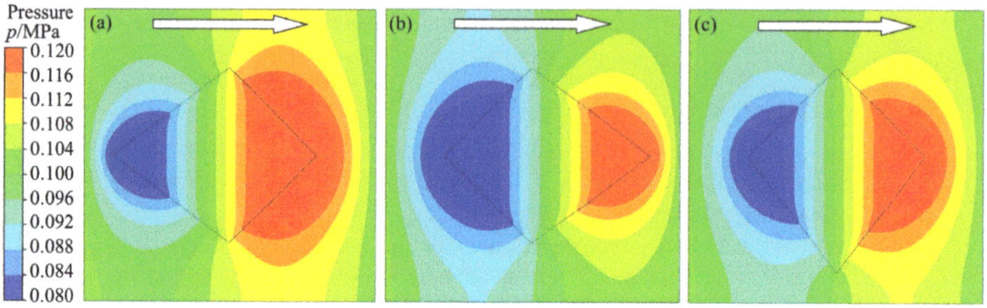

Figure 7. Pressure distribution on the bearing surface for different flow directions of the medium: (**a**) x direction, (**b**) $-x$ direction, and (**c**) y direction.

Table 1 presents the carrying capacity and friction coefficients for different flow directions. When the flow direction of the medium was following x, the maximum friction coefficient of the oil film bearing capacity was the smallest. In contrast, when the flow direction followed $-x$, the maximum friction coefficient of the oil film bearing capacity reached its largest value. The friction coefficients for the three flow directions were all lower than that of the non-textured surface (0.077), which demonstrated the high anti-wear performance of the textured surface.

Table 1. Bearing capacity and friction coefficient for different flow directions of the medium.

Direction of Flow	Bearing Capacity $F_z/\times10^{-2}$ N	Friction Coefficient f
x	10.5	0.070
$-x$	9.75	0.075
y	10.1	0.073

4. Analysis of the Numerical Simulation Results

4.1. The Effect of the Texture Area Rate of Rhombus-like Micro-Dimples on the Surface Lubrication Performance of Screw Pump Stator and Rotor Friction Pairs

To analyze the effect of the texture area rate on the surface lubrication performance of the friction pair surface, texture depth h was set to 50 μm, angle β between the texture center axis and the x-direction axis of the unit body was set to 0, and the ratio of the texture long axes a and c and the short axis b was fixed. The rhombus-like micro-dimple texture was selected for numerical calculation at area rates S_p of 5%, 8%, 12%, 16%, 21%, and 27%. Figure 8a shows the pressure distribution at the center axis of the carrying surface of the texture under different texture area rates. It can be seen that the pressure distribution followed the same trend for the different texture area rates. More precisely, it started as negative and then became positive, which indicated that the studied rhombus-like micro-dimple texture effectively generated a dynamic pressure lubrication effect. Figure 8b shows the surface friction coefficient and average pressure change curves under different texture area rates. When the texture area rate increased, the surface friction coefficient decreased, while the average pressure increased. When the texture area rate S_p was 27%, the friction coefficient reached its minimum value and the average pressure reached its maximum value, demonstrating an optimal lubrication performance.

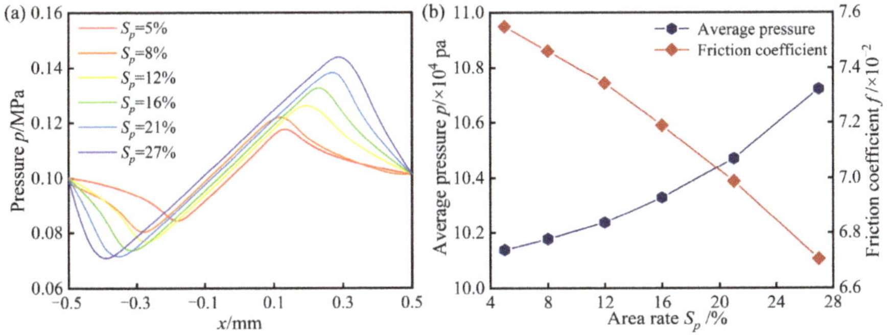

Figure 8. Curves of the friction coefficient and average pressure variation on the textured surface with different area rates: (**a**) pressure distribution at the center axis of the carrying surface, (**b**) friction coefficient and average pressure.

Figure 9 presents the oil film surface pressure distribution cloud diagrams at different texture area rates. It can be observed that a negative pressure region formed at the angle of the long side of the rhombus-like dimple texture, while a positive pressure region appeared at the angle of the short side. The area of positive pressure was larger than that of negative pressure, enhancing the hydrodynamic lubrication. When the texture area rate increased, the positive and negative pressure areas were expanded. When the texture area rate S_p was 27%, the difference between the positive and negative pressure areas reached its maximum value, yielding the best lubrication performance. This was because the increase in the texture area rate could make the stator and rotor friction pairs of the screw pump store more actual well fluids from shale oil wells, which affected the lubrication performance of the friction pairs and decreased the friction coefficient [49].

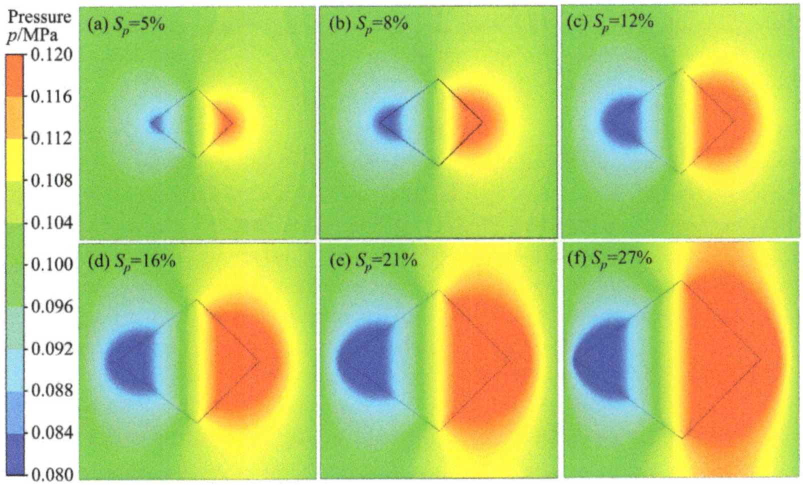

Figure 9. The pressure distribution cloud diagram of the textured surface under different area rates.

4.2. The Effect of the Texture Depth of Rhombus-like Micro-Dimples on the Surface Lubrication Performance of Screw Pump Stator and Rotor Friction Pairs

To analyze the effect of the texture depth on the surface lubrication performance of the friction pair, angle β between the central axis of the texture and the x-axis of the unit body was set to 0, and the texture area rate S_p was fixed at 27%. Numerical calculations were performed on rhombus-like micro-dimple textures at depths of 10, 20, 30, 40, 50, 60, 70, 80, and 90 μm. Figure 10a shows the pressure distribution at the center axis of the carrying surface for different texture depths. It can be seen that the amplitude of the maximum and minimum pressures reached its maximum value at a texture depth h of 20 μm and that the center axis gradually decreased when the texture depth h was greater than 20 μm. Figure 10b shows the surface friction coefficient and average pressure change curves at different texture depths. When the texture depth increased, the surface friction coefficient initially decreased and then increased, while the average pressure first increased and then decreased. When the texture depth h was 20 μm, the friction coefficient was minimized and the average pressure was maximized, resulting in optimal lubrication performance. The effects were attributed to the transition from the hydrodynamic lubrication effect in the texture to the mixed lubrication mechanism. When the texture depth was less than 20 μm, hydrodynamic lubrication dominated, resulting in a decrease in the friction coefficient and an increase in the bearing capacity. However, when the texture depth exceeded 20 μm, fluid resistance increased, leading to higher energy consumption and an increase in the friction coefficient.

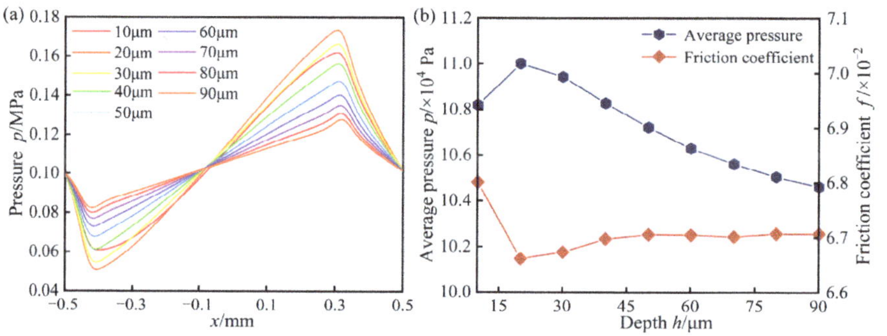

Figure 10. Curves of the friction coefficient and the average pressure variation in the textured surface at different depths: (**a**) pressure distribution at the center axis of the carrying surface, (**b**) friction coefficient and average pressure.

Figure 11 shows the pressure distribution on the oil film bearing surface at different texture depths. It can be observed that as the texture depth increased, the areas of both the positive and negative pressure regions increased first and then decreased. When the texture depth exceeded 50 μm, the hydrodynamic lubrication effect in the inlet and outlet regions of the texture diminished. This might be attributed to the texture depth being too large, the velocity gradient decreased, and the extracted fluid at the bottom of the texture unable to provide a support force in time [39]. The difference between the positive and negative pressure regions was at its largest when the texture depth h was 20 μm, which indicated that the surface lubrication performance was optimal at this depth.

Figure 12 illustrates the distribution of flow traces within the rhombus-like micro-dimple structure at varying texture depths. When the texture depth increased, the distribution of flow traces significantly changed. The vortex phenomenon started to appear inside the structure when the texture depth h was 30 μm. In general, the vortex phenomenon inside the structure was gradually strengthened when the texture depth h was more than 30 μm. Due to the vortex phenomenon inside the texture, when its depth increased, the amplitude of its pressure gradually decreased. When the texture depth was large, part of the kinetic energy generated by the oil film bearing surface was converted into the energy of the vortex inside the texture, which resulted in energy loss. In addition, the texture used to generate the dynamic pressure lubrication effect was weakened, and the surface lubrication performance was decreased [50]. Theoretically, the increase in the texture depth allowed the storage capacity of the extracted fluid in the screw pump to increase. However, the vortices generated within the texture decreased this storage capacity, which negatively affected the lubrication performance. Therefore, the texture depth should not exceed a certain limit. More precisely, it should approach the thickness of the oil film between the stator and rotor of the screw pump to reach optimal lubrication performance with the rhombus-like texture.

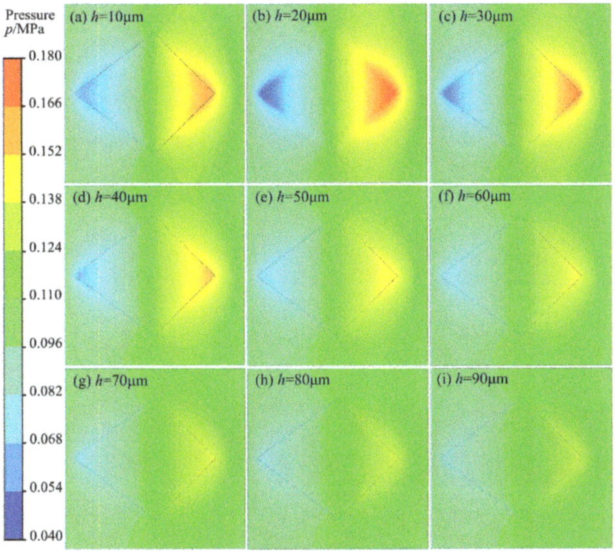

Figure 11. The pressure distribution cloud diagram of the textured surface at different area depths.

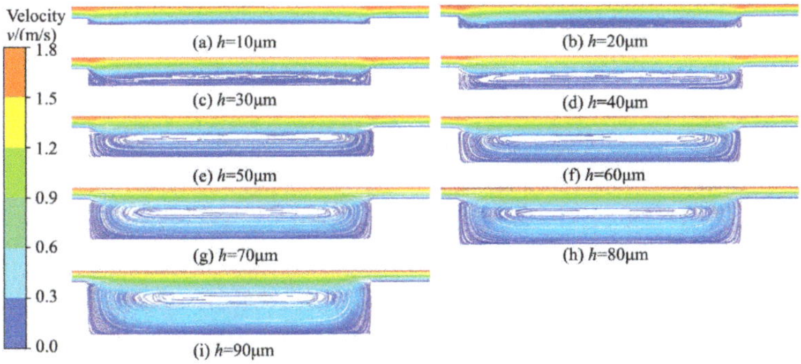

Figure 12. The internal flow trace diagram of texture under different depth conditions.

4.3. The Effect of the Rhombus-like Micro-Dimple Texture Shape on the Surface Lubrication Performance of Screw Pump Stator and Rotor Friction Pairs

When the long axes a and c of the rhombus-like dimple structure were constant, the variation in the ratio of the long axis a and the short axis b of the structure changed angle α between the long sides of the rhombus-like micro-dimple structure, leading to a variation in its shape. To analyze the effect of the texture shape on the surface lubrication performance of the friction surface, numerical calculations were performed for rhombus-like micro-dimple textures with long-side angles α of 48.5, 51.5, 55, 59, 63, 68, and 74°, while the texture area rate S_p, texture depth h, and angle β of the texture center axis and the x-direction axis of the texture unit were fixed at 27%, 20 μm, 0°, and 0°, respectively. Figure 13 shows the variation curves of the surface friction coefficient and the average pressure for different long-side pinch angles. When the long-side angle of the texture increased, the surface friction coefficient decreased, while the average pressure increased. For a long-side angle α of 74°, the surface friction coefficient reached its minimum value and the average pressure reached its maximum value, which indicated optimal lubrication performance. This was due to the variation in the angle between the long sides of the texture, which altered the texture geometry. This variation subsequently resulted in significant modification of the pressure distribution within the flow field.

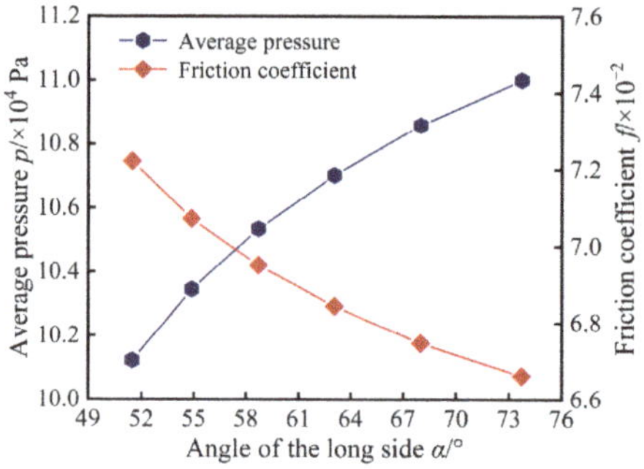

Figure 13. Curves of the friction coefficient and the average pressure variation in the textured surface with different long-side angles.

Figure 14 shows the pressure distribution cloud diagrams for different long-side angles on the friction surface. It can be observed that when the long-side angle of the texture increased, the area of positive pressure gradually decreased while that of negative pressure increased, which resulted in reduced net pressure on the texture structure and a diminished dynamic pressure lubrication effect. For a rhombus-like micro-dimple-textured structure with a long-side angle α of 74°, the difference between the positive and negative pressure areas was maximized, which indicated the best lubrication performance. This was attributed to the reduction in the long-side angle, which increased the short-side angle of the texture structure, leading to an increase in the wedge-convergence angle that generates the dynamic pressure lubrication effect. The narrowing area of the positive pressure zone decreased, and the widening area of the negative pressure zone increased, resulting in a decrease in net pressure. This change negatively affected the lubrication performance of the friction surface.

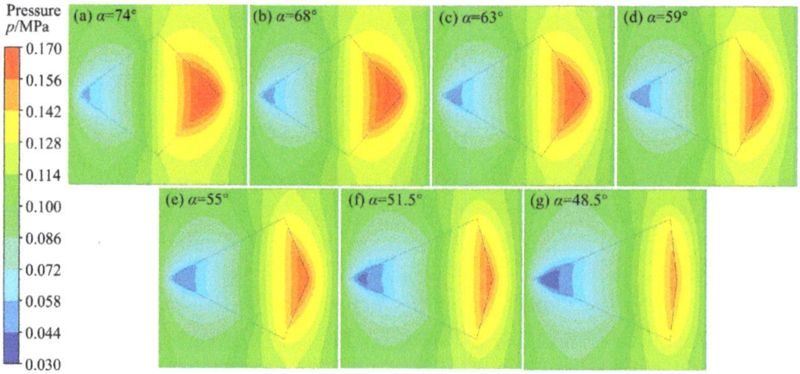

Figure 14. The pressure distribution cloud diagram of the textured surface with different long-side angles.

4.4. The Effect of the Rhombus-like Micro-Dimple Texture Angle on the Surface Lubrication Performance of Screw Pump Stator and Rotor Friction Pairs

To further analyze the effect of texture angle β on the surface lubrication performance of the friction surface, the texture area rate S_p, texture depth h, and angle α of the long side were, respectively, fixed at 27%, 20 μm, and 74°. Numerical calculations were carried out for the rhombus-like micro-dimple with angles β of the texture center axis and the x-direction axis of the unit body as 0, 15, 30, 45, 60, and 75°. Figure 15 shows the curves of the surface friction coefficient and the average pressure for the texture structure under different angle conditions. When the angle between the texture structure center axis and the x-direction axis of the unit body increased, the surface friction coefficient increased, the average pressure decreased, the surface friction coefficient of the texture friction surface was the smallest, and the average pressure was the largest. In addition, the lubrication performance was the best when angle β between the center axis of the texture structure and the x-direction axis of the unit body was 0°.

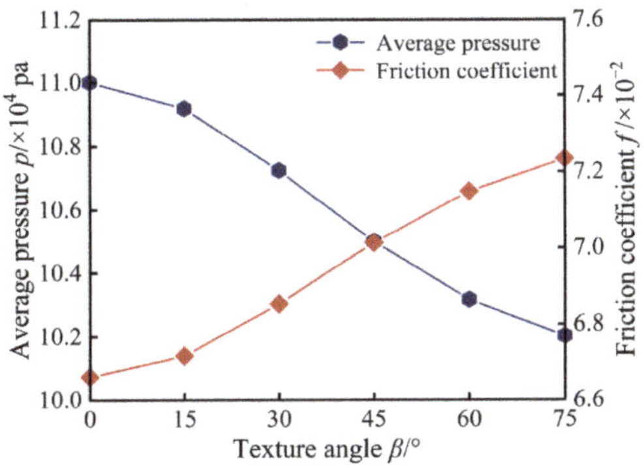

Figure 15. Curves of the friction coefficient and the average pressure variation in the textured surface with different angles.

Figure 16 illustrates the pressure distribution cloud diagrams for the friction surface under different angle conditions. It is evident that when the angle between the texture structure center axis and the x-direction axis increased, the area of positive pressure gradually decreased while that of negative pressure increased. This demonstrates that the net pressure generated on the bearing surface of the texture structure was gradually reduced, which resulted in a diminishing dynamic pressure lubrication effect. When angle β exceeded 45°, the positive pressure area became smaller than the negative pressure area. This resulted in the low lubrication performance of the friction surface. Conversely, for $\beta = 0°$, the difference between the positive and negative pressure areas was maximized, leading to the best lubrication performance of the friction surface. The reason was that the variations in angle β between the texture center axis and the x-direction axis of the unit affected the flow path and pressure distribution of the fluid on the bearing surface. When $\beta = 0°$, fluid flow tended to establish a stable positive pressure region. However, when β increased, the fluid flow direction changed, resulting in a reduction in the positive pressure area and an expansion of the negative pressure area. As a result, the overall pressure distribution of the fluid was not conducive to the generation of a hydrodynamic lubrication effect, leading to a decrease in the lubrication performance.

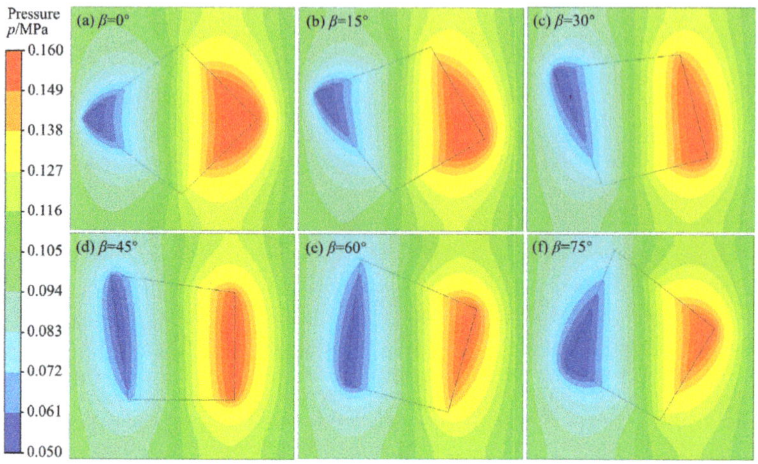

Figure 16. The pressure distribution cloud diagram of the textured surface at different angles.

4.5. The Effect of the Actual Well Fluid Viscosity of Shale Oil on the Lubrication Performance of the Rhombus-like Micro-Dimple Texture

The change in viscosity significantly influenced the lubrication performance of the texture surface [51]. To analyze the effect of the actual well fluid viscosity on the texture lubrication performance, numerical calculations were conducted on the rhombus-like dimple texture with a fixed area rate of 27%, depth of 20 μm, long-edge angle of 74°, and an angle of 0° under different viscosity conditions. Figures 17 and 18 illustrate the average pressure change of the textured surface and the pressure distribution of the bearing surface under different viscosities. The results indicated that as the viscosity of the actual well fluid increased, the average pressure on the texture surface and the bearing capacity also increased. From the pressure distribution, it was evident that both the positive and negative pressure regions were increased. Furthermore, with the increase in viscosity, the difference between the positive and negative pressure area grew larger, leading to higher net pressure generation and an enhanced dynamic pressure lubrication effect, consistent with the research results of Wang et al. [52]. This improvement could be attributed to the fact that higher viscosity facilitated oil film formation, strengthened the hydrodynamic lubrication effect, and, consequently, enhanced the bearing capacity.

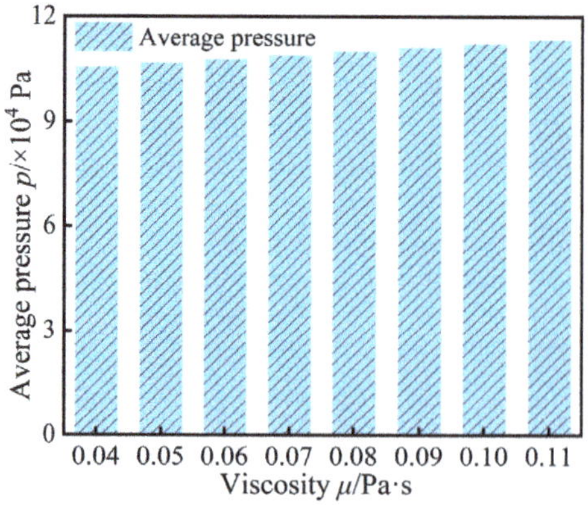

Figure 17. The average pressure change of the textured surface under different viscosities.

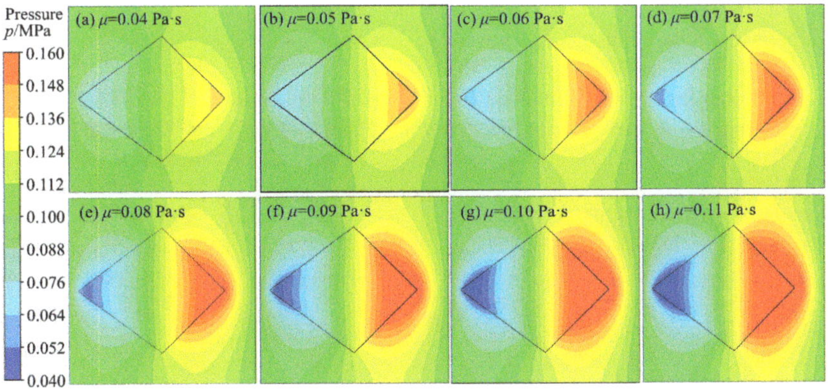

Figure 18. The pressure distribution cloud diagram of the bearing surface under different viscosities.

5. Conclusions

This study established a finite element calculation model for stator and rotor friction pairs with rhombus-like dimple textures and investigated the influence of texture parameters on lubrication performance. The following conclusions can be drawn:

(1) The geometrical parameters of the rhombus-like micro-dimple texture significantly affected the friction performance of the friction surfaces. When the texture area rate increased, the dynamic pressure lubrication performance of the friction pairs significantly increased. When the area rate increased, the lubricant reservoir within the texture expanded, which resulted in a decrease in the oil film friction coefficient and an increase in the average pressure and net load carrying capacity. The optimal lubrication effect was achieved at an area rate of 27%.

(2) When the texture depth increased, the oil film friction coefficient first decreased and then increased, while the average pressure first increased and then decreased. When the depth became greater than 30 μm, vortices started to form inside the structure. The intensity of these vortices increased with the increase in depth, which resulted in a more significant decrease in the dynamic pressure lubrication performance. The best lubrication performance was observed when the depth was close to 20 μm, which corresponded to the thickness of the oil film between the stator and rotor of the screw pump.

(3) When the angle of the long side of the structure increased, the dispersing wedge angle also increased, reducing the negative pressure zone area of the oil film, increasing the positive pressure zone area, decreasing the oil film friction coefficient, and increasing the net load carrying capacity.

(4) The angle between the texture center axis and the x-direction axis of the unit significantly affected the pressure distribution in the flow field. When this angle increased, the area of the negative pressure zone increased, while the positive pressure zone area decreased. This resulted in an increase in the friction coefficient and a decrease in the net bearing capacity. When the angle became greater than a certain threshold, the negative pressure area became smaller than the positive pressure area, which significantly reduced the oil film carrying capacity.

(5) The optimal parameters for the rhombus-like dimple texture were an area rate, depth, long-side angle, and angle of 27%, 20 μm, 74°, and 0°, respectively. When the oil film had the largest net bearing capacity and the smallest coefficient of friction, the dynamic pressure lubrication was the highest and the effect of wear reduction was the most significant. The results of the study provide a theoretical basis and design

guidance for improving the lubrication performance of the stator and rotor friction pairs in screw pumps.

Author Contributions: Conceptualization, X.S., X.L. and Y.S.; Methodology, X.S. and X.L.; Software, X.S.; Validation, X.S.; Formal analysis, X.S., X.L., C.L., Z.H., S.H. and X.N.; Investigation, X.S., X.L., C.L., Z.H., S.H., Y.S. and X.N.; Resources, X.L.; Data curation, X.S., Y.S. and X.N.; Writing—original draft, X.S.; Writing—review & editing, X.L.; Supervision, X.L., C.L., Z.H. and S.H.; Project administration, X.L., C.L., Z.H. and S.H.; Funding acquisition, X.L. All authors have read and agreed to the published version of the manuscript.

Funding: This study was supported by the National Natural Science Foundation of China (Nos. 52074161 and 52005281), the Taishan Scholar Project of Shandong Province (No. tsqn202211177), the Shandong Provincial Plan for the Introduction and Cultivation of Young Pioneers in Colleges and Universities (No. 2021-QingChuang-30613019), the Natural Science Foundation of Shandong Province (Nos. ZR2022ME173 and ZR2023QE011), and the Project of CNOOC Research Institute Ltd. (No. CCL2023RCPS2129RQN).

Data Availability Statement: If necessary, please contact the corresponding author.

Acknowledgments: The authors sincerely thank the anonymous reviewers for their insightful comments and suggestions.

Conflicts of Interest: Authors Zhongxian Hao and Shouzhi Huang were employed by PetroChina. The remaining authors declare that the research was conducted in the absence of any commercial or financial relationships that could be construed as a potential conflict of interest.

References

1. FestaBianchet, S.A.; Tyner, D.R.; Seymour, S.P.; Johnson, M.R. Methane venting at cold heavy oil production with sand (CHOPS) facilities is significantly underreported and led by high-emitting wells with low or negative value. *Environ. Sci. Technol.* **2023**, *57*, 3021–3030. [CrossRef] [PubMed]
2. Xu, J.; Deng, J.E.; Chen, Y.; Jia, L.X.; Xiao, Y.; Liu, W. Analysis of casing instability and factors influencing it in thermal production oil wells. *Chem. Technol. Fuels Oils* **2021**, *57*, 804–812. [CrossRef]
3. Ramos, A.M.; Brown, J.C.; Rojas, M.D.C.; Kuyuco, O.; Flores, J.G. Producing extra-heavy oil from the Orinoco Belt, Cerro Negro Area, Venezuela, using bottom-drive progressive cavity pumps. *SPE Prod. Oper.* **2007**, *22*, 151–155. [CrossRef]
4. Liu, X.F.; Liu, C.H.; Hao, Z.X.; Zheng, Y.; Zhang, K.; Wang, J.F.; Wei, S.B.; Hao, A.G.; Tao, J.L.; Cheng, H. Non-symmetric distributions of solids deposition for solid-water stratified flow in deviated tubing strings. *Pet. Sci.* **2023**, *20*, 3048–3061. [CrossRef]
5. Gusmanova, A.; Bekbayeva, R.; Koyshieva, A.; Abylai, K. Oil withdrawal technological advancement for multilayer field. *Int. J. Oil Gas Coal Technol.* **2022**, *30*, 265–282. [CrossRef]
6. Tavosi, S.; Alimardani, M.; Ghoreishy, R.H.M.; Mohammad, T. Fatigue failure modeling in rubber stator of downhole motors: Dependency to materials and geometry, and its application to life prediction. *Eng. Fail. Anal.* **2024**, *159*, 108072. [CrossRef]
7. Zheng, L.; Wu, X.D.; Han, G.Q.; Li, H.C.; Zuo, Y.; Zhou, D.K. Analytical model for the flow in progressing cavity pump with the metallic stator and rotor in clearance fit. *Math. Probl. Eng.* **2018**, *2018*, 3696930. [CrossRef]
8. Pan, L.; Tan, J.Z. Numerical investigation of contact stress between rotor and stator in a two-lead progressing cavity pump. *J. Pet. Sci. Eng.* **2015**, *134*, 176–185. [CrossRef]
9. Sun, W.X.; Wang, Y.Q.; Wang, K.; Wang, L. Research on three-dimensional modeling of multi-thread single screw pump rotor and stator based on UG. *Key Eng. Mater.* **2010**, *873*, 114–117. [CrossRef]
10. Guagliano, M. Fatigue failure of a bar of a twin-screw extruder for plastics. *Eng. Fail. Anal.* **2004**, *12*, 950–962. [CrossRef]
11. Borisova, K.E.; Ivanova, T.N.; Latypov, R.G. Study of screw pump stator and rotor working capacity to increase the output. *Procedia Eng.* **2017**, *206*, 688–691. [CrossRef]
12. Chen, J.; Wang, F.S.; Shi, G.C.; Cao, G.; He, Y.; Ge, W.T.; Liu, H.; Wu, H.A. Finite element analysis for adhesive failure of progressive cavity pump with stator of even thickness. *J. Pet. Sci. Eng.* **2015**, *125*, 146–153. [CrossRef]
13. Lv, X.R.; Wang, S.J.; Huo, X.Y. The effect of water content in crude oil on the tribological behavior of screw pump's stator rubber. *Chin. Sci. Bull.* **2014**, *59*, 2868–2874. [CrossRef]
14. Shi, C.S.; Zhu, X.H.; Chen, Y.K.; Tang, L.P. Failure analysis of general stator and uniform wall thickness stator. *Eng. Fail. Anal.* **2018**, *94*, 239–248. [CrossRef]

15. Bornillo, K.; Bovo, E.; Sorgato, M.; Lucchetta, G. Data-driven analysis of the effects of microtextured surfaces on friction reduction for plastic syringe applications. *Int. J. Adv. Manuf. Technol.* **2025**, *136*, 2583–2599. [CrossRef]
16. Xu, G.M.; Liu, P.; Liu, A.; Zhang, C.F.; Zhang, G.; Wu, Q.Q.; Lu, H.L. Mechanically stable superhydrophobic surfaces constructed by laser surface texturing and micro-arc oxidation of TC4 alloy prepared based on SLM. *Mater. Chem. Phys.* **2024**, *325*, 129791. [CrossRef]
17. Gu, C.X.; Cui, Y.M.; Zhang, D. Research on the optimal design approach of the surface texture for journal bearings. *Lubricants* **2024**, *12*, 111. [CrossRef]
18. Slawomir, W.; Waldemar, K.; Pawel, P. Selected methods and applications of anti-friction and anti-wear surface texturing. *Materials* **2021**, *14*, 3227. [CrossRef] [PubMed]
19. Pawel, P.; Waldemar, K.; Rafal, R. Surface texturing of cylinder liners: A review. *Materials* **2022**, *15*, 8629. [CrossRef]
20. Wang, J.F.; Xue, W.H.; Gao, S.Y.; Li, S.; Duan, D.L. Effect of groove surface texture on the fretting wear of Ti–6Al–4V alloy. *Wear* **2021**, *486*, 204079. [CrossRef]
21. Wei, H.; Chen, G.J.; Chen, Z.; Yu, Z.W.; Huang, J.S. Progress on bionic textured cutting tools: A review and prospects. *J. Bionic Eng.* **2023**, *21*, 19–55. [CrossRef]
22. Yang, X.D.; Chen, W.G.; Zhang, Y.; Li, Z.Y.; Yang, J.W.; Dai, B.G.; Zhang, J.H.; Li, D.Y.; Qiu, Z.L. Tribological properties of MAO coating deposited on aluminum base bearing alloy with different texture shapes. *Surf. Coat. Technol.* **2024**, *492*, 131191. [CrossRef]
23. Zhang, H.Y.; Liu, X.R.; Gong, J.Z.; Bai, S.Z.; Sun, K.; Jia, X.R. Thermohydrodynamic lubrication characteristics of piston rings in diesel engine considering transient heat transfer under the parameterized surface texture of cylinder liners. *Energies* **2023**, *16*, 7924. [CrossRef]
24. Jia, H.K.; Zhou, Z.Y.; Yin, B.F.; Zhou, H.Q.; Xu, B. Influence of microdimple on lubrication performance of textured plunger pump. *Ind. Lubr. Tribol.* **2021**, *73*, 563–571. [CrossRef]
25. Magnus, C.; Gulenc, I.T.; Rainforth, W.M. Ambient dry sliding friction and wear behaviour of laser surface textured (LST) Ti3SiC2 MAX phase composite against hardened steel and alumina. *Wear* **2022**, *490*, 204184. [CrossRef]
26. Wang, J.H.; Huang, D.J.; Wang, B.; Zhang, Y.; Wang, J. Study on the tribological behavior of laser surface texturing on silicon nitride ceramic under water lubrication. *Lubricants* **2025**, *13*, 21. [CrossRef]
27. Peng, S.J.; Huang, Z.Q.; Wang, M.L.; Xiong, C.; Chen, K. Study on the influence of surface texture parameters of oil production screw motor rotor on the tribological properties of its friction pair. *Processes* **2023**, *11*, 618. [CrossRef]
28. Wang, M.L.; Huang, Z.Q.; Qian, W.J.; Li, G. Study of the effect of laser textured rotors on the starting performance of metal–rubber mating pairs under different lubricating media environments. *Proc. Inst. Mech. Eng. Part J J. Eng. Tribol.* **2024**, *238*, 450–463. [CrossRef]
29. Paul Joshua, S.; Dinesh Babu, P. Friction and wear characteristics for laser surface textured rolling bearing steel: Effect of pattern density, depth, and area. *Trans. Indian Inst. Met.* **2022**, *75*, 2923–2930. [CrossRef]
30. Cui, Z.T.; Guo, Z.W.; Yuan, C.Q. Influence of different rhombic surface textures on the tribological performance of water-lubricated bearings. *Mater. Express* **2020**, *10*, 1452–1462. [CrossRef]
31. Otero, E.J.; Ochoa, G.E.; Vallinot, B.I.; Tanarro, E.C. Optimising the design of textured surfaces for reducing lubricated friction coefficient. *Lubr. Sci.* **2017**, *29*, 183–199. [CrossRef]
32. Guo, Q.G.; Zheng, L.; Zhong, Y.H.; Wang, S.K.; Ren, L.Q. Numerical simulation of hydrodynamic lubrication performance for continuous groove-textured surface. *Tribol. Int.* **2021**, *167*, 107411. [CrossRef]
33. Shen, Z.H.; Wang, F.C.; Chen, Z.G.; Ruan, X.P.; Zeng, H.H.; Wang, J.H.; An, Y.R.; Fan, X.L. Numerical simulation of lubrication performance on chevron textured surface under hydrodynamic lubrication. *Tribol. Int.* **2020**, *154*, 106704. [CrossRef]
34. Liu, W.L.; Ni, H.J.; Chen, H.L.; Wang, P. Numerical simulation and experimental investigation on tribological performance of micro-dimples textured surface under hydrodynamic lubrication. *Int. J. Mech. Sci.* **2019**, *163*, 105095. [CrossRef]
35. Hua, X.J.; Puoza, C.; Zhang, P.Y.; Yin, B.F.; Xie, X.; Din, J.L. Numerical simulation and experimental analysis of grease friction properties on textured surface. *Iran. J. Sci. Technol. Trans. Mech. Eng.* **2019**, *43*, 357–369. [CrossRef]
36. Wei, Y.; Yan, H.M.; Li, S.; Wang, X.W. Numerical and experimental study of a sector-shaped surface texture in friction reduction. *Tribol. Lett.* **2024**, *72*, 60. [CrossRef]
37. Li, Y.L.; He, Y.Y.; Luo, J.B. Multi-objective optimization of surface texture shape in fluid mechanical face seals using mass-conserving cavitation boundary condition. *Proc. Inst. Mech. Eng. Part J J. Eng. Tribol.* **2024**, *238*, 581–599. [CrossRef]
38. Wang, Z.Q.; Sun, L.T.; Han, B.; Wang, X.Q.; Ge, Z.W. Study on the thermohydrodynamic friction characteristics of surface-textured valve plate of axial piston pumps. *Micromachines* **2022**, *13*, 1891. [CrossRef]
39. Zhong, Y.H.; Zheng, L.; Gao, Y.H.; Liu, Z.N. Numerical simulation and experimental investigation of tribological performance on bionic hexagonal textured surface. *Tribol. Int.* **2019**, *129*, 151–161. [CrossRef]
40. Mao, Y.K.; Chen, W.G.; Li, Y.; Dai, Y.F.; Guo, W.X.; Zhang, J.B.; Wang, Y.H.; Wang, Z.X.; Li, D.Y.; Lin, N.M. Simulation and experimental study on tribological properties of herringbone texture on 42CrMo bearing alloy steel surface. *J. Mater. Eng. Perform.* **2024**, *34*, 1364–1377. [CrossRef]

41. Lv, X.R.; Wang, H.M.; Wang, S.J. Effect of swelling nitrile rubber in cyclohexane on its ageing, friction and wear characteristics. *Wear* **2015**, *328*, 414–421. [CrossRef]
42. Li, Y.F.; Nie, D.L.; Cai, Z.Q. The performance analysis of screw pump stator elastomers: Polyamide 6/hydrogenated nitrile blends: Mechanical, oil resistance and tribological properties. *Polym. Test.* **2023**, *128*, 108226. [CrossRef]
43. Qian, C.; Li, Y.L.; Zhao, J.; Wang, S.J.; He, E.Q. Thermal-oxidative aging and tribological properties of carbon nanotube/nitrile butadiene rubber composites with varying acrylonitrile content: Molecular dynamics simulations. *Polym. Eng. Sci.* **2023**, *63*, 1516–1527. [CrossRef]
44. Wang, Z.Q.; Xu, Y.F.; Hu, S.; Ji, H.; Yang, J. Research on lubrication mechanism with fluid–solid coupling of port plate pair in swash plate axial piston pump. *Proc. Inst. Mech. Eng. Part J J. Eng. Tribol.* **2020**, *234*, 515–527. [CrossRef]
45. Liu, X.; Shi, X.; Hao, Z.; Wei, S.; Sun, Y.; Niu, X.; Liu, C.; Li, M.; Li, Z. Effect of complex well conditions on the swelling and tribological properties of high-acrylonitrile stator rubber in screw pumps. *Polymers* **2024**, *16*, 2036. [CrossRef]
46. Liu, X.; Sun, Y.; Liu, C.; Shi, X.; Niu, X.; Zheng, G.; Wei, W.; Wei, S.; Huang, S. Lubrication characteristics of a warhead-type irregular symmetric texture on the stator rubber surfaces of screw pumps. *Lubricants* **2024**, *12*, 397. [CrossRef]
47. Liu, X.; Niu, X.; Liu, C.; Shi, X.; Sun, Y.; Hao, Z.; Huang, S.; Wang, Y.; Tao, H. Study on the tribological properties of multilayer concentric hexagonal laser texturing on rubber surfaces of screw pumps. *Materials* **2024**, *17*, 3708. [CrossRef]
48. Wang, J.H.; Yan, Z.J.; Shen, Z.Y.; Pan, X.X. Three-dimensional CFD analysis of the influence of surface texture morphology parameters on lubrication performance. *Lubr. Eng.* **2021**, *46*, 37–43. [CrossRef]
49. Schneider, J.; Braun, D.; Greiner, C. Laser textured surfaces for mixed lubrication: Influence of aspect ratio, textured area and dimple arrangement. *Lubricants* **2017**, *5*, 32. [CrossRef]
50. Sahlin, F.; Larsson, R.; Glavatskih, B.S.; Almqvist, T. Two-Dimensional CFD-Analysis of micro-patterned surfaces in hydrodynamic lubrication. *J. Tribol.* **2005**, *127*, 96–102. [CrossRef]
51. Tripathi, K.; Gyawali, G.; Joshi, B.; Amanov, A.; Wohn, S. Improved tribological behavior of grey cast iron under low and high viscous lubricants by laser surface texturing. *Mater. Perform. Charact.* **2017**, *6*, 24–41. [CrossRef]
52. Wang, L.L.; Zhao, X.T.; Guo, S.H.; Wang, M. Tribological properties of surface microtexture friction pairs under different lubrication conditions. *Adv. Mech. Eng.* **2019**, *11*, 1687814019881569. [CrossRef]

lubricants

Article

A Comprehensive Numerical Study of a Wedge-Shaped Textured Convergent Oil Film Gap

Raphael Scharf *, Michael Maier, Michael Pusterhofer and Florian Grün

Chair of Mechanical Engineering, Montanuniversität Leoben, Franz Josef-Straße 18, 8700 Leoben, Austria; michael.maier6@gmx.at (M.M.); michael.pusterhofer@unileoben.ac.at (M.P.); florian.gruen@unileoben.ac.at (F.G.)

* Correspondence: raphael.scharf@unileoben.ac.at

Abstract: The modification of surface geometries to reduce friction is an omnipresent topic of research. In nature, different low-friction surfaces, such as fish skins, exist. To transfer this knowledge to technical applications, for example, to journal or plain bearings, many numerical and experimental studies of textured surfaces have been performed. In this work, the influence of the geometric parameters (texture length l, width b, angle α and start position x_{start}) of a wedge-shaped texture on three different convergent oil film gaps was analyzed in full-film lubrication and compared with untextured oil film gaps. With the aid of a CFD (computational fluid dynamics) model, a comprehensive variation study was conducted, and the best-performing wedge-shaped texture was determined. The results show that an open texture at the inlet provides the largest improvement. Furthermore, it can be observed that the optimal relative texture width and absolute inlet height for the three investigated oil film gaps are similar. In contrast to the volume flow of the untextured geometry, the volume flow of the textured one is significantly higher, especially that perpendicular to the movement direction.

Keywords: surface texturing; wedge-shaped textures; hydrodynamic; finite element method; computational fluid dynamics

1. Introduction

To achieve climate targets, it is necessary to save energy and resources. One opportunity is a reduction in friction. Holmberg et al. [1] reported that 20% of all energy consumption is used to overcome friction. So, the reduction in friction is an omnipresent topic of research. In particular, nature demonstrates perfectly shaped surfaces, such as shark and fish skins [2] and bird wings [3], for drag reduction. Many researchers have tried to transfer these bio-inspired surfaces to technical applications and add additional indentations, called texture, to the surface, which should improve the tribological performance. The limiting factor for the geometry of the texture is the manufacturing process. Hence, simplified geometries, such as dimples, pockets, chevrons and so on, are used. Different surface texture manufacturing techniques exist, and according to [4], these methods can be divided into the following categories:

- Melting/vaporization (for example, electric discharge machining);
- Ablation (for example, laser surface texturing);
- Forced material removal (for example, micro-grinding);
- Dissolution (for example, chemical etching);
- Solidification (for example, micro-casting);
- Material addition (for example, chemical vapor deposition).

The advantages, disadvantages and applications of these methods are discussed in [4]. The quality of the generated surface is strongly influenced by the variable parameters of the manufacturing technique. In [5], the influence of the surface-texturing laser's pulse duration on the surface topography is depicted. Too long a pulse duration leads to micro-cracks, high-heat-affected zones and quite irregular surface profiles [5]. Another efficient manufacturing process is milling, and this process can be further optimized for texture manufacturing, for instance, by implementing spindle speed modulation, which leads to low costs for texture generation [6].

In [7,8], different textured surfaces for a parallel sliding contact were studied, and a significant drag reduction could be detected. Nevertheless, many technical applications are not parallel. Hence, Bei et al. and Rosenkranz et al. [9,10] examined the tribological performance of a pocket-, dot- and line-cross-like-shaped, textured convergent oil film gap and showed that the geometry of the oil film gap had a significant influence on the load-carrying capacity. Sharma et al. [11] performed a numerical study on a partially textured journal bearing and reported that texturing can enhance the performance for low and average journal eccentricities. The application of this knowledge to a convergent oil film gap demonstrated that partial surface texturing significantly enhances the performance of parallel sliding contacts, which is described in [7,8,12] and, for a convergent oil film gap with a low pitch angle α_0, in [10].

Different mechanisms underlying the operating principles of textured surfaces are discussed in the literature. Grützmacher et al. [13] published five different hypotheses about the performance enhancement of textured surfaces:

- Increase the hydrodynamic pressure;
- Supply the surfaces with additional lubricant;
- Store lubricant;
- Trap wear particles;
- Reduce the real contact area.

Besides these hypotheses, Hamilton et al. [14] reported that additional hydrodynamic pressure is a consequence of the existence of cavitation.

However, Babu et al. and Hu et al. [15,16] performed a detailed numerical and experimental analysis of different texture shapes and found that the geometrical parameters of every texture, such as the area density and texture height, have a strong influence on the enhancement of a textured surface. Besides the geometry itself, the position of the texture affects the tribological performance. According to Rosenkranz et al. [10], textures that are located at the inlet lead to a friction reduction when the pitch angle α_0 of a convergent oil film gap is low. Furthermore, Yagi et al. [17] found that an opened dimple on the inlet side increased the load-carrying capacity more than an enclosed dimple at the inlet position. Similar results can be observed for a journal bearing. The load-carrying capacity can be increased when the texture is located in a certain region in the convergent region, whereas textures that are placed in the region with the maximum pressure can only impair the performance [18].

As mentioned at the beginning, the surfaces of animals can inspire the development of surface texture. In [2] and Figure 1, a fish skin is depicted. The skin itself consists of different scales. The height profile of such a scale is also visualized in Figure 1. A simplification of the measured height profile, presented in the blue graph, leads to two wedges, presented in the dot-dashed turquoise graph. These simplified profiles served as inspiration for the development of a wedge-shaped texture.

The current paper focuses on a comprehensive study of such a texture on a convergent oil film gap. In [19,20], a wedge-shaped texture, among others, was examined under mixed lubrication. In contrast to van Liem et al. [19], who focused on the size and distribution of textures in mixed lubrication, Maier et al. [20] concentrated on the wear simulation of wedge-shaped surfaces. In both publications, specific geometries were chosen and analyzed. This leads to the question of the influence that the geometric parameters of a wedge-shaped texture have and which geometry will best enhance the performance. As in many technical applications, full-film lubrication is present, and this was not determined for wedge-shape textures by [19,20]. In this study, such a lubrication regime will be analyzed.

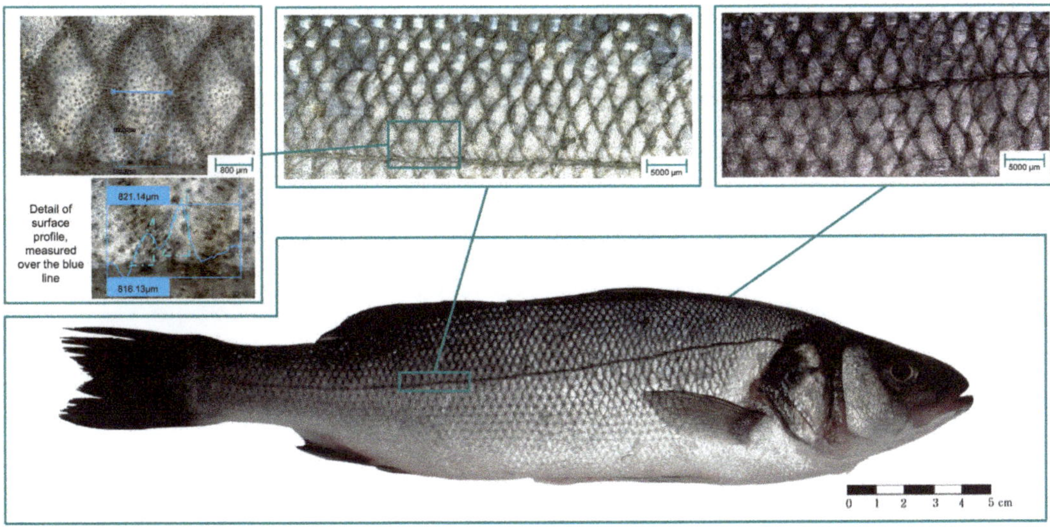

Figure 1. European seabass' skin—an inspirational source for textures.

2. Simulation Model

This section focuses on the methodical approach of the simulation model and its validation.

2.1. Mathematical Formulation of a Multiphase Flow

Numerical methods have the advantage, compared to experimental methods, of achieving time- and cost-efficient results. More than half of the research methods published in the field of surface texturing are simulation models [21]. In 77 % of cases, the Reynolds differential equation is solved [21]. Dobrica et al. [22] published a paper discussing the validity of the Reynolds equation in detail. This differential equation is valid for low Reynolds numbers Re, high dimple aspect ratios λ (the ratio between texture length l and texture height h) and small variations in lubricant thickness. The Reynolds number Re is the product of density ρ, velocity U and untextured oil film gap height h_0 divided by the kinematic viscosity ν [22]. In [23], the influence of inertia effects in textured lubricated sliding contacts was discussed noting that these effects become more pronounced at high Reynolds numbers. For this study, a maximum Reynolds number of about $Re \approx 10$ is expected, so the influence of inertia effects should be small. This means that the multiphase Stokes flow, which is presented in Equation (1), is applicable to this tribological system.

In this equation, $\vec{u}$ represents the velocity, p the pressure field, $\vec{f}$ represents the body force, ν the kinematic viscosity, ζ the second viscosity, $\boldsymbol{I}$ the identity matrix and ρ the density of the fluid. In the case of multiphase flow, ρ, ν and ζ are scalar fields. In addition to this equation, the continuity equation (cf. Equation (2)) is needed for time-independent density ρ. In [24], the mathematical inequality for the viscosity coefficients ν and ζ is presented. Both coefficients were set to the same values using this inequality and an existing simulation model, which is used for validation (cf. Section 2.2). For numerical implementation, the software package Netgen/NGsolve, based on a 3D finite element method (FEM), is used [25].

$$-\nabla\cdot\left\{\nu\left[\nabla\vec{u}+\left(\nabla\vec{u}\right)^T-\frac{2}{3}\left(\nabla\cdot\vec{u}\right)\boldsymbol{I}\right]+\zeta\left(\nabla\cdot\vec{u}\right)\boldsymbol{I}\right\}+\nabla p=\rho\vec{f} \tag{1}$$

$$-\nabla\cdot\left(\vec{u}\right)=0 \tag{2}$$

The multiphase Stokes flow (cf. Equation (1)) only needs to be solved if cavitation occurs; otherwise, the incompressible Stokes flow presented in Equation (3) can be solved numerically. Cavitation is determined by finding the minimum pressure and comparing it with the vapor pressure p_v. If the minimum pressure is less than p_v, cavitation occurs.

$$-\nu\,\Delta\vec{u} + \nabla p = \rho\,\vec{f} \tag{3}$$

In [26], the Merkle [27] cavitation model, among others, for a journal bearing was compared with the measured results in [28] and yielded favorable outcomes. This cavitation algorithm is also used in this study and presented in Equations (4)–(9). In this application, the stationary state is of interest, so the time derivative, which is generally a part of the transport equation (Equation (4)), is not required.

$$\nabla\cdot\left(\vec{u}\,\alpha_v\right) = \dot{m}^+ - \dot{m}^- \tag{4}$$

The source term, located on the right hand side of the transport equation (cf. Equation (4)), describes the vaporization and condensation process. α_v represents the vapor volume fraction and α_l represents the liquid volume fraction. The sum of both fractions must be 1, and the vapor and liquid volume fractions describe the volume of vapor V_v or liquid V_l, relative to the total volume (cf. Equation (7)).

$$\dot{m}^+ = Cc\frac{\rho}{\rho_l\rho_v}\frac{(1-\alpha_l)}{0.5U_\infty^2 t_\infty}\max(p-p_v,0) \tag{5}$$

$$\dot{m}^- = Cv\frac{\rho}{{\rho_v}^2}\frac{\alpha_l}{0.5U_\infty^2 t_\infty}\max(p_v-p,0) \tag{6}$$

The vaporization $\dot{m}^-$ and the condensation term $\dot{m}^+$ (cf. Equations (5) and (6)) depend on the condensation constant Cc, the vaporization constant Cv, vapor density ρ_v, liquid density ρ_l, input velocity U_∞, reference time t_∞ and vapor pressure p_v.

$$\alpha_v = \frac{V_v}{V_v + V_l} \tag{7}$$

In addition to the previously discussed equations, it is essential to incorporate Equation (8) to compute the density field ρ and Equation (9) to characterize the kinematic viscosity field (ν) when dealing with a multiphase flow.

$$\rho = \rho_v\,\alpha_v + \rho_l\,(1-\alpha_v) \tag{8}$$

$$\nu = \nu_v\,\alpha_v + \nu_l\,(1-\alpha_v) \tag{9}$$

2.2. Comparison of Different Simulation Models

To compare the self-programmed simulation model implemented in this study with an established software tool, OpenFOAM-8 [29] and its interPhaseChangeFoam solver were used. Since mesh generation for a wedge-shaped texture in OpenFOAM is quite complex, the geometry of a block-shaped texture on a convergent oil film gap, as depicted in Figure 2, was created. The bottom surface of the geometry is moving with a velocity $U = 10$ m/s, while the top surface and the texture surface are static. A Dirichlet boundary condition of $p_{abs} = 1.013$ bar was applied to the lateral surfaces. Additional boundary conditions for the liquid volume fraction were required. The lateral surface has a pressure of 1.013 bar, resulting in a liquid volume fraction α_l of 1 at these surfaces. The fluid parameters and parameters for the Merkle cavitation algorithm are presented in Section 2.3. A grid study was conducted for both models, and the results in the equilibrium state were analyzed. Figure 3 depicts the pressure distribution for different y-coordinate positions for both simulation models. Despite the differences in mathematical approach (finite volume method (FVM) in OpenFOAM and finite element method (FEM) in NGSolve) and element types (hexahedrons in OpenFOAM and tetrahedrons in NGSolve), similar results can be achieved. The comparison of the lifting and drag forces, calculated according to Section 2.3, showed a difference of about 6%. These small deviations between the different simulation models can be explained by different numerical methods and potentially by the fact that the NGSolve model does not include the inertia effects of the multiphase flow. Nevertheless, the use of this simulation model is valid because this variational study requires a large number of simulations (more than 50,000) to understand the various influencing geometric parameters. Neglecting the inertia effects significantly reduces computational times, as the nonlinear term of the Navier–Stokes equation is disregarded. As shown in the Results section, the improvement in tribological performance is much greater than the difference between both simulation models.

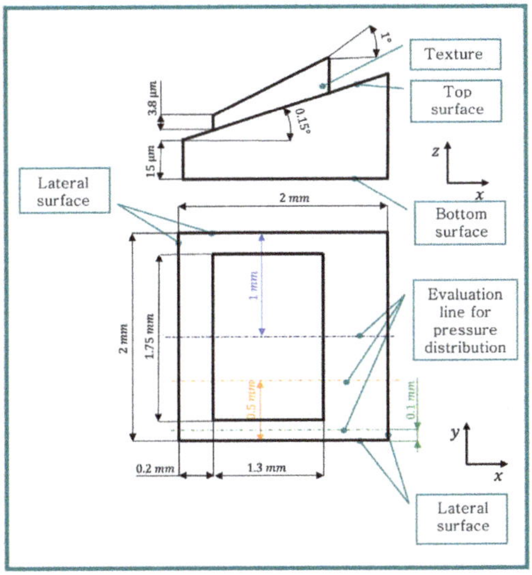

Figure 2. Geometry for validating the simulation model (unscaled).

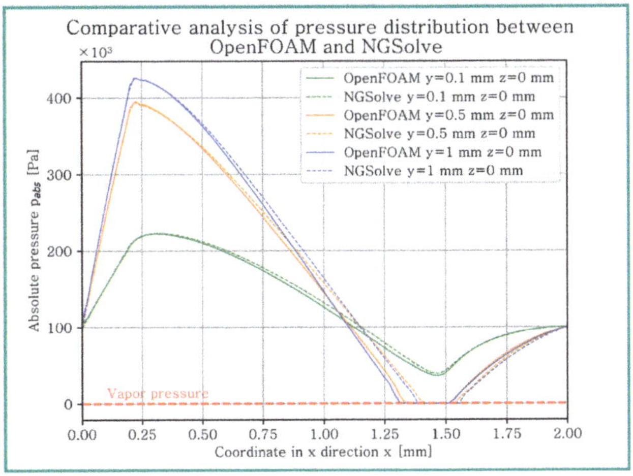

Figure 3. Comparative analysis between OpenFOAM and NGSolve.

2.3. Simulation Methodology for Wedge-Shaped Textured Surfaces

In Figure 4, the oil film with a global length l_0, width b_0 and pitch angle α_0 is visible. It was assumed that the surface was smooth, and the roughness of a surface was not included in this simulation model, unlike [30,31], who used flow factors to describe the real surface topography. Within this study, the performance of the texture geometry parameters themselves should be analyzed in detail and the effects of surface roughness should be neglected.

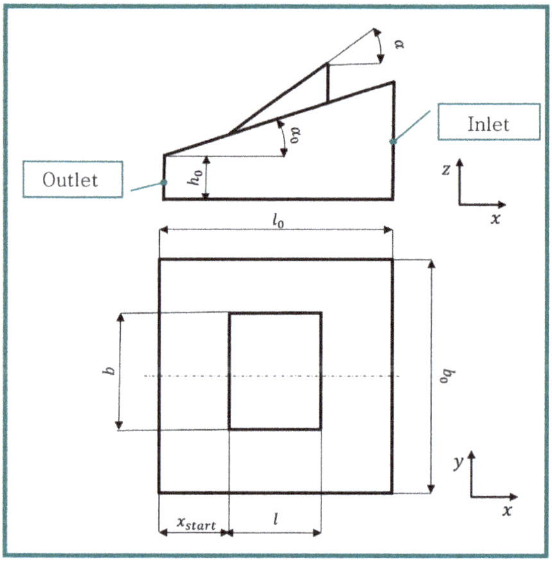

Figure 4. Geometry of textured surface (unscaled).

Three different global geometries were analyzed in detail, and the parameters themselves are presented in Table 1. Two criteria were important for the selection of these parameters. On the one hand, the tribological performance should be enhanced, and on the other hand, the computation time has to be quite short, so that it is possible to conduct a huge number of simulations in an adequate time. Therefore, a low α_0 and a small global geometry were selected (cf. [10]). The parameters α_0 and h_0 were kept constant to vary as few parameters as possible and to focus this study on the influencing parameters of the texture parameters themselves.

Table 1. Geometry parameters.

Designation	l_0 [mm]	b_0 [mm]	α_0 [°]	h_0 [µm]
Geometry I	2	2	0.15	15
Geometry II	3	3	0.15	15
Geometry III	4	4	0.15	15

The geometrical parameters of the texture, length l, width b angle α and start position x_{Start} were varied in a defined interval, with the requirement that the texture does not overlap the oil film gap. The texture angle α was varied between 0.2° and 2.2°. The step size of the geometry parameters between each simulation was 0.1 mm or 0.2 mm for l and b, depending on the width b_0 and length l_0 of the oil film gap. A step size of 0.2° for α and 0.1 mm for x_{Start} was selected. For only a few diagrams, some additional simulations, where a smaller step size was chosen, were added. The texture start position was varied in a defined interval $x_{Start} = [0.1, 0.2, \ldots 0.8]$ mm for all geometries. So, all in all, 9000 simulations for geometry I, 20,000 for geometry II and 20,000 for geometry III were conducted.

In Figure 5, the boundary conditions of an unscaled model are depicted. A constant sliding velocity of $U = 10$ m/s is defined on the bottom surface. This velocity is typical for technical applications in hydrodynamic lubrication and can also be tested on a pin-on-disc test rig, as can be observed in [32]. The lateral surface of the texture, the top surface of the oil film gap and the texture itself are stationary. On the lateral surface of the oil film gap, the ambient pressure p_0 is applied. Additionally, a liquid volume fraction of $\alpha_l = 1$ is defined on this surface. The boundary condition of zero gradient is applied to all the other surfaces that are not mentioned. Table 2 shows the oil and cavitation parameters used for the simulation model, as presented in [33,34].

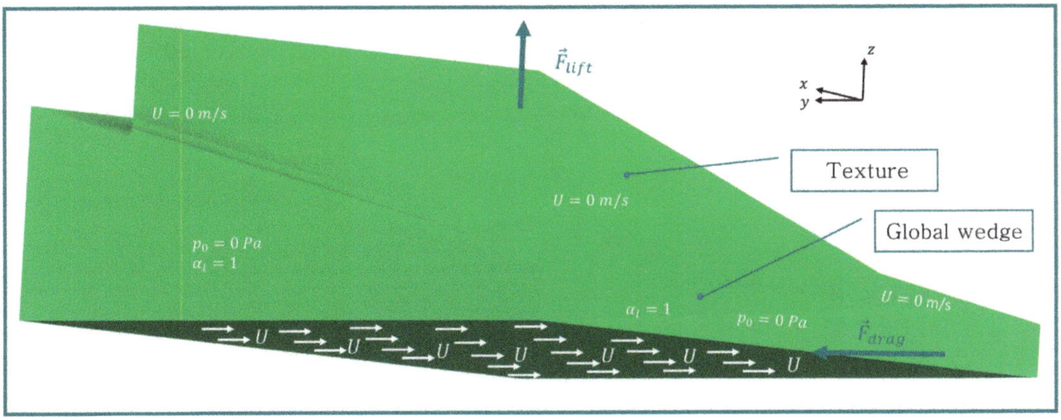

Figure 5. Boundary conditions for velocity $\vec{u}$, pressure p and liquid volume fraction α_l of simulation model, as well as lifting force $\vec{F}_{lift}$ and drag force $\vec{F}_{drag}$ (unscaled).

Table 2. Fluid and cavitation model parameters.

Parameters	
Density oil liquid ρ_l	857 kg/m^3
Density oil vapor ρ_v	0.13 kg/m^3
Dynamic viscosity liquid η_l	22.4 mPas
Dynamic viscosity vapor η_v	2×10^{-2} mPas
Condensation coefficient C_C	33.3
Vaporization coefficient C_v	1.55×10^{-3}
Vapor pressure p_v	165 Pa
Input velocity U_∞	10 m/s
Reference time t_∞	l_0/U_∞

The geometry was discretized using second order 3D tetrahedron-shaped elements. To achieve sufficient precision, simulations with different maximum mesh sizes were conducted. These analyses were used to identify an adequate mesh size where only small changes, less than 1%, could be detected in the results. Based on these results, a maximum mesh size of 0.025 mm was selected for discretizing the geometries. A direct solver called the unsymmetric multi-frontal package UMFPACK was used to solve the partial differential equations.

In postprocessing, the lifting force $\vec{F}_{lift}$, which is equivalent to the load that can be absorbed by the oil film gap, can be calculated by integrating the pressure field on the bottom surface (cf. Equation (10)). To calculate the drag force $\vec{F}_{drag}$, the shear stresses τ_x are integrated over the bottom surface. To validate the model, an analytical solution according to [35] for an untextured convergent oil film gap was used to calculate the lifting force $\vec{F}_{lift}$ and the drag force $\vec{F}_{drag}$. The comparison showed similar results.

$$\vec{F}_{lift\cdot} = \int_A p_{(x,y,z=0)} \cdot \vec{n}\, dA \tag{10}$$

Within this model, the lifting force $\vec{F}_{lift}$ and drag force $\vec{F}_{drag}$ can be calculated for different textured geometries. Since different geometries result in different lifting—$\vec{F}_{lift}$ and friction forces $\vec{F}_{drag}$—it is necessary to combine these two physical quantities for comparison with an untextured convergent oil film gap. This leads to the performance enhancement ratio (PER) according to Sharma et al. [11] (cf. Equation (11)). The two parameters are combined into one to simplify the visualization. If PER is greater than one, the performance is enhanced; otherwise, it is impaired. However, it should be noted that a reduction in friction- $\vec{F}_{drag}$ and lifting force $\vec{F}_{lift}$ can lead to a performance enhancement ratio PER greater than one.

$$\mathrm{PER} = \frac{\dfrac{\left|\vec{F}_{lift,textured}\right|}{\left|\vec{F}_{lift,untextured}\right|}}{\dfrac{\left|\vec{F}_{drag,\ texured}\right|}{\left|\vec{F}_{drag,untextured}\right|}} \tag{11}$$

3. Results

Overall, numerous simulations were conducted to provide an overview of the results. Scatter plots are quite useful for this purpose, while standard plots are commonly used to investigate the influence of different parameters. This section will also focus on a detailed analysis of the velocity and pressure fields of selected texture geometries.

3.1. Analysis of Variation Parameters Depending on PER

Figure 6a–h depict the results of geometry I for different start values x_{start}. Each point in each plot represents one simulation for a specific texture geometry with the parameters l, b, α and x_{start}. The color of each point indicates the performance enhancement ratio (PER). The lower limit of the color bar for Figures 6–8 was set as one, allowing the visualization to focus on geometries that improve tribological performance, by increasing lifting force $\vec{F}_{lift}$, reducing friction or both. The upper limit for these colored plots was determined by finding the maximum PER. For geometry I, it is 2.65, which is achievable at a texture length $l = 1.8$ mm, width $b = 1.7$ mm, angle $\alpha = 1.2°$ and at a start position $x_{start} = 0.2$ mm. Figures 7 and 8 present the same information as Figure 6, but for geometry II and III. The maximum improvement for geometry II can be achieved with a texture length $l = 2.8$ mm, width $b = 2.6$ mm, angle $\alpha = 0.8°$ and a start position $x_{start} = 0.2$ mm. For geometry III, the maximal PER is 1.67 at a texture length $l = 3.7$ mm, width $b = 3.4$ mm, angle $\alpha = 0.6°$ and a start position $x_{start} = 0.3$ mm. All geometries were simulated with the same start positions, and a selection of these results is visible in Figures 6–8. The simulation effort for geometry II and III is higher, compared to geometry I. The results show that the best performance is achieved with low texture start positions x_{start}, and consequently for geometry II's start positions $x_{start} = 0.1$ mm and $x_{start} = 0.2$ mm a smaller step size of 0.1 mm for texture length l and width b was defined (cf. Figure 7a) in contrast to Figure 7b–d with a step size of 0.2 mm.

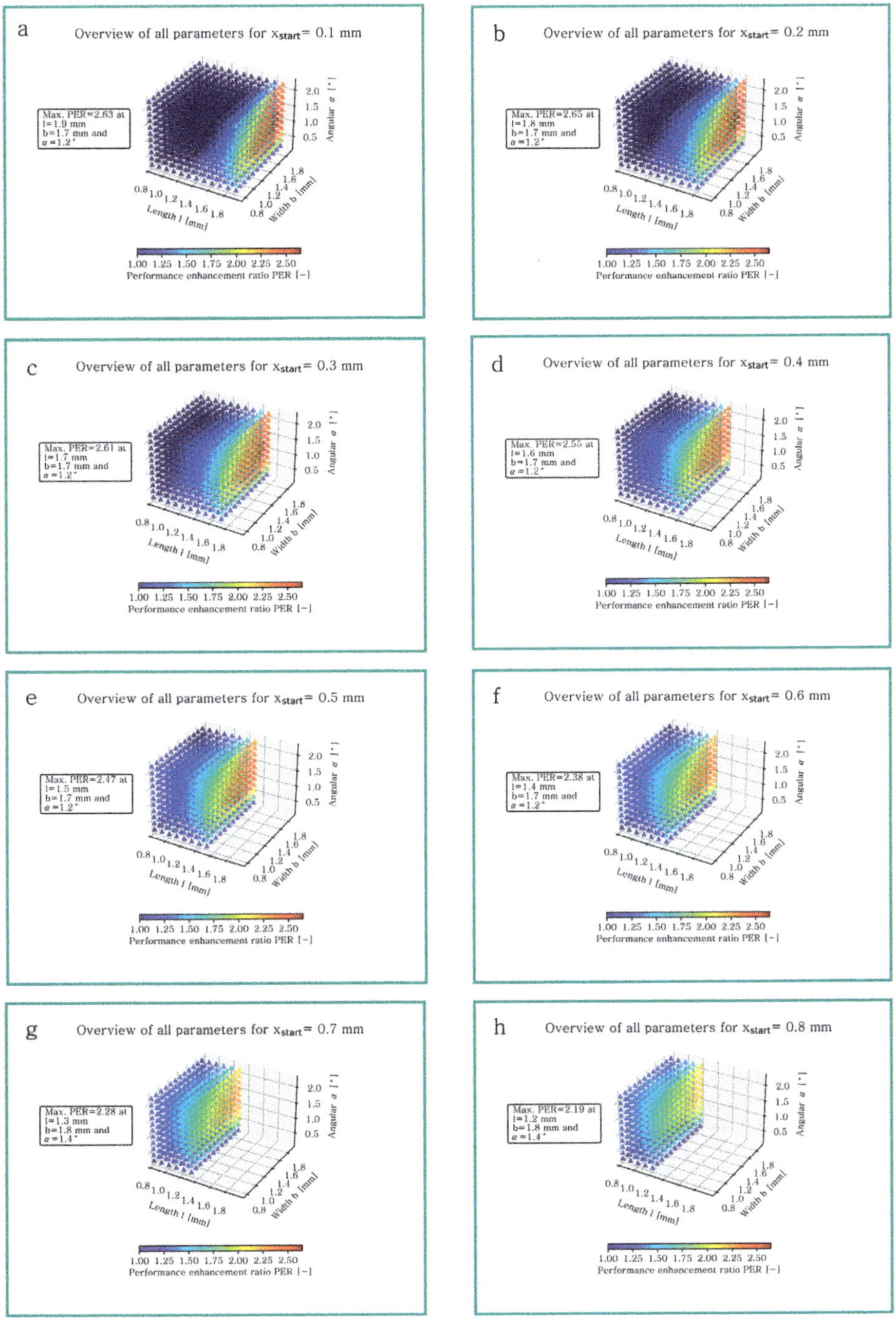

Figure 6. Overview of all simulations of geometry I for (**a**) $x_{start} = 0.1$ mm, (**b**) $x_{start} = 0.2$ mm, (**c**) $x_{start} = 0.3$ mm, (**d**) $x_{start} = 0.4$ mm, (**e**) $x_{start} = 0.5$ mm, (**f**) $x_{start} = 0.6$ mm, (**g**) $x_{start} = 0.7$ mm and (**h**) $x_{start} = 0.8$ mm.

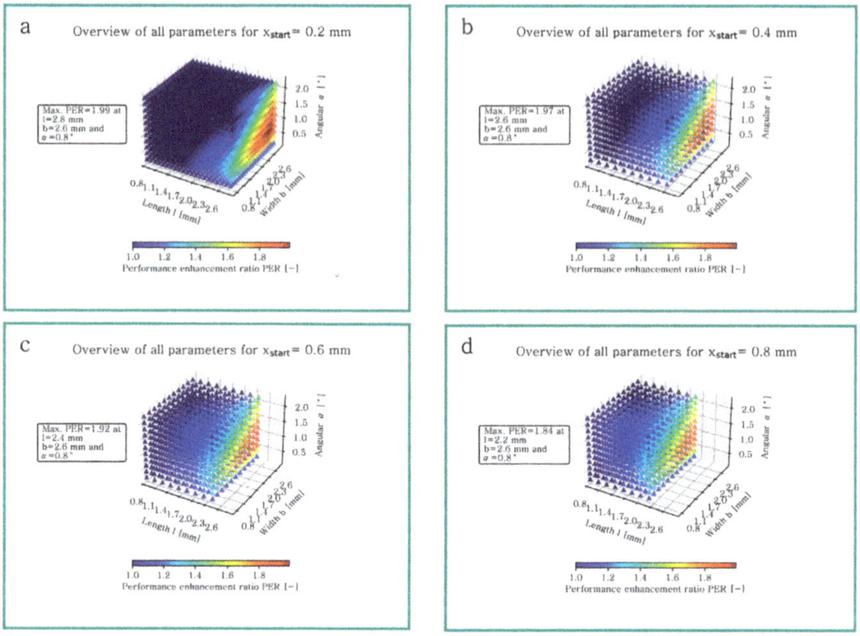

Figure 7. Overview of all simulations of geometry II for (**a**) x_{start} = 0.2 mm, (**b**) x_{start} = 0.4 mm, (**c**) x_{start} = 0.6 mm and (**d**) x_{start} = 0.8 mm.

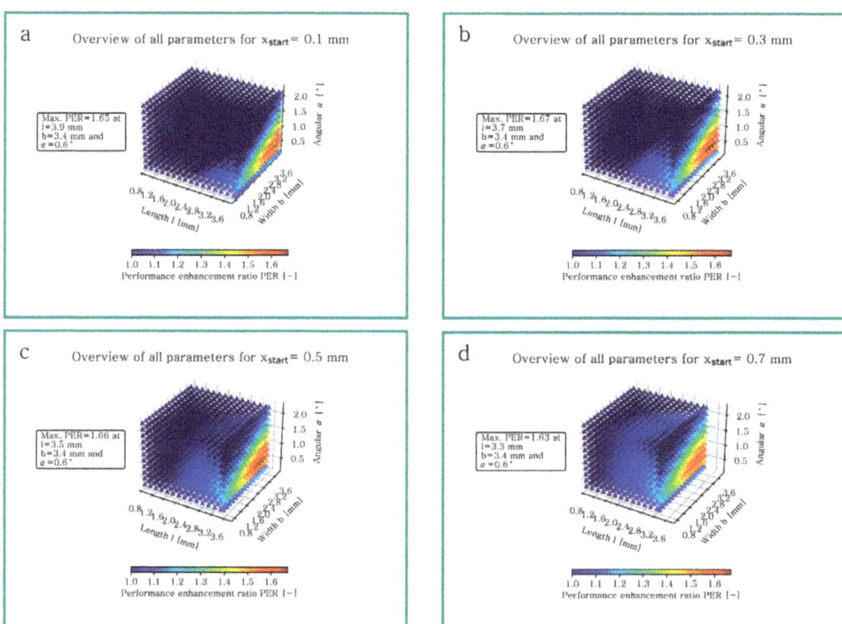

Figure 8. Overview of all simulations of geometry III for (**a**) x_{start} = 0.1 mm, (**b**) x_{start} = 0.3 mm, (**c**) x_{start} = 0.5 mm and (**d**) x_{start} = 0.7 mm.

In Table 3, the performance enhancement ratio (PER) for the best operating textures for each geometry is shown. Additionally, the ratio, describing the improvement in lifting PER_{lift} and drag force PER_{drag}, is illustrated. It is evident for all geometries that the lifting force increases in comparison to the untextured convergent oil film gap. Furthermore, the drag force of these textured geometries decreases compared to the untextured geometry. Additionally, it is noticeable that as the global length l_0 and width b_0 increase, PER_{lift} in particular decreases.

Table 3. Maximal performance enhancement ratios of best operating geometries.

	PER [−]	$PER_{lift} = \frac{\lvert \vec{F}_{lift,textured} \rvert}{\lvert \vec{F}_{lift,untextured} \rvert}$ [−]	$PER_{drag} = \frac{\lvert \vec{F}_{drag,\ texured} \rvert}{\lvert \vec{F}_{drag,untextured} \rvert}$ [−]
Geometry I	2.65	2.18	0.82
Geometry II	1.99	1.64	0.82
Geometry III	1.67	1.43	0.87

Although the maximum PER for all geometries is different, similar trends can be observed. The best performance for every geometry can be achieved, with textures where the sum of texture length l and start position x_{start} is equivalent to the global length l_0, resulting in an open configuration at the inlet side. An investigation of the texture width b shows that tribological performance can be enhanced with increasing texture width b to a local maximum, beyond which the PER decreases. The influence of the texture angle α is also depicted in Figures 6–8. As the global length and width increase, the optimum texture angle α decreases and the range for performance enhancement becomes smaller.

The detailed influence of the different parameters is illustrated in Figures 9–11. In each figure, three parameters were held constant while one was varied. The constant parameters were based on the values of the simulation model that yielded the maximum enhancement.

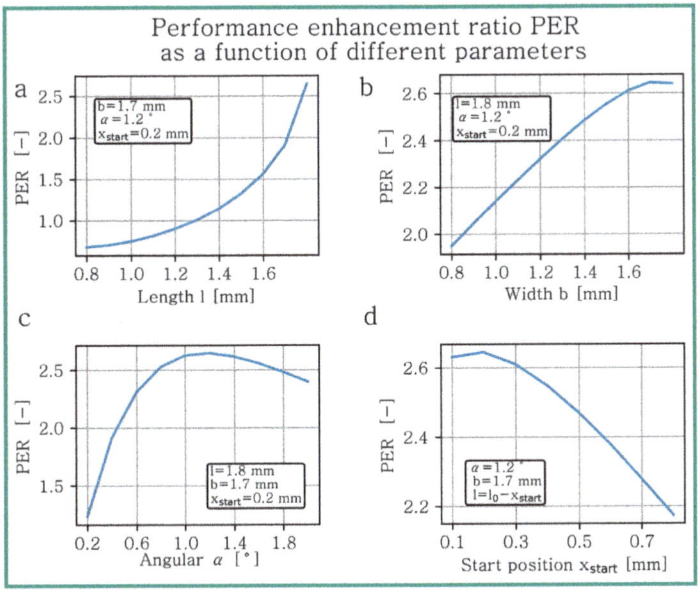

Figure 9. Performance enhancement ratio (PER) for geometry I as a function of texture (**a**) length l, (**b**) width b, (**c**) angle α and (**d**) start position x_{start}.

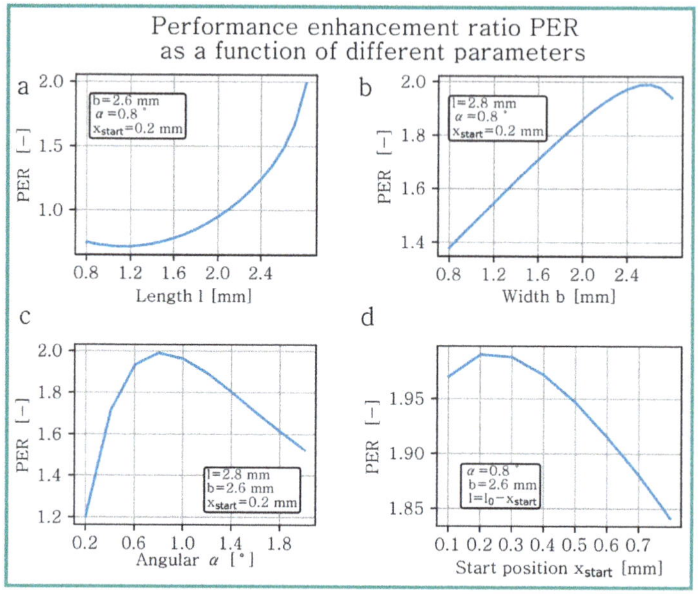

Figure 10. Performance enhancement ratio PER for geometry II as a function of texture (**a**) length l, (**b**) width b, (**c**) angle α and (**d**) start position x_{start}.

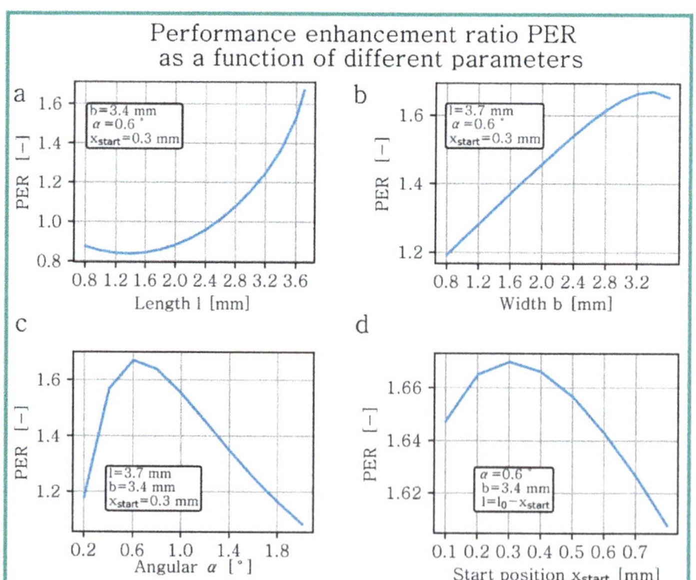

Figure 11. Performance enhancement ratio PER for geometry III as a function of texture (**a**) length l, (**b**) width b, (**c**) angle α and (**d**) start position x_{start}.

Although the absolute values of the PER are different, the same trends for geometry I, II and III can be observed. As already mentioned, an open texture against the movement direction improves the performance the most. An increasing texture length l leads to a rising PER, as shown in Figures 9a, 10a and 11a. For small texture lengths l, a significant deterioration can be noticed. In Figure 9a, it is visible that for texture lengths $l > 1.3$ mm the performance is enhanced. This corresponds to a percentage length l/l_0 of 65%. Similar analogies can be observed for geometry II and III. For $l > 2.1$ mm (geometry II) and $l > 2.6$ mm (geometry III), the PER is greater than one, which corresponds to a percentage length l/l_0 of 70% and 65%.

The PER depending on texture width b is presented in Figures 9b, 10b and 11b. For all geometries, the curve progression is similar and the maximal PER can be achieved at a related width b/b_0 of 85%, 86.7% and 85.0% for geometry I, II and III. An analysis of the influence of α showed that with increasing global geometry, the maximum enhancement is located at a lower texture angle α (cf. Figures 9c, 10c and 11c). For geometry I, the best improvement can be achieved with a texture angle $\alpha = 1.2°$, for geometry II $\alpha = 0.8°$ and for geometry III $\alpha = 0.6°$.

In Figures 9d, 10d and 11d, the PER as a function of the start position x_{start} is presented. In these figures, the texture length l of each simulation point is different, because it was assumed that the texture is open at the inlet side. So, the texture length l was calculated by subtracting x_{start} from l_0. An optimum relative start position x_{start}/l_0 can be reached at 10.0%, 6.7% and 7.5% for geometry I, II and III.

In summary, as can be seen from the figures discussed, small variations in texture geometry parameters have a significant effect on tribological performance.

For additional analysis, the pressure and velocity field of the best performing texture geometries were investigated.

3.2. Pressure and Velocity Field Analysis

In Figures 12–14, the pressure distribution and the calculated volume flow Q for an untextured and best performing textured oil film gap are presented. The small variations that violate volume continuity can be attributed to rounding and numerical errors. The qualitative characteristic of the pressure field for all textured and untextured geometries is similar. Compared to the untextured geometry, the maximum pressure of the textured geometry is located closer to the outlet. Additionally, the texture increases the pressure, as shown in Figures 12b, 13b and 14b. In addition to the pressure field, the volume flow Q between the textured and untextured geometries differs, especially the volume flow Q perpendicular to the direction of movement is significantly larger, which is visually represented by the width of the arrows. This means that 14.9%, 16.5% and 17.4% of the inlet flow are flowing out there for the textured geometry I, II and III. In contrast, only 6.9%, 9.4% and 11.6% of the inlet flow exit on the faces perpendicular to the movement direction for the untextured geometries I, II and III.

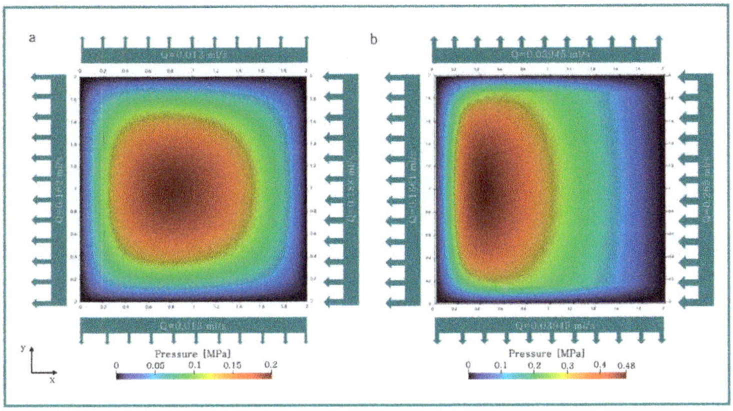

Figure 12. Pressure distribution and volume flow of geometry I (**a**) untextured and (**b**) textured ($l = 1.8$ mm, $b = 1.7$ mm, $\alpha = 1.2°$, $x_{start} = 0.2$ mm).

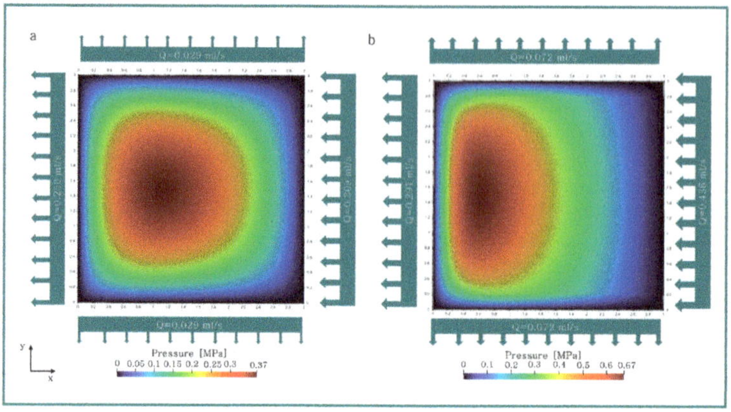

Figure 13. Pressure distribution and volume flow of geometry II (**a**) untextured and (**b**) textured ($l = 2.8$ mm, $b = 2.6$ mm, $\alpha = 0.75°$, $x_{start} = 0.2$ mm).

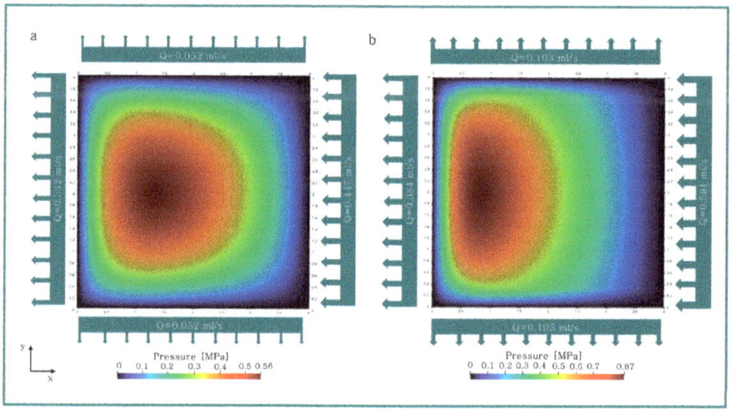

Figure 14. Pressure distribution and volume flow of geometry III (**a**) untextured and (**b**) textured ($l = 3.7$ mm, $b = 3.4$ mm, $\alpha = 0.6°$, $x_{start} = 0.3$ mm).

In Figures 15–17, the velocity field of the untextured geometries I, II, III and the best performing textures of these convergent oil film gaps are visible. The texture depicted in these pictures had the highest performance enhancement ratio of all conducted simulations. Since the real height of an oil film gap in comparison to its horizontal and vertical size is small, the z-coordinates of these figures were scaled by a factor of eight. The size of the flow vectors is constant, and the magnitude of the velocity is represented by the color. It is apparent in all slices B-B that the velocity magnitude is larger on the right side. Since all slices were created parallel to the bottom surface with a displacement of 10 μm, the velocity increases towards the right side where the oil film gap is higher. This can be explained by the simple shear case, where the velocity is linear over the oil film gap height and the flow due to pressure is not present. An increasing height consequently leads to a higher velocity for the same vertical position. This can also be observed in the outer regions of the textured geometry, where there is a sudden change in velocity between the crossover of the textured and untextured region. Compared to the untextured geometries, the magnitude of the velocity of the textured geometries is significantly higher, as seen in Figures 15b, 16b and 17b, due to the increased oil film gap height caused by the texture. Additionally, the y-coordinates of the textured flow vectors are higher compared to the untextured geometry, causing an increase in volume flow Q on the faces normal to the directional movement (cf. Figures 12b, 13b and 14b). Slice A-A of Figures 15a, 16a and 17a shows an expected velocity contribution where the fluid on the bottom surface is moving with U, while the top surface is stagnant. On the right side of the textured geometry slices (cf. Figures 15b, 16b and 17b), a backstreaming of the oil with a low magnitude can be observed. As the global geometry increases, the length of the backstreaming region decreases.

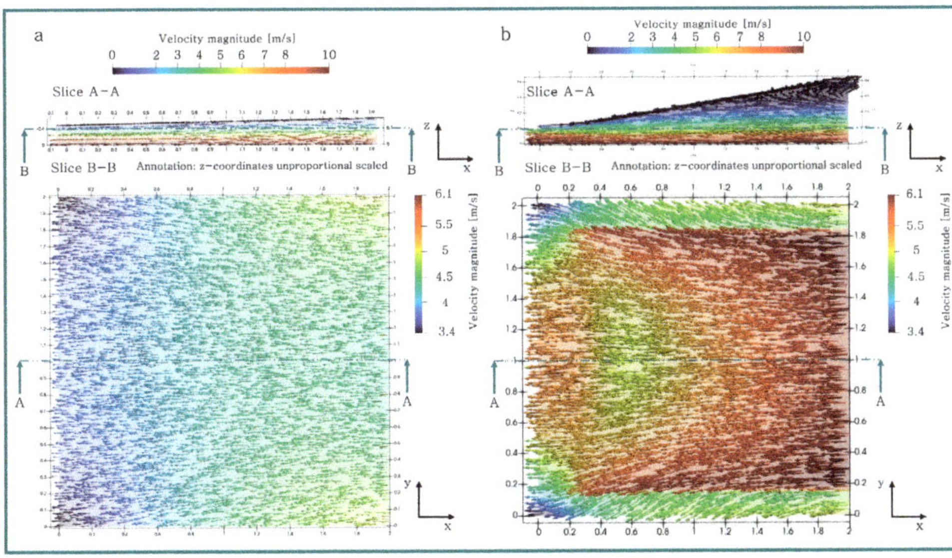

Figure 15. Velocity field of geometry I (**a**) untextured and (**b**) textured (l = 1.8 mm, b = 1.7 mm, $\alpha = 1.2°$, x_{start} = 0.2 mm).

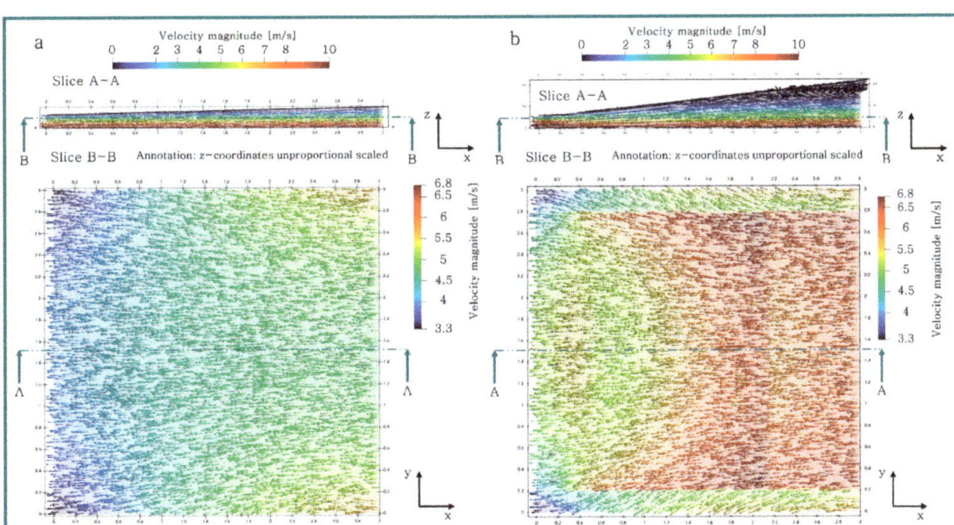

Figure 16. Velocity field of geometry II (**a**) untextured and (**b**) textured (l = 2.8 mm, b = 2.6 mm, $\alpha = 0.8°$, x_{start} = 0.2 mm).

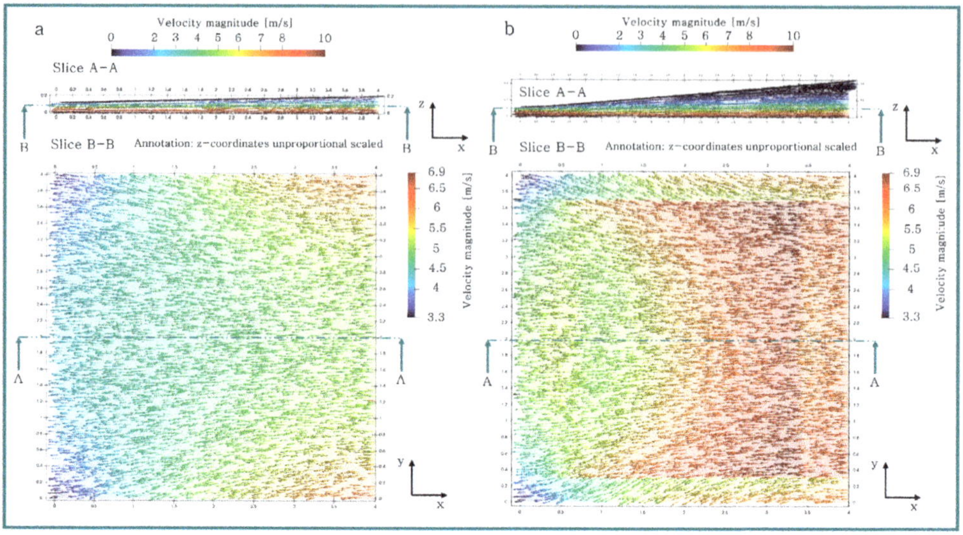

Figure 17. Velocity field of geometry III (**a**) untextured and (**b**) textured ($l = 3.7$ mm, $b = 3.4$ mm, $\alpha = 0.6°$, $x_{start} = 0.3$ mm).

4. Discussion

The principal mechanisms of textured surfaces, such as the increase in hydrodynamical pressure and the supply of surfaces with additional lubricant as published by Grützmacher et al. [13], were also observed in this study. The increase in pressure and volume flow Q can be seen in Figures 12–14. The volume flow Q also increases, when the texture is not open against the movement of direction, probably due to an increasing velocity at the inlet. This phenomenon is illustrated in Figure 18, which shows an untextured convergent oil film gap in the center, surrounded by different textures. It is qualitatively presented that the volume flow at the inlet for all four textures increases compared to the untextured oil film gap. For an open texture, the relative inlet surface area $A_{in,rel}$ is higher compared to closed textures or untextured lubrication gaps. The average velocity magnitude $\left|\overline{v_{in,rel}}\right|$ on this surface is lower than the average magnitude $\left|\overline{v_{in,rel}}\right|$ of a smooth oil film cap, resulting in an increased volume flow at the inlet Q_{in} due to the larger surface area $A_{in,rel}$. These observations for an opened texture are presented in Figures 13 and 14. The three principal mechanisms of surface textures, trap wear particles, reduce the real contact area and store lubricant, according to Grützmacher et al. [13], and so cannot be described with this simulation model.

The hypothesis that cavitation is the reason for additional pressure in some cases [14], or that cavitation can increase the load-carrying-capacity [36], is disproved for the best performing texture geometries, as the minimum pressure for these textures and many others was not lower than the vapor pressure p_V. In Figure 18, the pressure distribution for different textures is qualitatively presented and it can be observed that for an at-the-inlet opened texture, the minimal pressure is not lower than the ambient pressure. Additionally, it is evident that in some cases, cavitation can negatively impact performance by decreasing the lifting force $\vec{F}_{lift}$, when the pressure falls below the ambient pressure. However, it is important to note that textures where cavitation occurs and enhance tribological performance do exist. For the texture, which is analyzed in detail in this study, the highest performing enhancement ratio (PER) can be achieved and cavitation does not occur.

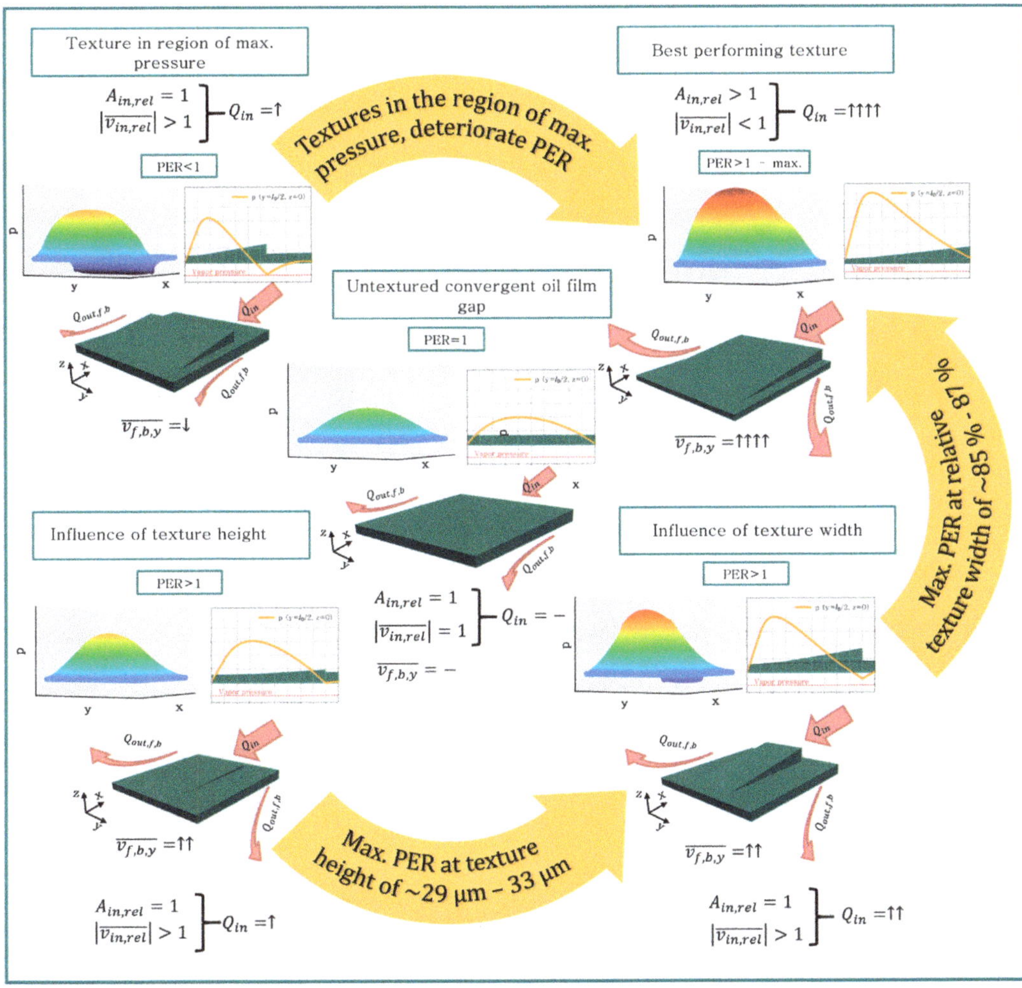

Figure 18. Qualitative visualization of influencing parameters in wedge-shaped textures.

In Figures 12–14, it is depicted that textures increase the volume flow $Q_{out,f,b}$ perpendicular to the movement direction. Figure 18 similarly presents this qualitatively, showing that the average magnitude of the velocity on the front and back surface $\overline{v_{f,b,y}}$ can be increased, compared to the untextured lubricant gap. This results in an increasing volume flow $Q_{out,f,b}$ perpendicular to the movement direction. However, it should be noted that this is not a universally valid mechanism, because an infinitely expanded untextured oil film gap is also providing a higher lifting force, in contrast to a finitely expanded one [35]. Therefore, the additional inflow of the textured oil film gaps potentially improves performance, which is one operational principle according to Grützmacher et al. [13].

As can be seen in Figures 9a, 10a and 11a and Figure 18, an open texture at the inlet noticeably increases the PER. Similarly, in [17], a parallel sliding surface with multiple dimples exhibits a similar behavior. This open texture may be providing additional lubrication, thereby enhancing the tribological performance.

In Figures 9a, 10a and 11a, it can be observed that for texture lengths l less than $l < 1.3$ mm (geometry I), $l < 2.1$ mm (geometry II) and $l < 2.6$ mm (geometry III), the performance enhancement ratio is less than one. When compared to the pressure field of the untextured geometries, cf. Figures 12a, 13a and 14a, it is evident that the area of maximum pressure is located within these lengths. This suggests that textures located only in the area of maximum pressure negatively impact performance (cf. Figures 12–14). This phenomenon is also illustrated in the top left corner of Figure 18, showing the pressure distribution of an oil film gap, where the texture is located in the region of the maximal pressure of a smooth lubrication gap. Due to cavitation, the lifting force $\vec{F}_{lift}$ is reduced, despite the higher maximal pressure compared to a smooth wedge. In summary, it can be concluded that textures located in the region of maximal pressure have a negative impact on tribological performance. Liang et al. [18] also published similar results for partially textured journal bearings.

Bei et al. [9] published that a maximal load-carrying capacity for a convergent oil film gap and pocket-shaped textures could be achieved at a dimensionless texture width $b/b_0 = 70\%$, which is not very different from the values found in this study. Here, the optimal related width b/b_0 is between 85% and 87% (cf. Figure 18). In [9], the texture is defined through the parameters length, width and height. In this work, the texture height h, (also referred to as texture depth in the literature) is not constant over the texture length l. This led to the conclusion that for this variation study, a texture angle α was used instead of a texture height h.

According to Bei et al. [9], the load-carrying capacity first decreases, then increases and finally decreases over the texture width and depth. Within this work, it was observed that there is an initial increase followed by a decrease in performance, depending on texture width b and angle α.

In [37], the influence of square-shaped dimple depths is discussed in detail and the best performance could be achieved with a dimple depth of $h = 30$ µm. In [37], nearly the same density ρ and doubled dynamic viscosity η as in this study were used. To compare this value with the results of this variation study, trigonometric calculations are necessary to calculate the maximal height of the texture. For geometry I, the highest PER could be acquired for a maximal texture height of $h = 33$ µm, for geometry II $h = 31.8$ µm and for geometry III $h = 29.1$ µm, as visualized in Figure 18. These values are in the same range as those mentioned by Ullah et al. [37]. Another interesting observation regarding the texture height h could be made in this study. The absolute oil film gap height at the inlet, which is the sum of the maximal texture height and the maximal height of the untextured convergent oil film gap, is nearly equivalent for all geometries (51.6 µm for geometry I, 53.2 µm for geometry II and 54.5 µm for geometry III). The conclusion of Ullah et al. [37] that a decreasing texture height h improves performance cannot be observed at all in this study. It can be shown that with increasing texture height, the PER first increases and then decreases. One possible reason why this behavior was detected in this study is that in comparison to Ullah et al. [37], simulations with a very small texture angle α, which corresponds to small heights, were also conducted.

In general, it can be observed that the geometry parameters of the texture strongly influence the performance of a lubricated sliding oil film gap. A comparison with the general tolerances outlined in ISO 2768 reveals that for nominal lengths less than 3 mm, a deviation of ± 0.1 mm for the medium class and ± 0.05 mm for the fine class is acceptable [38]. For angle dimension deviations for nominal lengths up to 10°, a deviation of $\pm 1^\circ$ is permissible [38]. Comparing these tolerances with Figures 9–11 shows that the texture angle α and the start position x_{start} would change significantly with these tolerances, while the texture length l and width b would not be as strongly influenced.

5. Conclusions

Within this work, a comprehensive numerical study of a wedge-shaped texture on three different convergent oil film gaps was conducted and the influence of the geometry parameters was analyzed in detail. The following conclusion for the already mentioned global geometries can be summarized:

- The performance of a wedge-shaped texture is strongly influenced by the geometry parameters. An increasing texture length l improves the tribological behavior and the maximal PER can be achieved with an open texture at the inlet. This means that the texture increases the area of the face where the fluid advects.
- Depending on the dimensions of the oil film gap, a relative texture width b/b_0 of 85% to 87% leads to the best performance.
- With increasing dimensions of the oil film gap, the optimum texture angle α decreases. The best performance can be achieved if the oil film gap at the inlet side has an absolute height of about $52 - 55$ µm, more or less independent of the geometry of the oil film gap.
- A relative start position of the texture x_{start}/l_0, referring to the length of the oil film gap l_0, between 6.7% and 10% enhances the performance in the best way.
- The texture increases the fluid velocity perpendicular to the movement direction, leading to an increasing volume flow perpendicular to the movement direction.
- The texture also induces additional pressure.
- Textures that are only located in the area of maximal pressure deteriorate the tribological performance.

Author Contributions: Conceptualization, R.S., M.M. and M.P.; methodology, R.S.; software, R.S.; validation, R.S. and M.P.; formal analysis, R.S.; investigation, R.S.; resources, R.S.; data curation, R.S.; writing—original draft preparation, R.S.; writing—review and editing, M.M., M.P. and F.G. visualization, R.S.; supervision, M.P. and F.G.; project administration, F.G.; funding acquisition, F.G. All authors have read and agreed to the published version of the manuscript.

Funding: This research received no external funding.

Data Availability Statement: Data are contained within the article.

Conflicts of Interest: The authors declare no conflicts of interest.

References

1. Holmberg, K.; Erdemir, A. The impact of tribology on energy use and CO_2 emission globally and in combustion engine and electric cars. *Tribol. Int.* **2019**, *135*, 389–396. [CrossRef]
2. Tian, G.; Zhang, Y.; Feng, X.; Hu, Y. Focus on Bioinspired Textured Surfaces toward Fluid Drag Reduction: Recent Progresses and Challenges. *Adv. Eng. Mater.* **2022**, *24*, 2100696. [CrossRef]
3. Dutta, P.; Nagar, O.P.; Sahu, S.K.; Savale, R.R.; Raj, R.G. Aerodynamic analysis of bionic winglet- slotted wings. *Mater. Today Proc.* **2022**, *62*, 6701–6707. [CrossRef]
4. Arslan, A.; Masjuki, H.H.; Kalam, M.A.; Varman, M.; Mufti, R.A.; Mosarof, M.H.; Khuong, L.S.; Quazi, M.M. Surface Texture Manufacturing Techniques and Tribological Effect of Surface Texturing on Cutting Tool Performance: A Review. *Crit. Rev. Solid State Mater. Sci.* **2016**, *41*, 447–481. [CrossRef]
5. Pou-Álvarez, P.; Riveiro, A.; Nóvoa, X.; Fernández-Arias, M.; del Val, J.; Comesaña, R.; Boutinguiza, M.; Lusquiños, F.; Pou, J. Nanosecond, picosecond and femtosecond laser surface treatment of magnesium alloy: Role of pulse length. *Surf. Coat. Technol.* **2021**, *427*, 127802. [CrossRef]
6. Huang, F.; Jin, X. Surface texture generation using high-feed milling with spindle speed modulation. *Precis. Eng.* **2021**, *72*, 13–24. [CrossRef]
7. Bijani, D.; Deladi, E.L.; Akchurin, A.; de Rooij, M.B.; Schipper, D.J. The Influence of Surface Texturing on the Frictional Behaviour of Parallel Sliding Lubricated Surfaces under Conditions of Mixed Lubrication. *Lubricants* **2018**, *6*, 91. [CrossRef]
8. Liu, Y.; Zhang, H.; Dai, S.; Dong, G. Designing a bioinspired scaly textured surface for improving the tribological behaviors of starved lubrication. *Tribol. Int.* **2022**, *173*, 107594. [CrossRef]
9. Bei, G.; Ma, C.; Wang, X.; Sun, J.; Ni, X. Study on Tribological Characteristics of Textured Surface under Convergent Oil Film Gap. *Lubricants* **2022**, *10*, 183. [CrossRef]
10. Rosenkranz, A.; Costa, H.L.; Profito, F.; Gachot, C.; Medina, S.; Dini, D. Influence of surface texturing on hydrodynamic friction in plane converging bearings—An experimental and numerical approach. *Tribol. Int.* **2019**, *134*, 190–204. [CrossRef]
11. Sharma, S.; Jamwal, G.; AWASTHI, R.K. Numerical study on steady state performance enhancement of partial textured hydrodynamic journal bearing. *Ind. Lubr. Tribol.* **2019**, *71*, 1055–1063. [CrossRef]
12. Dobrica, M.B.; Fillon, M.; Pascovici, M.D.; Cicone, T. Optimizing surface texture for hydrodynamic lubricated contacts using a mass-conserving numerical approach. *Proc. Inst. Mech. Eng. Part J J. Eng. Tribol.* **2010**, *224*, 737–750. [CrossRef]
13. Grützmacher, P.G.; Profito, F.J.; Rosenkranz, A. Multi-Scale Surface Texturing in Tribology—Current Knowledge and Future Perspectives. *Lubricants* **2019**, *7*, 95. [CrossRef]

14. Hamilton, D.B.; Walowit, J.A.; Allen, C.M. A Theory of Lubrication by Microirregularities. *J. Basic Eng.* **1966**, *88*, 177–185. [CrossRef]
15. Babu, P.V.; Ismail, S.; Ben, B.S. Experimental and numerical studies of positive texture effect on friction reduction of sliding contact under mixed lubrication. *Proc. Inst. Mech. Eng. Part J J. Eng. Tribol.* **2021**, *235*, 360–375. [CrossRef]
16. Hu, D.; Guo, Z.; Xie, X.; Yuan, C. Effect of spherical-convex surface texture on tribological performance of water-lubricated bearing. *Tribol. Int.* **2019**, *134*, 341–351. [CrossRef]
17. Yagi, K.; Sugimura, J. Performance of Balancing Wedge Action in Textured Hydrodynamic Pad Bearings. *J. Lubr. Technol.* **2017**, *139*, 248. [CrossRef]
18. Liang, X.; Liu, Z.; Wang, H.; Zhou, X.; Zhou, X. Hydrodynamic lubrication of partial textured sliding journal bearing based on three-dimensional CFD. *Ind. Lubr. Tribol.* **2016**, *68*, 106–115. [CrossRef]
19. van Liem, N.; Zhenpeng, W.; Renqiang, J. Effect of shape/size and distribution of microgeometries of textures on tribo-performance of crankpin bearing. *Proc. Inst. Mech. Eng. Part J J. Eng. Tribol.* **2022**, *236*, 421–433. [CrossRef]
20. Maier, M.; Pusterhofer, M.; Grün, F. Multiscale Wear Simulation in Textured, Lubricated Contacts. *Coatings* **2023**, *13*, 697. [CrossRef]
21. Gropper, D.; Wang, L.; Harvey, T.J. Hydrodynamic lubrication of textured surfaces: A review of modeling techniques and key findings. *Tribol. Int.* **2016**, *94*, 509–529. [CrossRef]
22. Dobrica, M.B.; Fillon, M. About the validity of Reynolds equation and inertia effects in textured sliders of infinite width. *Proc. Inst. Mech. Eng. Part J J. Eng. Tribol.* **2009**, *223*, 69–78. [CrossRef]
23. Cupillard, S.; Glavatskih, S.; Cervantes, M.J. Inertia effects in textured hydrodynamic contacts. *Proc. Inst. Mech. Eng. Part J J. Eng. Tribol.* **2010**, *224*, 751–756. [CrossRef]
24. Shibata, Y.; Enomoto, Y. Global Existence of Classical Solutions and Optimal Decay Rate for Compressible Flows via the Theory of Semigroups. In *Handbook of Mathematical Analysis in Mechanics of Viscous Fluids*; Giga, Y., Novotný, A., Eds.; Springer International Publishing: Cham, Switzerland, 2018; pp. 2085–2181.
25. Netgen/NGSolve Finite Element Tool. Available online: https://docu.ngsolve.org/latest/index.html (accessed on 10 January 2024).
26. Concli, F. Equilibrium of a journal bearing: A simplified CFD-analytical coupled approach. In *Advances in Fluid Mechanics XIII*; WIT Transactions on Engineering Sciences: Southampton, UK, 2020; pp. 13–25.
27. Merkle, C.L.; Feng, J.; Buelow, P.E. Computational modeling of sheet cavitation. In Proceedings of the 3rd International Symposium on Cavitation, Grenoble, France, 7–10 April 1998.
28. Gao, G.; Yin, Z.; Jiang, D.; Zhang, X. Numerical analysis of plain journal bearing under hydrodynamic lubrication by water. *Tribol. Int.* **2014**, *75*, 31–38. [CrossRef]
29. OpenFoam. Available online: https://www.openfoam.com/ (accessed on 8 October 2023).
30. Pusterhofer, M.; Bergmann, P.; Summer, F.; Grün, F.; Brand, C. A Novel Approach for Modeling Surface Effects in Hydrodynamic Lubrication. *Lubricants* **2018**, *6*, 15. [CrossRef]
31. Maier, M.; Pusterhofer, M.; Summer, F.; Grün, F. Validation of statistic and deterministic asperity contact models using experimental Stribeck data. *Tribol. Int.* **2021**, *165*, 107329. [CrossRef]
32. Chen, Y.M.; Rigaut, B.; Armanet, F. Wear behaviour of partially stabilized zirconia at high sliding speed. *J. Eur. Ceram. Soc.* **1990**, *6*, 383–390. [CrossRef]
33. Savio, A.; Cianferra, M.; Armenio, V. Analysis of Performance of Cavitation Models with Analytically Calculated Coefficients. *Energies* **2021**, *14*, 6425. [CrossRef]
34. Hong, B.; Lv, S.; Liu, Q.; Hou, J.; Ji, Y.; Gao, Z.; Gao, J.; Hu, J. A Simple Method to Evaluate the Vapor Pressure of Transformer Oil at Various Temperatures. In Proceedings of the 2018 7th International Conference on Energy and Environmental Protection (ICEEP 2018), Shenzhen, China, 14–15 July 2018; Advances in Engineering Research. Atlantis Press: Amsterdam, The Netherlands, 2018; Volume 2018.
35. Lang, O.R.; Steinhilper, W. *Gleitlager: Berechnung und Konstruktion von Gleitlagern Mit Konstanter und Zeitlich Veränderlicher Belastung; mit 6 Arbeitsblättern*, 1978th ed.; Springer: Berlin, Germany, 2014.
36. Meng, F.M. On influence of cavitation in lubricant upon tribological performances of textured surfaces. *Opt. Laser Technol.* **2013**, *48*, 422–431. [CrossRef]
37. Ullah, M.Z.; Rizwan, M.; Raza, A.; Ahmed, A.; Abid, M. Effect of dimple shape and depth on tribological performance of textured surface. In Proceedings of the 2021 International Bhurban Conference on Applied Sciences and Technologies (IBCAST), Islamabad, Pakistan, 12–16 January 2021; IEEE: Piscataway Township, NJ, USA, 2021; pp. 719–725.
38. *DIN ISO 2768-1:1991-06*; General Tolerances; Tolerances for Linear and Angular Dimensions without Individual Tolerance Indications. Identical with ISO 2768-1:1989; International Organization for Standardization: Geneva, Switzerland, 1991.

lubricants

Article

Deterministic Simulation of Surface Textures for the Piston Ring/Cylinder Liner System in a Free Piston Linear Engine

Felipe Kevin Correia Luz [1], Francisco J. Profito [2], Marcelo Braga dos Santos [3], Samuel A. N. Silva [3] and Henara Lillian Costa [1,3,*]

1 School of Engineering, Universidade Federal do Rio Grande, Campus Carreiros, Rio Grande 96203 900, Brazil; felipekevin.fkcl@furg.br
2 Department of Mechanical Engineering, Polytechnic School, University of São Paulo, São Paulo 17033 360, Brazil; fprofito@usp.br
3 Laboratory of Tribology and Materials, Universidade Federal de Uberlândia, Uberlandia 38408 100, Brazil; marcelo.bragadossantos@ufu.br (M.B.d.S.); snunesilva@ufu.br (S.A.N.S.)
* Correspondence: henaracosta@furg.br

Abstract: The use of synthesis gas (SYNGAS) from waste gasification has been pointed out as a key strategy to help the energy transition. However, SYNGAS' low calorific power is considered a difficult obstacle to its technological use in internal combustion engines. To overcome this, a novel free-piston linear motor has been proposed to pave the way for the use of SYNGAS in the mobility sector. Surface texturing has vast potential to reduce friction losses in this system. This study utilizes a deterministic numerical model to investigate the mixed lubrication performance of a textured piston ring/cylinder liner conjunction in a free piston engine. The model considers the simultaneous solution of the lubrication and asperity contact problems at the roughness scale, including texturing features on the cylinder surface. The numerical model employs the Reynolds equation with mass-conserving cavitation to calculate the inter-asperity fluid pressure. The rough contact model utilizes the Hertz theory for elastic contact to calculate the contact pressure at each asperity between the piston liner surface and the admitted smooth and rigid ring surface. Surface texturing demonstrated remarkable effectiveness, particularly in the hydrodynamic lubrication regime, with a maximum friction reduction of 38.5% observed for an area coverage of 50%. This was accompanied by a notable shift in the transition from the boundary to the mixed lubrication regime. The textured surfaces exhibited consistent efficiency in reducing fluid pressure and shear stress as the coverage of the textured areas increased. The incorporation of dimples on these surfaces played a crucial role by augmenting the lubricant storage capacity while concurrently reducing the real shear and contact areas. This study offers valuable insights into the nuanced friction-reducing mechanisms of surface textures, illuminating their influence on the coefficient of friction and the formation of lubricant films across various lubrication regimes.

Keywords: free piston linear engine; surface texturing; lubrication; deterministic simulation; piston ring system

1. Introduction

In view of the new challenges that the energy transition imposes [1], the search for mobility systems with lower friction losses is often linked to individual or combined actions of (i) development and use of materials and coatings that provide lower friction [2,3]; (ii) development and use of new lubricants, for example, new additive packages and/or lower viscosity lubricants [4,5]; and (iii) changes in the topography of the surfaces in contact, aiming at operating in lubrication regimes that result in less friction [6–8].

As the transition to electric vehicles will be slow and not occur at the same pace or proportions in all regions, low-carbon renewable fuels will still be needed for many years to come to eliminate fossil fuel use (which leads to a high carbon footprint) [9–11].

In the search for sustainable energy sources, the synthesis of methane gas from urban solid residues emerges as a viable option. However, the synthesis gas (SYNGAS) derived from this source exhibits a low calorific capacity [12], which presents a challenge when considering its applicability in conventional internal combustion engines [13]. Despite this limitation, purified methane gas carries potential as a substitute for current fossil fuels in specific scenarios [14,15].

A key strategy to tap into this potential involves enhancing engine efficiency. Traditional internal combustion engines suffer from various energy losses, with the primary source of dissipation occurring within the kinematic chain of the connecting rod-crank mechanism [16]. A free-piston linear engine is a feasible solution to reduce such losses and elevate engine efficiency [17]. This engine design can improve thermal cycle efficiency by up to 56% [18,19], promoting a transformative shift in the landscape of internal combustion engines. As an ingenious departure from traditional designs, this innovative engine configuration distinguishes itself by liberating the piston from the mechanical shackles of a crankshaft, presenting a paradigm shift in the pursuit of enhanced efficiency, reduced mechanical intricacies, and heightened adaptability across a spectrum of applications [17,18,20].

Unlike their conventional counterparts, free-piston linear engines redefine the dynamics of power generation by eliminating the traditional linkage between pistons and crankshafts. Instead, these engines harness the inherent back-and-forth motion of a piston within a cylinder, a motion unbridled by the constraints of a mechanical connection to a crankshaft (see Figure 1). The combustion process unfolds within a dedicated combustion chamber, where the expanding gases exert force on the piston, propelling it outward. This liberated reciprocating motion is the cornerstone for directly generating electrical power or performing mechanical work, circumventing the need for a conventional crankshaft-driven mechanism.

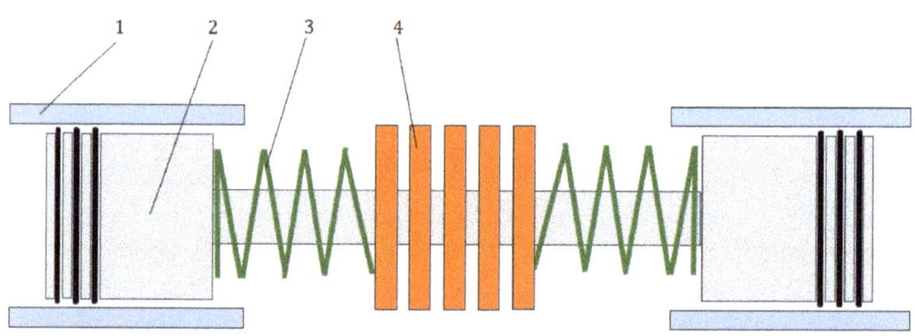

Figure 1. Schematics of a free-piston linear engine with opposed pistons. 1—cylinder liner; 2—piston; 3—springs; 4—stator.

The intrinsic freedom of movement within a free-piston linear engine brings forth a myriad of advantages, ranging from heightened operational efficiency to a streamlined mechanical architecture. Designed to optimize efficiency and reduce mechanical complexity, these engines find application in diverse fields such as power generation, automotive propulsion, and hybrid systems. The appeal of free-piston linear engines lies not only in their capacity to revolutionize energy conversion processes, but also in their adaptability to compact and lightweight configurations, presenting an enticing solution for specific niche applications [21].

A substantial source of energy loss in internal combustion engines arises from friction in the piston/cylinder liner system [22]. Specifically, frictional losses are mainly attributed to the interaction between the cylinder liner surface and the piston rings during engine operation. The alteration in surface textures offers a practical avenue for optimizing lubrication, diminishing friction, and controlling wear [23,24].

The reduction in frictional losses can be analyzed through the interaction of the tribological pair, namely the cylinder liner and the piston rings (see Figure 2). Approximately

50 to 75% of frictional losses occur at the interface between the piston rings and the cylinder liner surface [25]. The lubrication regimes (Figure 3) in the contact between the piston rings and the liner surface are (i) boundary lubrication at the top and bottom dead centers, characterized by significant asperity contact between the surfaces with little or almost no presence of lubricant; (ii) mixed lubrication regime, with reduced asperity contact and the presence of a lubricant film, that partially separates some of the surface asperities as the velocity increases; and (iii) hydrodynamic lubrication regime, with a full lubricant film separating the surfaces in the middle of the stroke (mid-stroke), distinguished by the absence of asperity contact and the formation of a complete film of lubricating oil [25]. In the piston-ring system, the most significant frictional forces are concentrated at the top and bottom dead center positions [26]. However, due to the low velocities at these positions, they account for only a minor fraction of the total frictional losses [27]. The majority of frictional power losses in internal combustion engines are situated within the hydrodynamic lubrication regime in the mid-stroke, where the piston reaches its maximum velocity, leading to increased shear forces in the lubricating oil and, thus, higher viscous losses [28].

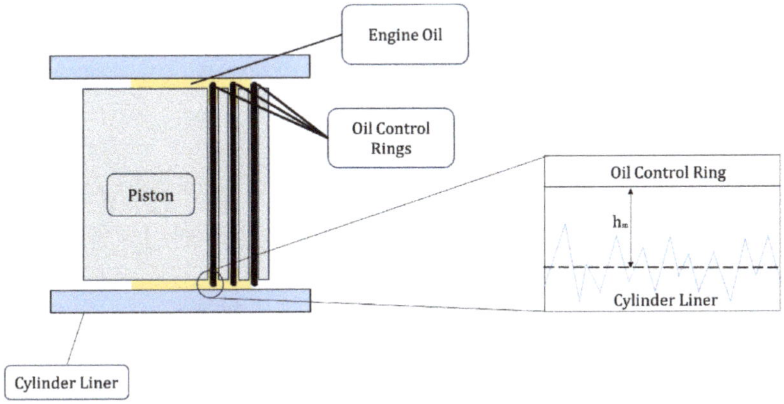

Figure 2. Schematics of the interaction between the cylinder liner and the piston rings.

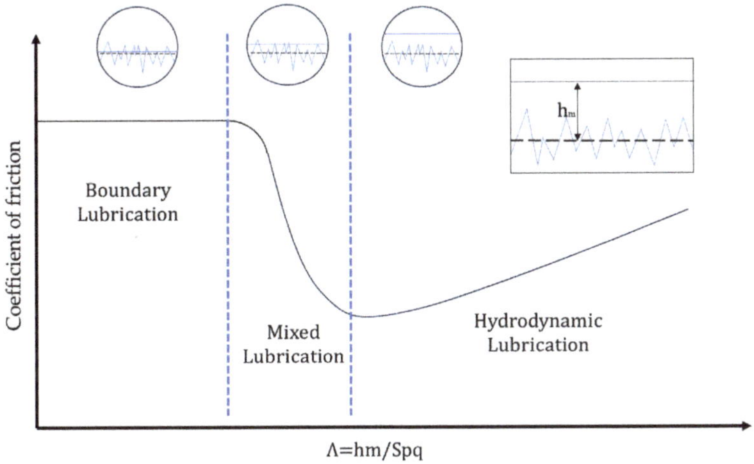

Figure 3. Stribeck curve describing the correlation between the coefficient of friction and the dimensionless film thickness or lambda ratio ($\Lambda = h_m/Spq$).

The surface finish of piston cylinder liners significantly influences the friction coefficient, the load-carrying capacity between the tribological pair of cylinder liners and piston rings, and the durability of the sealing rings [29,30]. Plateau honing is a surface finishing process in cylinder liners that removes the highest peaks of the liner surface, resulting in a reduced surface roughness, imparting a surface texture characterized by deep valleys that serve as lubricant reservoirs [31,32]. In the mid-stroke, the hydrodynamic lubrication film generation can be facilitated by controlled cavitation within the valleys created by the honing grooves, effectively acting as micro-hydrodynamic bearings. Simultaneously, the relatively smooth plateau regions between the grooves are responsible for the robust load-carrying capacity.

While the honing process yields advantageous surface features for optimizing performance, recent studies have indicated that a combination of various texture levels can be particularly effective in reducing friction coefficient and wear. Of particular interest is the integration of honing textures with spherical pockets, commonly referred to as dimples. The behavior of the dimples is contingent upon their geometric attributes, including diameter, depth, and the density of the textured region [33]. Deep dimples, for instance, serve as reservoirs for lubricants, aiding in maintaining the lubricant film. Additionally, they can retain wear debris, thereby mitigating wear-related issues. In contrast, shallow dimples function as micro-hydrodynamic bearings, augmenting the film thickness and loading capacity while concurrently diminishing the coefficient of friction [34]. The combination of honing textures with dimples thus presents an intriguing avenue for further investigation, as it offers the potential to optimize surface performance through a tailored approach that considers the specific geometric properties of these dimples [25].

Numerical simulation is a powerful tool to help optimize the performance of textured surfaces in the ring/cylinder liner system, but it is very challenging due to the complexities involved [35]. In the mixed lubrication regime, the models should account for both lubrication and asperity contact [36,37].

Modeling the mixed lubrication behavior of piston ring/cylinder liner conjunctions involves describing the coupled effects of lubricant flow and asperity interactions [38,39]. These multiphysics phenomena can be considered using deterministic models that account for the whole surface topography [35,37,40] or averaging and homogenization models [41].

In this context, this study employs a deterministic numerical simulation approach to investigate various surface texturization parameters to optimize the tribological performance of textured liners for a free-piston linear engine. This investigation aims to assess the impact of surface textures on the friction behavior of the piston ring/cylinder liner conjunction, providing insights into potential strategies for enhancing engine performance that might enable the successful use of waste gas as fuel.

2. Deterministic Mixed Lubrication Model

Evaluating the frictional behavior of piston ring/cylinder liner systems across the various lubrication regimes necessitates a comprehensive approach and well-suited models. While acknowledging that several system properties can vary under real operating conditions, introducing effects such as temperature variation, system dynamics, mechanical deformation, or the influence of surface topography on frictional losses increases both the computational costs and the complexity of the simulation analysis. To address these difficulties while preserving the most significant mixed lubrication effects and the detailed influence of surface features in a computationally efficient simulation framework, the deterministic mixed lubrication model proposed by Profito, et al. [42] was used to investigate the lubrication performance of the textured liner. The model is based on the integrated solution of hydrodynamic and asperity contact problems in the same simulation framework using the full-scale representation of the cylinder liner topography.

The hydrodynamic pressure is calculated using the steady-state isothermal Reynolds equation with the Elrod–Adams p–θ mass-conserving cavitation model, which can be written as follows [42]:

$$\frac{\partial}{\partial x}\left(\frac{\rho h^3}{12\mu}\frac{\partial p_h}{\partial x}\right)+\frac{\partial}{\partial y}\left(\frac{\rho h^3}{12\mu}\frac{\partial p_h}{\partial z}\right)=\frac{\partial}{\partial x}\left(\frac{\rho h U\theta}{2}\right), \quad (1)$$

with the complementarity boundary conditions for cavitation [42]:

$$(p_h-p_{cav})\cdot(1-\theta)=0 \Rightarrow \begin{cases} p_h>p_{cav}, & \theta=1 & \text{in } \mathcal{D}^+ \\ p_h=p_{cav}, & 0\le\theta<1 & \text{in } \mathcal{D}^- \\ p_h=p_{cav}, & & \text{on } C \end{cases}, \quad (2)$$

In the above equations, p_h is the hydrodynamic pressure, θ is the lubricant film fraction that represents the fluid saturation in the cavitation regions, ρ and μ are the lubricant density and viscosity, respectively, h is the local lubricant gap that considers the full-scale geometric features of the surface topography, and U is the sliding velocity of the contacting surfaces. The liner surface was assumed to be rough and deformable, and the piston ring countersurface was admitted to be smooth, flat, and rigid. Furthermore, D^+ and D^- denote the pressured and cavitated regions, respectively, and C the cavitation boundaries. The hydrodynamic shear stress (τ_h) in the sliding direction acting on the ring surface is calculated as follows [42]:

$$\tau_h=\frac{h}{2}\frac{\partial p_h}{\partial x}+\mu\theta\left(\frac{U}{h}\right). \quad (3)$$

The mean contact pressure at each asperity of the rough liner surface interacting with the smooth rigid ring surface is calculated using an asperity-based elastic, perfectly plastic model based on Hertz's theory for the contact between parabolical-shaped bodies. The surface hardness of the softer material is assumed to limit the plastic deformation in the contact zone. The expression of the mean contact pressure (p_{asp_s}) for each asperity interaction is as follows [42]:

$$p_{asp_s}=\begin{cases}\frac{4E_s^*}{3\pi}\left(\frac{w_s}{\beta_s}\right)^{\frac{1}{2}} & ,p_{asp_s}\le H_V \\ H_V & ,\ p_{asp_s}>H_V\end{cases}, \quad (4)$$

where w_s is the contact interference, β_s is the local asperity radius of curvature, E_s^* is the combined elastic modulus of the asperity, and H_V is the hardness of the softer material. β_s is determined by approximating the geometry of each asperity with a best-fit parabolical cap using the neighboring points of the highest contact point; this allows for the contact pressure at each asperity to be distributed over the points that compound the asperity. The asperity contact shear stress (τ_{asp_s}) is calculated based on the Coulomb–Amontons law for sliding friction, as follows [42]:

$$\tau_{asp_s}=\mu_{BL}p_{asp_s}, \quad (5)$$

where μ_{BL} is the boundary friction coefficient usually determined experimentally.

In order to reduce the computational effort of the numerical solutions, the rough liner surface is segmented into slices of the same width as the smooth ring surface to emulate the ring contact area. For each slice, deterministic simulations are performed for a constant lubricant viscosity and sliding velocity and a range of imposed separations (h_m) between the rigid surface and the reference plane of the rough surface (see more details in Section 3.1). Subsequently, the governing equations of the hydrodynamic and asperity contact problems described above are solved quasi-statically for each imposed separation. The p–θ Reynolds equation (Equations (1) and (2)) is solved numerically using the hybrid-type Element-based Finite Volume Method (EbFVM) proposed by Profito, Giacopini, Zachariadis and Dini [42] to solve lubrication problems, automatically ensuring mass conservation even in conditions with inter-asperity cavitation, as are the cases investigated in this work.

The deterministic simulations provide the distributions of the hydrodynamic pressure, hydrodynamic shear stress, asperity contact pressure, and asperity shear stress for each surface slice and imposed separations. In many situations, it is convenient to evaluate the global effect of these field results as a function of the dimensionless film thickness $\Lambda = h_m/Spq$ using the following average quantities:

$$
\begin{aligned}
&P_h(\Lambda) = \frac{1}{N}\sum_{i=1}^{N} p_{h_i}, \qquad \overline{\tau}_h(\Lambda) = \frac{1}{N}\sum_{i=1}^{N} \tau_{h_i}, \\
&P_{asp}(\Lambda) = \frac{1}{N}\sum_{i=1}^{N} p_{asp_i}, \qquad \overline{\tau}_{asp}(\Lambda) = \frac{1}{N}\sum_{i=1}^{N} \tau_{asp_i}, \qquad \text{COF}(\Lambda) = \frac{\overline{\tau}_h(\Lambda)+\overline{\tau}_{asp}(\Lambda)}{P_h(\Lambda)+P_{asp}(\Lambda)},
\end{aligned}
\tag{6}
$$

where N is the number of mesh points of each surface slice, and P_h, $\overline{\tau}_h$, P_{asp}, $\overline{\tau}_{asp}$, and COF are the average hydrodynamic pressure, hydrodynamic shear stress, asperity contact pressure, asperity contact shear stress and coefficient of friction, respectively.

3. Methodology

The analysis process is summarized via the flowchart in Figure 4, outlining each sequential step and its key attributes. The analysis begins with measurements of surface topography, followed by adjustments to isolate the impact of surface roughness and texture. The subsequent stages highlight how the proposed surface textures influence the system lubrication performance using the results obtained with the deterministic mixed lubrication model.

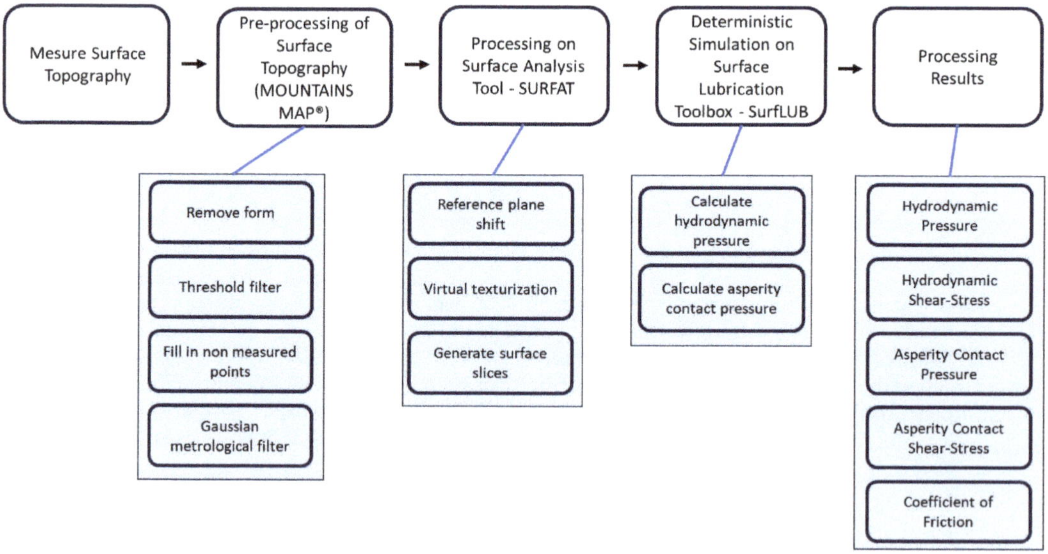

Figure 4. Flowchart of the methodology applied in the conducted analysis.

3.1. Surface Processing

The surfaces of commercial grey cast iron honed liner samples were measured using a Taylor Robson CCI laser interferometer. These measurements employed the stitching technique, allowing for the sampling of a large area.

The measured surfaces were initially processed using the MounstainsMap® Imaging Topography software version 9.0 from DigitalSurf. The treatment of the surfaces involved the following steps: (i) surface levelling; (ii) form removal, using a third-degree polynomial fitting; (ii) application of a threshold to remove measurement outliers, eliminating surface height data that fell outside the range of 0.20% to 99.80% according to the Abbot–Firestone curve; (iii) filling in non-measured points; and (iv) application of a robust Gaussian metrological filter with a cut-off of 800 μm to separate surface roughness and waviness.

The surface topography data were subsequently processed using the in-house Matlab toolbox SURFAT ('Surface Analysis Toolbox') to prepare the surface topographies for the numerical simulations. SURFAT is a general-purpose surface analysis toolbox developed at the Laboratory of Surface Phenomena (LFS) of the University of São Paulo (USP). SURFAT divided the rough liner surfaces into slices to facilitate data processing and reduce the computational cost of the numerical simulations. Each slice has the same width of the ring face (1000 μm in this work) to reproduce the ring contact area on the liner. An overlap of 50 μm was applied to emulate the ring's sliding motion effect along the stroke, resulting in 97 slices. A computational mesh of 376 × 251 was used to discretize each slice for the numerical simulations.

The honing grooves and surface textures cause a shift in the position of the reference plane of featured surfaces, which no longer coincides with the plateau region, where the actual contact takes place. A plateau segmentation method based on the Gaussian Mixture Model clustering method was used to overcome such reference plane shift, as illustrated in the Results section. This correction, proposed in a recent work, considers the roughness of the plateau region for adjusting the location of the reference plane. This adjustment is necessary because the surface separations and the lubricating film thickness are defined in the numerical model from the reference plane of the plateau, which is more convenient for mixed lubrication analysis [43].

Within the SURFAT program, a MATLAB code extension was developed to create virtual texturing so that spherical dimples could be added to measured surfaces of the commercial honed liners. The input parameters for creating the dimples include the dimple radius (*Dr*), the distance between the centers of the dimples (*Ds*), and the ratio between the dimple's depth and radius (*Dd*). Surfaces with different textured area densities were generated using the parameters shown in Table 1 for numerical simulations. These virtual surfaces enabled the assessment of the impact of textured area density on the surface properties and lubrication performance, which are crucial for optimizing tribological systems.

Table 1. Virtual texturing parameters.

Coverage Area Percentage	*Dr* (μm)	*Ds* (μm)	*Dd*
Original Untextured Surface	-	-	-
5%	100	792.66	0.03
10%	100	560.50	0.03
15%	100	457.65	0.03
20%	100	396.33	0.03
30%	100	323.60	0.03
50%	100	250.66	0.03

The simulations were conducted using the in-house computational code SURFLUB (SURFace LUBrication), which solves the governing equations of the deterministic mixed lubrication model described in Section 2. SURFLUB was also developed at the Laboratory of Surface Phenomena (LFS) of the University of São Paulo (USP). The analysis involved the imposition of different lambda ratios ($\Lambda = h_m / Spq$) under steady-state conditions, where h_m is the imposed separation between the reference plane of the piston liner surface roughness and the smooth and rigid theorical surface of the piston's oil control ring, and *Spq* is the standard deviation of the roughness in the plateau region of the cylinder liner surface. The variation in the imposed Λ values aimed to use a larger number of points in the regions of thinner films, as described in Table 2, where the variation step increases as the imposed film becomes thicker.

Table 2. Step increments for the variation in imposed lambda values (Λ).

Λ range	0.01–0.05	0.06–0.10	0.10–0.50	0.50–1.00	1.00–5.00
Variation step	0.01	0.02	0.05	0.10	0.25

The lubricant properties were selected for a commercial SAE 30 oil. The mechanical properties of the contacting materials were obtained from the literature. All other relevant parameters for the numerical simulations are summarized in Table 3.

Table 3. Simulation parameters.

Sliding velocity (U)	2.82 $m \cdot s^{-1}$
Lubricant dynamic viscosity (μ)	0.01 Pa·s
Lubricant specific mass (ρ)	0.82 g/cm^3
Young's Modulus (E)	120 GPa
Hardness (H)	420 HV
Boundary coefficient of friction (μ_{BL})	0.14

3.2. Outlier Removal

A post-processing treatment was applied to the averaged simulation results (see Section 2) to remove outliers using Tukey's method. The detection and removal of outliers are crucial procedures in statistical data analysis. Tukey's method provides an effective approach to identifying and addressing these extreme observations, beginning by calculating the Interquartile Range (IQR) of the dataset. The IQR is the difference between the third quartile (Q3) and the first quartile (Q1): IQR = Q3 − Q1. Next, the upper and lower limits are established to identify outliers. These limits are typically determined as follows: lower limit: calculated as Q1 − 1.5 × IQR; upper limit: calculated as Q3 + 1.5 × IQR; values that fall below the lower limit or above the upper limit are considered outliers. Once the limits are defined, it is possible to identify and, if necessary, remove outliers from the dataset [44].

4. Results and Discussion

4.1. Original Surface Processing

The results of the original surface pre-processing are shown in Figure 5, where level, form and waviness have been adequately removed. This step is fundamental to ensure that only the effect of the surface roughness is accounted for in the numerical simulations. The surface roughness parameters calculated after the pre-processing are summarized in Table 4. These parameters are widely used for the statistical characterization of the surface topography of cylinder liners [31] and are defined by the standard ISO 25178-2 [45]. As the surface of the ring is assumed to be smooth and rigid in the numerical model, only the liner topography is considered during surface processing. As defined in Section 2, the standard deviation of the plateau roughness, *Spq*, is used to calculate the dimensionless separation (Λ) between the surfaces.

Table 4. Roughness parameters of the simulated surfaces for different coverage area percentages.

Parameters	Original Surface	Coverage Area Percentage					
		5%	10%	15%	20%	30%	50%
Spq (μm)	0.25	0.25	0.25	0.25	0.25	0.26	0.28
Ssk (μm)	−2.06	−2.02	−1.79	−1.63	−1.46	−1.18	−0.68
Sku (μm)	7.58	7.53	6.28	5.49	4.74	3.90	2.94
Std (deg)	28.24	151.80	151.80	151.80	151.80	151.80	151.80
Spk (μm)	0.22	0.22	0.22	0.22	0.22	0.22	0.23
Sk (μm)	0.68	0.71	0.76	0.80	0.86	1.03	2.00
Svk (μm)	1.99	2.24	2.52	2.67	2.80	2.93	2.23

Table 4. *Cont.*

Parameters	Original Surface	Coverage Area Percentage					
		5%	10%	15%	20%	30%	50%
Smrk1 (%)	5.61	5.35	4.95	4.61	4.28	3.46	1.19
Smrk2 (%)	71.18	69.54	67.26	65.33	63.20	58.80	57.55
Vv ($\mu m^3/\mu m^2$)	0.61	0.68	0.78	0.85	0.94	1.13	1.52

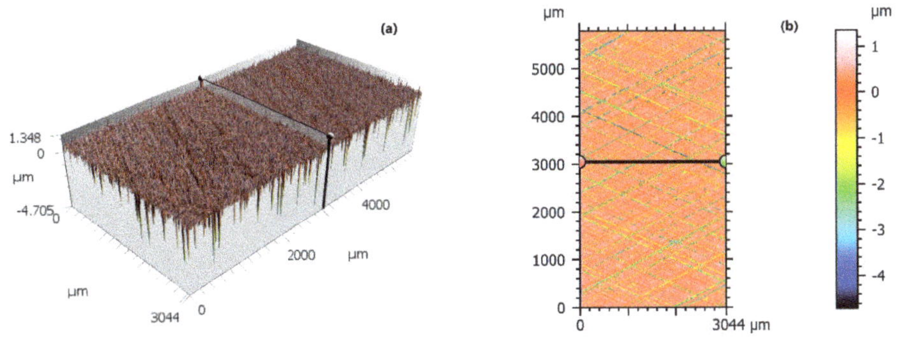

Figure 5. Cylinder liner topography; (**a**) 3D representation of the surface roughness; (**b**) 2D representation of the surface roughness.

The preparation for the numerical simulation of the surface topography uses the Gaussian Mixture Model method to identify clusters in the height distributions (Figure 6c). The mean height of the cluster with the highest frequency is used to shift the reference plane to the plateau to compensate for its deviation caused by the honed grooves (Figure 6e). This shift in the reference plane is exemplified by the selected profiles shown in Figure 6d, where blue is the profile of the original surface, and red is the profile of the shifted surface.

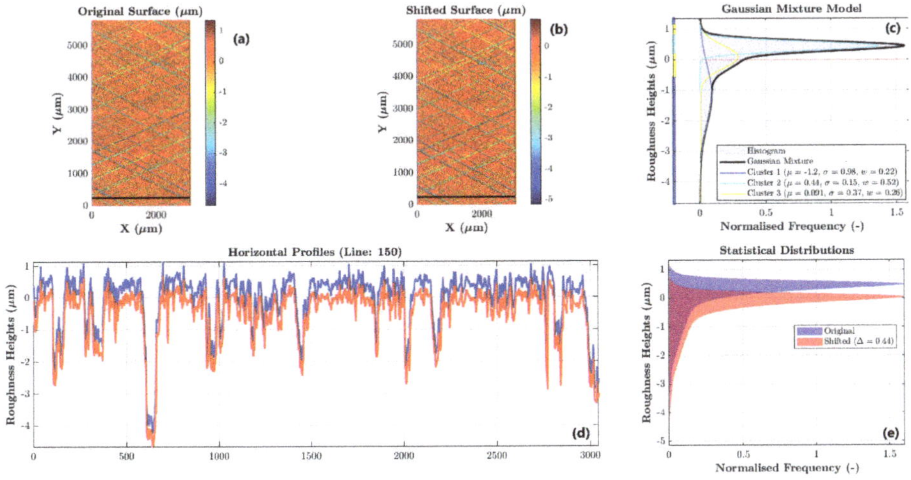

Figure 6. Mean plane shift using a Gaussian Mixture Model: (**a**) top view of the original surface; (**b**) top view of the shifted original surface; (**c**) Gaussian Mixture Model used for surface clustering; (**d**) line profiles comparing original (blue) and shifted condition (red); (**e**) comparison of height asperity distributions of the original and shifted surfaces.

After shifting the reference plane, virtual texturing was applied to the original surface, resulting in the textured surfaces shown in Figure 7. The virtual texturing of the pre-processed surface does not change the position of the reference plane.

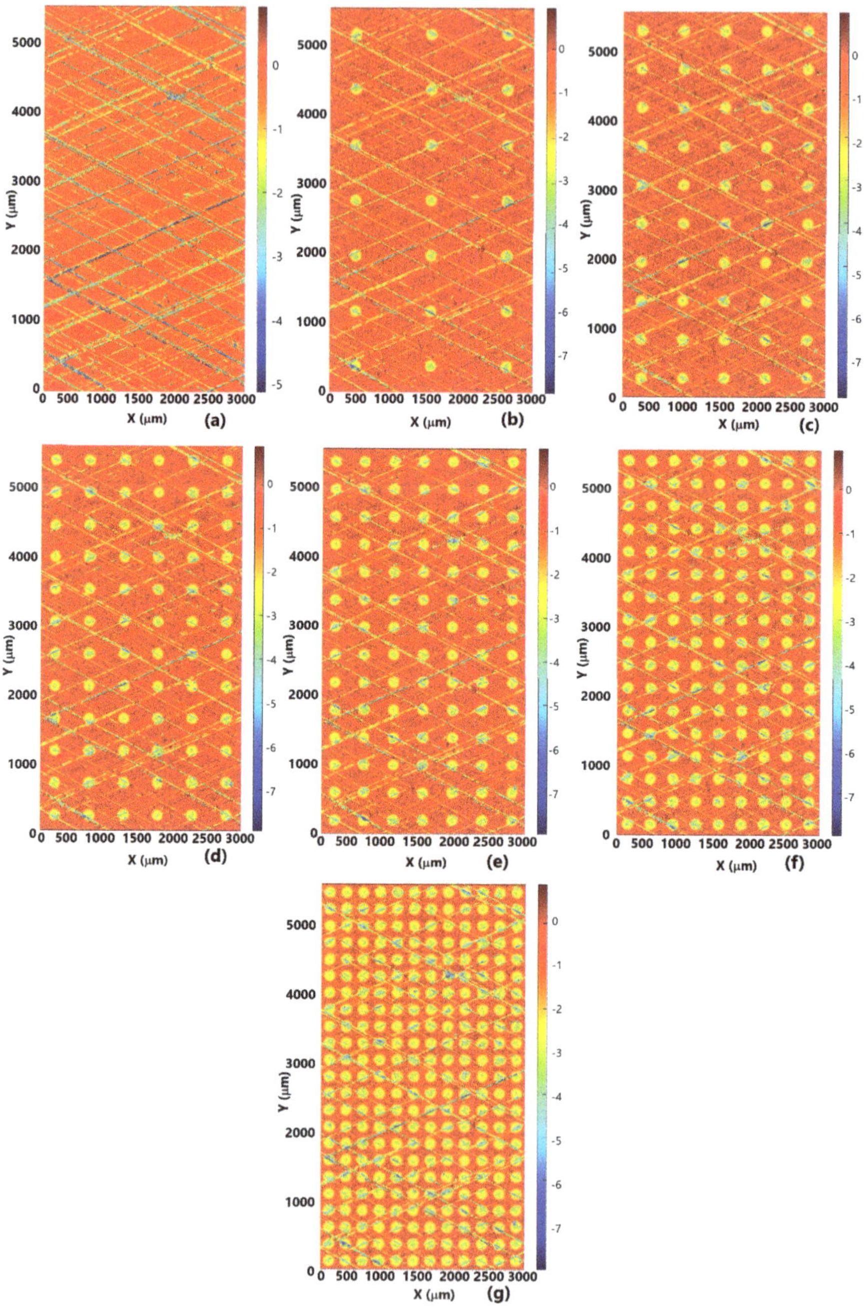

Figure 7. Surfaces for the numerical simulations with different area coverage percentages of the textures: (**a**) original surface; (**b**) 5%; (**c**) 10%; (**d**) 15%; (**e**) 20%; (**f**) 30%; (**g**) 50%.

4.2. Field Results

Representative examples of the film thickness, hydrodynamic pressure, film fraction (cavitation), and contact pressure distributions are depicted in Figures 8 and 9 for the first slices of the original and textured (50%) surfaces, respectively. Similar field results for textured surfaces with all the area coverages can be found in the Supplementary Materials Figure S1. In all these figures, (a) corresponds to the dimensionless separation $\Lambda = 0.5$, (b) $\Lambda = 2.5$, and (c) $\Lambda = 4.5$. Therefore, the selected field results relate to distinct lubrication regimes, enabling a comprehensive evaluation of the surface texture behavior across the boundary, mixed, and hydrodynamic lubrication regimes.

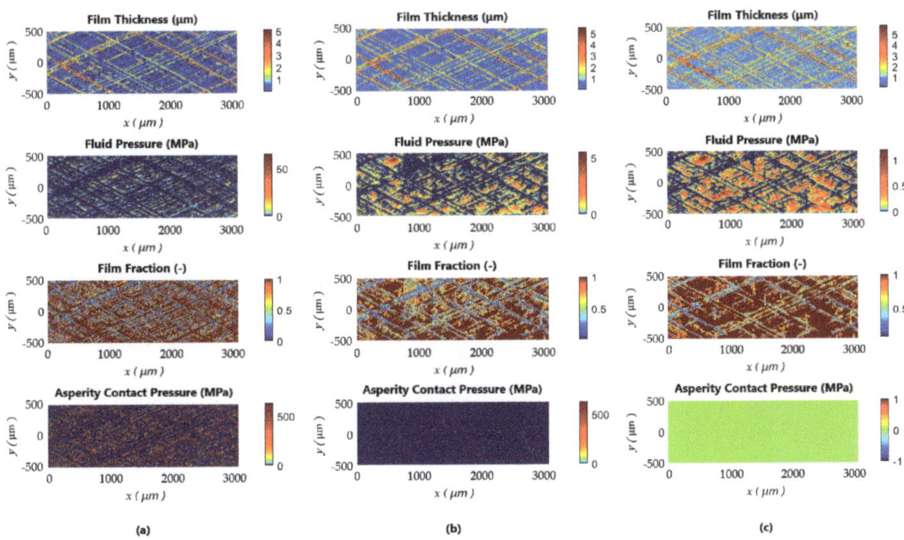

Figure 8. Field results of the original surface for: (**a**) $\Lambda = 0.5$; (**b**) $\Lambda = 2.5$; and (**c**) $\Lambda = 4.5$.

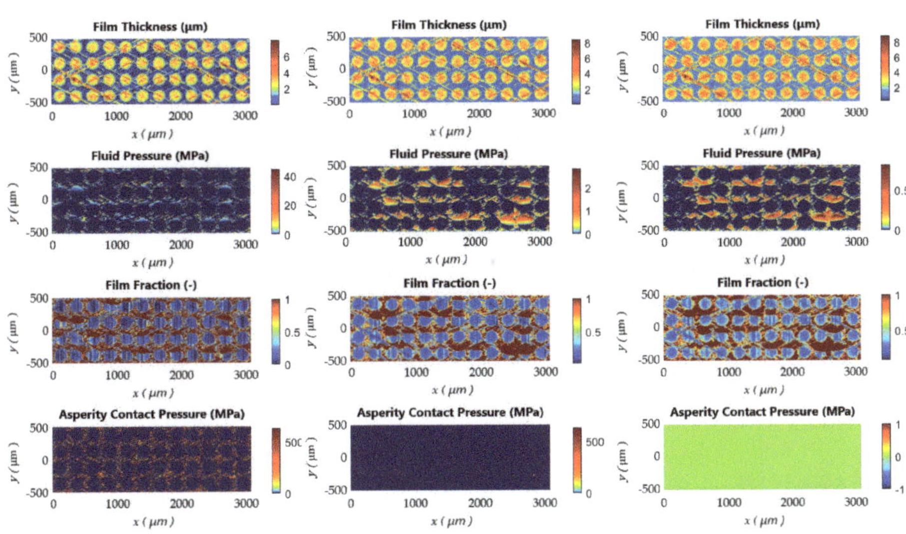

Figure 9. Field results of the textured surface (50%) for: (**a**) $\Lambda = 0.5$; (**b**) $\Lambda = 2.5$; and (**c**) $\Lambda = 4.5$.

4.3. Average Results

The average results of the numerical simulations allow for an overall performance comparison of the evaluated topographies in terms of hydrodynamic pressure support (Figure 10a) and shear stress (Figure 10b), as well as regarding the asperities contact pressure (Figure 10c), shear stress (Figure 10d) and total coefficient of friction (Figure 10e).

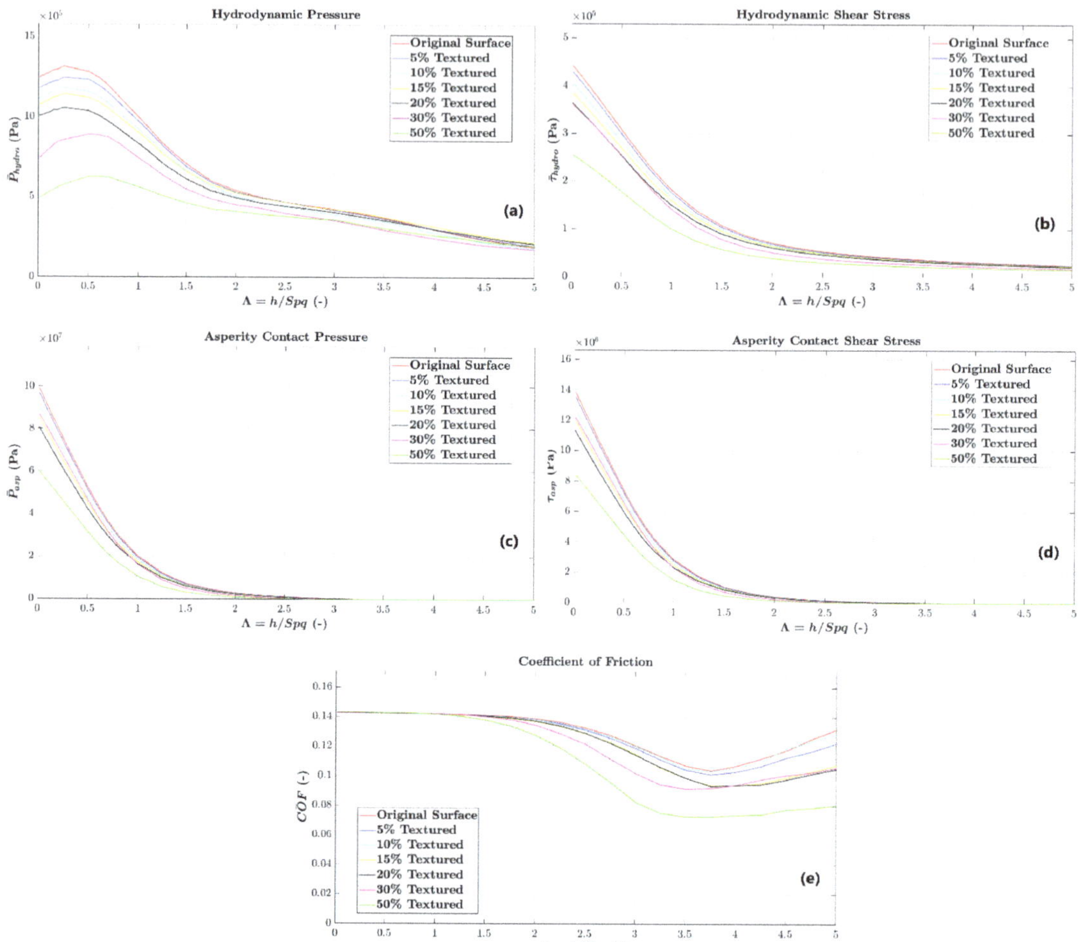

Figure 10. Average results of the deterministic simulations: (**a**) hydrodynamic pressure; (**b**) hydrodynamic shear stress; (**c**) asperity contact pressure; (**d**) asperity contact shear stress; (**e**) coefficient of friction.

As the percentage of the textured area increases, a discernible decrease in hydrodynamic pressure is evident, as depicted in Figure 10a, specifically within the dimensionless film thickness associated with the boundary lubrication regime. An analysis of the fluid pressure field results in Figures 8 and 9 reveals near-zero pressure on the dimples despite the presence of cavitated fluid, as indicated by the film fraction field results. Notably, the cavitated fluid does not contribute to a significant micro-hydrodynamic bearing effect capable of elevating fluid pressure on the plateau regions.

This observed behavior persists across varying percentages of textured areas, with the 50% textured area displaying the lowest hydrodynamic pressure within the boundary lubrication regime. As the transition to the mixed and hydrodynamic lubrication regime occurs, the hydrodynamic pressure curves begin to converge.

Despite the heightened presence of fluid associated with larger percentages of textured areas, a noteworthy reduction in hydrodynamic shear stress is observed in the boundary and mixed lubrication regimes, as illustrated in Figure 10b. This reduction is attributed to a simultaneous decrease in the fluid shear rate over the textures and in the effective viscosity due to the texture-induced cavitation, as discerned from both the hydrodynamic shear stress data and the field results for film fraction. Equation (3) underscores the impact of the local film thickness (h) and film fraction (θ) on the hydrodynamic shear stress, offering insights into the observed decrease in these results.

The predicted reductions in asperity contact pressure (Figure 10c) and shear stress (Figure 10d) resulted from a diminished real contact area due to increased textured area percentages, as can be observed with the diminishing upper support coverage parameters in Table 3 for the *Smrk2* parameter with the increase in textured area coverage. Despite this, it is essential to note that in the deterministic model employed, the boundary friction coefficient was imposed as 0.14 (Table 3), leading to the convergence of the predicted coefficient of friction (Figure 10e) for all textured area percentages for the dimensionless film thickness associated with the boundary lubrication regime.

The curves representing the coefficient of friction (Figure 10e) exhibit a notable leftward shift for the transition point from the boundary to the mixed lubrication regime as the textured area percentage increased, with the 50% textured area percentage entering the mixed lubrication regime for the lowest Λ values. Furthermore, as the textured area increased, there was a reduction in the increase rate for the coefficient of friction during the hydrodynamic lubrication regime.

Surface textures with 5% and 10% textured area percentages displayed remarkably similar behavior across all three lubrication regimes, as did surfaces with 15% and 20% textured area percentages. This trend suggests a consistent response among these specific textured area percentages, highlighting their comparable performance throughout different lubrication conditions.

The field results in Figures 8 and 9 show the roughness and texture-induced hydrodynamic pressure generation, especially near the honed grooves and dimples. Furthermore, the influence of inter-asperity cavitation (fluid fraction field) within the honed grooves and textures on the local lubricant flow and its impact on the fluid pressure buildup is noticeable.

Analyzing the field results for $\Lambda = 0.5$ (boundary lubrication regime), spikes in fluid pressure are evident. However, while the hydrodynamic pressure is high at some localized points, an insufficient amount of fluid limits its significant influence on the overall coefficient of friction across all surface topographies. Furthermore, as expected for this lubrication regime, most frictional losses originate from the asperity interactions [46].

The contact between surfaces diminishes in the mixed lubrication regime ($\Lambda = 2.5$), and the lubricant can flow more easily through the interface. Therefore, the hydrodynamic micro-bearing effect starts increasing the pressure on the plateau region. In the honing grooves and inside dimples, the cavitated fluid pushes out, elevating the pressure on the plateau downstream of the convergent portions of the surface features [47]. Furthermore, the surface topographies with increased area coverage yielded lower friction due to reduced contact and hydrodynamic shear stress over the dimples.

In the hydrodynamic lubrication regime, the contact effect of surface asperities is eliminated, as shown in the contact pressure distributions for all simulated surface topographies for $\Lambda = 4.5$. The film fraction parameter also indicates the integrity of the lubricant film in the plateau region. The increase in the area coverage of the textures directly reduces fluid pressure due to decreased plateau areas. A larger textured area reduces plateau regions with increased pressure from the hydrodynamic bearing effect, lowering the average hydrodynamic pressure, as shown in Figure 6. The global friction coefficient followed a similar trend to the mixed lubrication regime for all simulated surface topographies, with friction decrease now solely governed by the hydrodynamic shear stress reduction mechanism.

The rise in void volume (Vv) due to an increased textured area signifies an augmented capacity for lubricant storage in surface topography, helping to maintain the lubricant film. Despite the larger amount of lubricant due to textures, there is a reduction in hydrodynamic shear stress across all lubrication regimes with increased area coverage. Furthermore, the increase in the area coverage of the textures influences the transition point from the boundary to the mixed lubrication regime (see Figure 10) and significantly reduces the coefficient of friction in both mixed and hydrodynamic lubrication regimes. The combination of reduced plateau contact area and increased void volume significantly impacts the global coefficient of friction, with the textured surface with 50% area coverage demonstrating the best performance.

5. Conclusions

This work used a deterministic mixed lubrication model to evaluate the frictional behavior across diverse lubrication regimes for the proposed surface textures. The results underscore the advantageous impact of the hydrodynamic bearing effect induced by fluid cavitation within surface features, leading to a notable reduction in the global coefficient of friction as the textured area increases.

Remarkably, the surface with a 50% textured area exhibited optimal performance, enabling its transition into the mixed lubrication regime with a lower film thickness (Λ) than other surface textures. In the hydrodynamic lubrication regime, this surface demonstrated a marginal increase in the global coefficient of friction compared to the other surface topographies.

These findings provide valuable insights for optimizing tribological performance, highlighting the pivotal role of surface topography in influencing lubrication regimes and frictional behavior across a broad spectrum of engineering applications.

Supplementary Materials: The following supporting information can be downloaded at: https://www.mdpi.com/article/10.3390/lubricants12010012/s1, Figure S1: Field results at different Λ ratios for textured surfaces with textured area coverages varying from 5% to 30%.

Author Contributions: Conceptualization, H.L.C. and M.B.d.S.; methodology, F.J.P.; software, F.J.P., S.A.N.S. and F.K.C.L.; formal analysis, H.L.C. and F.J.P.; investigation, F.K.C.L.; resources, M.B.d.S. and H.L.C.; writing—original draft preparation, F.K.C.L.; writing—review and editing, H.L.C. and F.J.P.; visualization, F.K.C.L. and S.A.N.S.; supervision, H.L.C.; project administration, M.B.d.S.; funding acquisition, M.B.d.S. and H.L.C. All authors have read and agreed to the published version of the manuscript.

Funding: The authors would like to thank FINEP for funding project 01.21.0144.00 (0751/21). HLC acknowledges financial support from CNPq (project 305453/2017-3), as well as CNPq/INCT CT Trib (project 406654/2022-0).

Data Availability Statement: Original data can become available upon request directly to the authors.

Conflicts of Interest: The authors declare no conflicts of interest. The funders had no role in the design of the study; in the collection, analyses, or interpretation of data; in the writing of the manuscript, or in the decision to publish the results.

References

1. Almohaimeed, S.A.; Suryanarayanan, S.; O'Neill, P. Reducing carbon dioxide emissions from electricity sector using demand side management. *Energy Sources Part A-Recovery Util. Environ. Eff.* **2021**, 1–21. [CrossRef]
2. Moreira, H.C.R.; Oliveira, J.A.B.; Carreño, N.L.V.; Silva, R.; Garcia, I.T.S.; Costa, H.L.; Alano, J.H. Effects of niobium pentoxide nanoparticles on the tribological properties of electrodeposited ZnNi coatings. *Surf. Topogr.-Metrol. Prop.* **2022**, *10*, 024003. [CrossRef]
3. Cesur, I.; Akgündüz, M.; Çelik, H.A.; Çay, Y.; Ergen, G. Tribological Analysis and Optimization of Ring-Cylinder Couple Coated with Different Materials. *Arab. J. Sci. Eng.* **2023**, 1–12. [CrossRef]

4. Prajapati, D.K.; Ahmad, D.; Katiyar, J.K.; Prakash, C.; Ajaj, R.M. A numerical study on the impact of lubricant rheology and surface topography on heavily loaded non-conformal contacts. *Surf. Topogr.-Metrol. Prop.* **2023**, *11*, 035006. [CrossRef]
5. Hei, D.; Zheng, M.R.; Liu, C.; Jiang, L.T.; Zhang, Y.F.; Zhao, X.W. Study on the frictional properties of the top ring-liner conjunction for different-viscosity lubricant. *Adv. Mech. Eng.* **2023**, *15*, 16878132231155002. [CrossRef]
6. Rodrigues, T.A.; Arencibia, R.V.; Costa, H.L.; da Silva, W.M. Roughness analysis of electrochemically textured surfaces: Effects on friction and wear of lubricated contacts. *Surf. Topogr.-Metrol. Prop.* **2020**, *8*, 024011. [CrossRef]
7. Atulkar, A.; Pandey, R.K.; Subbarao, P.M.V. Synergistic effect of textured piston ring and nano-lubricant on performance parameters and emissions of IC engine. *Surf. Topogr.-Metrol. Prop.* **2021**, *9*, 035009. [CrossRef]
8. Mohammadpour, M.; Morris, N.J.; Leighton, M.; Rahnejat, H. Effect of surface topography upon micro-impact dynamics. *Surf. Topogr.-Metrol. Prop.* **2016**, *4*, 014001. [CrossRef]
9. Costa, H.L.; Cousseau, T.; Souza, R.M. Current Knowledge on Friction, Lubrication, and Wear of Ethanol-Fuelled Engines-A Review. *Lubricants* **2023**, *11*, 292. [CrossRef]
10. Udeagha, M.C.; Ngepah, N. Striving towards carbon neutrality target in BRICS economies: Assessing the implications of composite risk index, green innovation, and environmental policy stringency. *Sustain. Environ.* **2023**, *9*, 2210950. [CrossRef]
11. Kime, S.; Jacome, V.; Pellow, D.; Deshmukh, R. Evaluating equity and justice in low-carbon energy transitions. *Environ. Res. Lett.* **2023**, *18*, 123003. [CrossRef]
12. Zheng, J.J.; Impeng, S.; Liu, J.; Deng, J.; Zhang, D.S. Mo promoting Ni-based catalysts confined by halloysite nanotubes for dry reforming of methane: Insight of coking and H2S poisoning resistance. *Appl. Catal. B-Environ. Energy* **2024**, *342*, 123369. [CrossRef]
13. Pawananont, K.; Leephakpreeda, T. Experimental investigation and optimal combustion control of untreated landfill gas via fuzzy logic rule knowledge based approach. *Waste Manag.* **2021**, *121*, 383–392. [CrossRef] [PubMed]
14. Zhang, X.Y.; Khandavilli, M.; Gautam, R.; Alabbad, M.; Li, Y.; Chatakonda, O.; Kloosterman, J.W.; Middaugh, J.; Sarathy, S.M. An updated functional-group-based approach to modeling the vacuum residue oil gasification kinetics. *Fuel* **2024**, *357*, 129759. [CrossRef]
15. Tsipis, E.V.; Matveev, D.V.; Sharafutdinov, A.U.; Yalovenko, D.V.; Samoilov, A.V.; Fedotov, Y.S.; Dyakina, M.S.; Zhigacheva, D.V.; Agarkov, D.A.; Bredikhin, S.I.; et al. Performance of SOFCs using model waste gases: A case study. *Fuel* **2024**, *358*, 130129. [CrossRef]
16. Xu, S.Y.; Nguyen, V.; Wang, X.J.; Zhou, H.X. Sensitivity Analysis of the Geometrical Dimensions of the Crankpin Bearing on the Tribological Property of an Engine. *Sae Int. J. Engines* **2022**, *15*, 367–380. [CrossRef]
17. Yang, F.Y.; Feng, H.H.; Wu, L.M.; Zhang, Z.Y.; Wang, J.Y. Performance prediction and parameters optimization of an opposed-piston free piston engine generator using response surface methodology. *Energy Convers. Manag.* **2023**, *295*, 117633. [CrossRef]
18. Hung, N.B.; Lim, O. A review of free-piston linear engines. *Appl. Energy* **2016**, *178*, 78–97. [CrossRef]
19. Wu, J.P.; Nicholas, P.; Peter Van, B.; Goldborough, S. *Homogeneous Charge Compression Ignition Free Piston Linear Alternator*; University of North Texas Libraries, UNT Digital Library: Denton, TX, USA, 1998.
20. Guo, C.D.; Zuo, Z.X.; Feng, H.H.; Roskilly, T. Advances in free-piston internal combustion engines: A comprehensive review. *Appl. Therm. Eng.* **2021**, *189*, 116679. [CrossRef]
21. Guo, C.D.; Wang, Y.H.; Tong, L.; Feng, H.H.; Zuo, Z.X.; Jia, B.R. Research on Piston Dynamics and Engine Performances of a Free-Piston Engine Linear Generator Coupling with Various Rebound Devices. *Energies* **2023**, *16*, 6570. [CrossRef]
22. Holmberg, K.; Andersson, P.; Erdemir, A. Global energy consumption due to friction in passenger cars. *Tribol. Int.* **2012**, *47*, 221–234. [CrossRef]
23. Atulkar, A.; Pandey, R.K.; Subbarao, P.M.V. Role of textured piston rings/liners in improving the performance behaviours of IC engines: A review with vital findings. *Surf. Topogr.-Metrol. Prop.* **2021**, *9*, 023002. [CrossRef]
24. Li, S.J.; Wu, Z.P.; Dong, B.W.; Luo, W.Y.; Song, H.L.; Guo, H.T.; Zhou, A.Q. Improvement of lubrication performance of sliding pairs with multi-depth groove textures based on genetic algorithm. *Surf. Topogr.-Metrol. Prop.* **2023**, *11*, 025011. [CrossRef]
25. Wakuri, Y.; Hamatake, T.; Soejima, M.; Kitahara, T. Piston ring friction in internal combustion engines. *Tribol. Int.* **1992**, *25*, 299–308. [CrossRef]
26. dos Santos Filho, D.; Tschiptschin, A.P.; Goldenstein, H. Effects of ethanol content on cast iron cylinder wear in a flex-fuel internal combustion engine–A case study. *Wear* **2018**, *406–407*, 105–117. [CrossRef]
27. Morris, N.; Rahmani, R.; Rahnejat, H.; King, P.D.; Howell-Smith, S. A Numerical Model to Study the Role of Surface Textures at Top Dead Center Reversal in the Piston Ring to Cylinder Liner Contact. *J. Tribol.-Trans. Asme* **2016**, *138*, 021703. [CrossRef]
28. Johansson, S.; Nilsson, P.H.; Ohlsson, R.; Rosén, B.-G. A Novel Approach to Reduction of Frictional Losses in a Heavy-Duty Diesel Engine by Reducing the Hydrodynamic Frictional Losses. *Adv. Tribol.* **2017**, *2017*, 1–17. [CrossRef]
29. Miao, C.W.; Guo, Z.W.; Yuan, C.Q. An experimental study on tribological properties and air tightness of co-textured cylinder liner-piston ring on an engine tester. *Surf. Topogr.-Metrol. Prop.* **2021**, *9*, 015005. [CrossRef]
30. Leighton, M.; Rahmani, R.; Rahnejat, H. Surface-specific flow factors for prediction of friction of cross-hatched surfaces. *Surf. Topogr.-Metrol. Prop.* **2016**, *4*, 025002. [CrossRef]
31. Pawlus, P.; Reizer, R. Functional importance of honed cylinder liner surface texture: A review. *Tribol. Int.* **2022**, *167*, 107409. [CrossRef]

32. Reizer, R.; Pawlus, P.; Wieczorowski, M. Simulation of plateau-honed cylinder liner surface texture creation using superimposition approach. *Precis. Eng.-J. Int. Soc. Precis. Eng. Nanotechnol.* **2023**, *82*, 10–24. [CrossRef]
33. Lu, P.; Wood, R.J.K. Tribological performance of surface texturing in mechanical applications—A review. *Surf. Topogr.-Metrol. Prop.* **2020**, *8*, 043001. [CrossRef]
34. Rosenkranz, A.; Szurdak, A.; Gachot, C.; Hirt, G.; Mücklich, F. Friction reduction under mixed and full film EHL induced by hot micro-coined surface patterns. *Tribol. Int.* **2016**, in press. [CrossRef]
35. Liao, K.; Liu, Y.; Kim, D.; Urzua, P.; Tian, T. Practical challenges in determining piston ring friction. *Proc. Inst. Mech. Eng. Part J J. Eng. Tribol.* **2012**, *227*, 112–125. [CrossRef]
36. Chu, N.R.; Jackson, R.L.; Ghaednia, H.; Gangopadhyay, A. A Mixed Lubrication Model of Piston Rings on Cylinder Liner Contacts Considering Temperature-Dependent Shear Thinning and Elastic-Plastic Contact. *Lubricants* **2023**, *11*, 208. [CrossRef]
37. Chen, H.; Liao, K.; Tian, T. A Numerical and Experimental Study of Twin-land Oil Control Ring Friction in Internal Combustion Engines Part 2. *SAE Int. J. Engines* **2012**. [CrossRef]
38. Tomanik, E. Modelling the hydrodynamic support of cylinder bore and piston rings with laser textured surfaces. *Tribol. Int.* **2013**, *59*, 90–96. [CrossRef]
39. Profito, F.J.; Tomanik, E.; Zachariadis, D.C. Effect of cylinder liner wear on the mixed lubrication regime of TLOCRs. *Tribol. Int.* **2016**, *93*, 723–732. [CrossRef]
40. Liu, Y.; Li, Y.; Tian, T. Development and Application of Ring-Pack Model Integrating Global and Local Processes. Part 2: Ring-Liner Lubrication. *SAE Int. J. Engines* **2017**, *10*, 1969–1983. [CrossRef]
41. Söderfjäll, M.; Almqvist, A.; Larsson, R. A model for twin land oil control rings. *Tribol. Int.* **2016**, *95*, 475–482. [CrossRef]
42. Profito, F.J.; Giacopini, M.; Zachariadis, D.C.; Dini, D. A General Finite Volume Method for the Solution of the Reynolds Lubrication Equation with a Mass-Conserving Cavitation Model. *Tribol. Lett.* **2015**, *60*, 18. [CrossRef]
43. Silva, S.A.N. A Comparison of Texturized Surfaces for Ring-Cylinder through Deterministic Simulation. Master's Thesis, Universidade Federal de Uberlândia, Uberlândia, Brazil, 2022.
44. Schwertman, N.C.; Owens, M.A.; Adnan, R. A simple more general boxplot method for identifying outliers. *Comput. Stat. Data Anal.* **2004**, *47*, 165–174. [CrossRef]
45. *ISO-25178-2*; Geometrical Product Specifications (GPS): Surface Texture: Areal—Part 2: Terms, Definitions and Surface Texture Parameters. ISO: Geneva, Switzerland, 2021.
46. Scaraggi, M.; Mezzapesa, F.P.; Carbone, G.; Ancona, A.; Tricarico, L. Friction Properties of Lubricated Laser-MicroTextured-Surfaces: An Experimental Study from Boundary- to Hydrodynamic-Lubrication. *Tribol. Lett.* **2013**, *49*, 117–125. [CrossRef]
47. Vlădescu, S.-C.; Olver, A.V.; Pegg, I.G.; Reddyhoff, T. Combined friction and wear reduction in a reciprocating contact through laser surface texturing. *Wear* **2016**, *358–359*, 51–61. [CrossRef]

lubricants

MDPI

Article

Multi-Field Coupling Numerical Analysis and Experimental Validation of Surface-Textured Metal Seals in Roller Cone Bits

Yi Ma [1,2,*], Ziang Li [1], Ziyang Yuan [1], Xiangkai Meng [1,2], Xudong Peng [1,2,*] and Jinbo Jiang [1,2]

1 College of Mechanical Engineering, Zhejiang University of Technology, Hangzhou 310023, China; 2112102181@zjut.edu.cn (Z.L.); yzy810369918@163.com (Z.Y.); mengxk@zjut.edu.cn (X.M.); jiangjinbo@zjut.edu.cn (J.J.)

2 Engineering Research Center of Process Equipment and Remanufacturing, Ministry of Education, Hangzhou 310023, China

* Correspondence: myant@zjut.edu.cn (Y.M.); xdpeng@zjut.edu.cn (X.P.)

Abstract: Metal seals play a crucial role in ensuring the dependable functioning of the bearing system in roller cone bits. Due to the intricate nature of downhole conditions, the microstructure design of sealing end faces, specifically the surface texture, is rarely used in metal seals. This study focuses on examining the applicability of surface texture technology in metal seals for cone bits to enhance the lubrication characteristics and friction and wear properties of liquid film between end faces. A multi-field coupling model of surface-textured metal seals was established and experimentally verified. The liquid film carrying the capacity and sealing performance of five different shapes of surface-textured end faces (circle, ellipse I (horizontal), ellipse II (vertical), triangle, and chevron) were investigated under varying rotational speeds and environmental pressures. The influences of texture parameters, including depth and area ratio, on the lubrication characteristics and friction reduction effects of surface-textured metal seals were investigated, and the microscopic wear morphology characteristics of the sealing end faces were analyzed. The results show that surface textures can effectively improve the liquid film state and comprehensive performance of metal seals. Specifically, triangle textures significantly enhance the hydrodynamic pressure effect and weaken the abrasive wear and adhesive wear of the end faces. However, when the environmental pressure is $p \geq 30$ MPa, the surface textures gradually lose the advantages of dynamic lubrication but can still reduce friction. The proper texture depth and area ratio can achieve zero leakage (obtained theoretically) and minimal friction in metal seals.

Keywords: roller cone bits; metal seals; surface texture; multi-field coupling; hydrodynamic performance; friction and wear

1. Introduction

Roller cone bits are essential rock-breaking drilling tools in the field of geological exploration engineering [1,2]. Metal seals are extensively employed as vital components in the bearing sealing system of these bits, owing to their simple structure, compact size, and ability to withstand high pressure [3]. The performance and longevity of roller cone bits are directly influenced by the effectiveness of these metal seals [4]. However, when subjected to varying downhole temperature and pressure conditions, as well as external mudflow [5,6], the metal seals are susceptible to a range of issues, including an inadequate opening force between end faces, an excessive temperature rise, and severe wear [7,8]. Previous studies have shown that the utilization of surface texture technology can enhance the load-bearing capacity of the liquid film on end faces and the anti-wear performance of conventional mechanical seals [9–11]. Nevertheless, the feasibility and effectiveness of surface texture technology to enhance the performance of metal seals in downhole conditions remain uncertain. Therefore, clarifying the liquid film state and comprehensive performance of surface-textured metal seals is necessary for the applicability study of texture in multi-physics drilling environments.

In recent years, some researchers have conducted numerical simulations and macrostructural optimization of metal seals in roller cone bits. Xiong and Salant [12,13] and Ma et al. [14,15] established steady-state and dynamic multi-field coupling numerical models for first- and second-generation metal seals, respectively, and analyzed the changes in liquid film characteristics, temperature field distribution, and leakage rate with operating parameters. Although few literatures have studied surface-textured metal seals, the mature research methods and results of surface texture in bearings and mechanical seals can serve as valuable insights for investigating the multi-field coupling of surface-textured metal seals. Etsion et al. [16,17] developed an analytical model for micropore-textured mechanical seals, demonstrating that optimally textured seals can significantly reduce friction and heat generation. Brunetiere and Tournerie [18] and Yang et al. [19] considered the cavitation phenomenon, asperity contact, and viscosity–temperature effect of mechanical seals and numerically studied the differences in the liquid film bearing capacity and frictional state of end faces with different texture parameters or shapes. The findings revealed a strong correlation between the leakage of sealing end faces and the density of surface texture. Moreover, the study demonstrated that triangle textures are more effective for enhancing the hydrodynamic performance of the mechanical seals. Qiu and Khonsari [20] and Dingui et al. [21] conducted friction and performance tests on laser-surface-textured rings or mechanical seals, respectively. They found that the size and depth of dimples have a significant impact on the load-carrying capacity and friction coefficient, and the interaction between roughness and surface textures on the behavior of mechanical seals cannot be ignored. However, there is a scarcity of studies that investigate the multi-field coupling simulation and experimental research of surface-textured metal seals. Consequently, the applicability of surface textured technology in the metal seals in cone bits needs to be solved urgently.

This study aimed to enhance the lubrication and sealing performance of metal seals in roller cone bits by applying surface texture technology. A multi-field coupling numerical model of surface-textured metal seals was established to reveal the mechanism of textures on the friction reduction and lubrication enhancement of end faces. The influences of key operating parameters and surface texture parameters on the liquid film characteristics and sealing performance of metal seals were analyzed. To validate the accuracy of the numerical results, a comprehensive test bench of surface-textured metal seals was constructed, and the interface temperature rise and wear morphology on the end faces of metal seals were tested and compared. The results offer novel insights into the application of surface textures with varying shapes and parameters on metal seals in intricate downhole environments.

2. Numerical Model

2.1. Geometric Model

The metal seals specifically developed for roller cone bits were adopted as the research objective, as shown in Figure 1. Metal seals, categorized as convergent cone seals, consist primarily of metal sealing rings (rotor and stator) and auxiliary rubber seals (O-ring and backup ring) [22,23]. The rotor is installed within the cone and rotates with the cone during drilling. The stator is elastically supported by the O-ring and backup ring, which are fixed using the palm journal, ensuring a close fit between the end faces of the stator and rotor. The lubricant oil and drilling mud serve as the media on the inner and outer diameter sides of metal seals, respectively [12,24]. The cavities on both sides of the metal seals maintain a specific pressure difference (Δp) to prevent drilling mud from invading the bearing system. The structural and operating parameters of the metal seals are shown in Table 1.

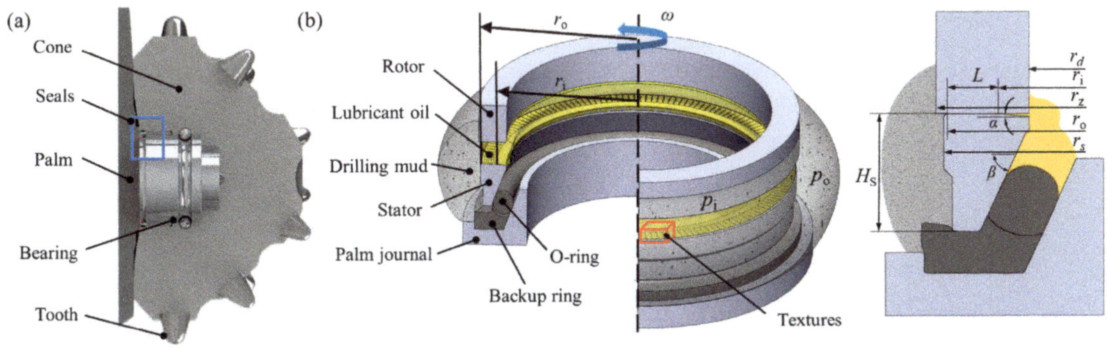

Figure 1. Geometric model of roller cone bit and metal seals. (**a**) Roller cone bit; (**b**) Metal seals.

Table 1. Structural and operating parameters of metal seals.

Structural Parameters	Values	Operating Parameters	Values
Inner diameter of stator r_d (mm)	29	Environmental pressure p_o (MPa)	3~69
Outer diameter of rotor r_s (mm)	34.5	Pressure difference Δp (MPa)	0.5
Sealing inner diameter r_i (mm)	31	Environmental temperature T_o (°C)	8~180
Sealing outer diameter r_o (mm)	34.3	Rotational speed n (rpm)	100~500
Wedging angle α (°)	5	Density of lubricant oil (kg/m^3)	861
Incline angle β (°)	65	Density of drilling mud (kg/m^3)	1742
Surface roughness σ (μm)	0.2	Viscosity of lubricant oil (Pa·s)	0.002~0.189
Dry friction coefficient f_0	0.08	Viscosity of drilling mud (Pa·s)	0.02~0.03

The sealing end face of the stator can be divided into three zones: the wedge angle zone (radial range of r_d~r_i) on the inner diameter side, the sealing zone (radial range of r_i~r_o) in the middle, and the chamfering zone (radial range of r_o~r_s) on the outer diameter side. Within the sealing zone of the stator, five typical micro-texture shapes, namely, circle, ellipse I (horizontal), ellipse II (vertical), triangle, and chevron [25,26], are adopted, as shown in Figure 2. The micro-textures with the same size are uniformly distributed in the sealing zone in a circular array of T columns. Each column comprises N_t micro-textures along the radius with a depth, h_p, and an area ratio, S_p. The texture parameters are shown in Table 2.

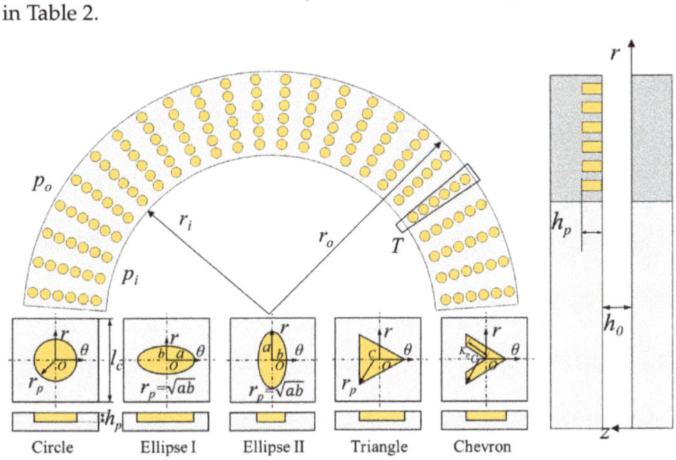

Figure 2. Surface texture patterns on the stator.

Table 2. Structural parameters of textures.

Type	3-D Shape	Definition [25]	Parameter	Value
Circle		$S_p = \frac{\pi r_p^2}{l_c^2}$	Texture depth h_p (μm)	2.6~8.6
ellipse I		$S_p = \frac{\pi ab}{l_c^2}$	Texture control unit l_c (μm)	550
ellipse II		$S_p = \frac{\pi ab}{l_c^2}$	Column number T	300
Triangle		$S_p = \frac{3\sqrt{3}r_p^2}{4l_c^2}$	Texture area ratio S_p	0.1~0.4
Chevron		$S_p = \frac{3\sqrt{3}(1-K_p)r_p^2}{4l_c^2}$	Texture quantity N_t	6

2.2. Multi-Field Coupling Model

2.2.1. Fluid Lubrication Analysis

The following assumptions are made in the fluid lubrication analysis: the fluid flow on the sealing end faces is laminar, the influence of inertial force and volume force can be ignored, the pressure gradient along the thickness direction of the liquid film is unaffected, and the sealing medium is a Newtonian fluid.

Considering the influence of roughness on the sealing performance, the average flow model [27,28] is introduced, and the Reynolds equation can be expressed as

$$\frac{1}{r}\frac{\partial}{\partial\theta}\left(\frac{\phi_\theta h^3}{12\mu}\frac{\partial p}{\partial\theta}\right)+\frac{\partial}{\partial r}\left(\frac{\phi_r r h^3}{12\mu}\frac{\partial p}{\partial r}\right) = \frac{\omega}{2}r\left(\phi_c\frac{\partial h}{\partial\theta}+\sigma\frac{\partial\phi_s}{\partial\theta}\right) \tag{1}$$

where r and θ are the radial coordinate and the circumferential coordinate, respectively, μ is the viscosity of the sealing medium, p is the liquid film pressure, ω is the rotational angular velocity of the rotor, and h is the liquid film thickness, which can be described respectively for the textured zone and nontextured zone [29,30] as follows:

$$h = \begin{cases} h_0 + h_p + \delta \ \text{ Textured zone} \\ h_0 + \delta \ \text{ Non-textured zone zone} \end{cases} \tag{2}$$

where h_0 is the sealing gap, h_p is the texture depth and δ is the deformation of the sealing end faces.

ϕ_r and ϕ_θ are pressure flow factors, expressed as

$$\phi_r = \phi_\theta = 1 - Ce^{-\gamma H} \tag{3}$$

where H is the film thickness ratio ($H = h/\sigma$), σ is the comprehensive root mean square roughness of the end face, and C and γ are constants.

ϕ_s is the shear flow factor, expressed as

$$\phi_s = \left(\frac{\sigma_1^2 - \sigma_2^2}{\sigma^2}\right)\Phi_s \tag{4}$$

where σ_1 and σ_2 are the standard deviations of the surface roughness of the rotor and stator, respectively, and Φ_s can be calculated as [28]

$$\begin{cases} \Phi_s = 1.899 \cdot H^{\alpha_1} e^{-\alpha_2 H + \alpha_3 H^2} & (H \leq 5) \\ \Phi_s = 1.126 \cdot e^{-\alpha_4 H} & (H > 5) \end{cases} \tag{5}$$

where α_1, α_2, α_3, and α_4 are constants.

ϕ_c is the contact factor, which can be expressed as

$$\phi_c = \frac{1}{2}[1 + \mathrm{erf}(\frac{h}{\sqrt{2}\sigma})] \tag{6}$$

Based on the Jakobsson–Floberg–Olsson (JFO) cavitation boundary condition [31,32], the liquid film density ratio ($\Phi = \rho/\rho_c$) is introduced to consider the cavitation effect, where ρ is the density of the sealing medium and ρ_c is the density of the sealing medium at the cavitation zone. Equation (1) can be written as

$$\frac{1}{r}\frac{\partial}{\partial \theta}\left(\phi_\theta h^3 \frac{\partial p}{\partial \theta}\right) + \frac{\partial}{\partial r}\left(\phi_r r h^3 \frac{\partial p}{\partial r}\right) = 6\mu\omega r\left(\phi_c \frac{\partial (h\Phi)}{\partial \theta} + \sigma \frac{\partial \phi_s}{\partial \theta}\right)\begin{cases} \Phi = 1, p > p_c & \text{Complete liquid film zone} \\ 0 < \Phi < 1, p < p_c & \text{Cavitation zone} \end{cases} \tag{7}$$

where p_c is the cavitation pressure.

In this study, the Reynolds Equation (7) is discretized using the finite difference method. The computational domain in the polar coordinate system is partitioned into structured control volume grids, and nodes denoted as (i, j), (i − 1, j), (i, j − 1) are generated at the junction of each control volume grid, as shown in Figure 3a. The nodes along the r direction and θ direction are referred to as radial nodes and circumferential nodes, respectively. Each column on the sealing end faces is divided into N_r nodes along the radial direction and divided into the same unit node length along the circumferential direction to obtain fully structured grids [33,34]. According to the number of textures N_t, each column can be regarded as a composition of N_t evenly distributed units along the radial direction. Given the initial film thickness h_0 and texture shape criterion r_p, the center of the unit is taken as the origin to determine whether the position of each node in each unit is within the texture boundary. Thus, the initial grid nodes of the textured column are generated. Furthermore, the grid refinement and independence verification of the computational domain are performed to enhance the texture boundary treatment. The numerical models of each texture column with the number of radial nodes N_r ranging from 50 to 450 are selected for independence verification. The results show that the error of the numerical results is within 0.5% when the number of radial nodes N_r reaches 300. Therefore, this study adopts the numerical model with 300 radial nodes to conduct the simulation, considering the calculation time and accuracy, as shown in Figure 3b.

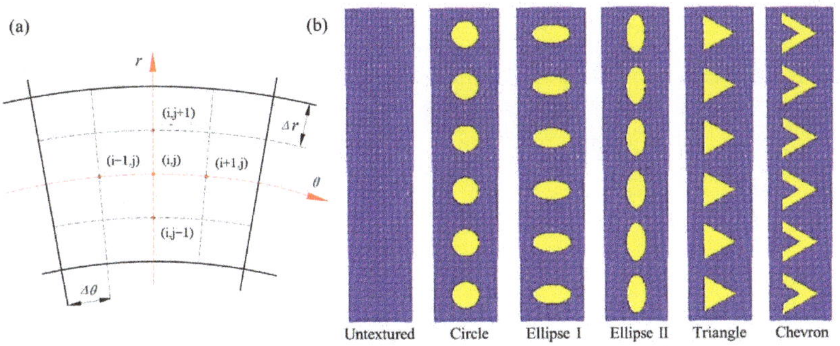

Figure 3. Control volume nodes and grids. (**a**) Control volume nodes; (**b**) Control volume grids.

After discretization, the Reynolds Equation (7) can be written as

$$\mathrm{A}p_{\mathrm{i+1,j}} + \mathrm{B}p_{\mathrm{i-1,j}} + \mathrm{C}p_{\mathrm{i,j+1}} + \mathrm{D}p_{\mathrm{i,j-1}} - \mathrm{E}p_{\mathrm{i,j}} - \mathrm{F}\Phi_{\mathrm{i,j}} + \mathrm{G}\Phi_{\mathrm{i-1,j}} - \mathrm{H} = 0 \quad (8)$$

where A, B, C, D, E, and F are coefficients, expressed as

$$\begin{array}{l} \mathrm{A} = \frac{1}{r_{\mathrm{i,j}}\Delta\theta^2}k_{\mathrm{i+\frac{1}{2},j}} = \frac{1}{r_{\mathrm{i,j}}\Delta\theta^2}(\phi_\theta h^3)_{\mathrm{i+\frac{1}{2},j}},\ \mathrm{B} = \frac{1}{r_{\mathrm{i,j}}\Delta\theta^2}k_{\mathrm{i-\frac{1}{2},j}} = \frac{1}{r_{\mathrm{i,j}}\Delta\theta^2}(\phi_\theta h^3)_{\mathrm{i-\frac{1}{2},j}}, \\ \mathrm{C} = \frac{1}{r_{\mathrm{i,j}}\Delta\theta^2}k_{\mathrm{i,j+\frac{1}{2}}} = \frac{1}{\Delta r^2}(\phi_r r h^3)_{\mathrm{i,j+\frac{1}{2}}},\ \mathrm{D} = \frac{1}{r_{\mathrm{i,j}}\Delta\theta^2}k_{\mathrm{i,j-\frac{1}{2}}} = \frac{1}{\Delta r^2}(\phi_r r h^3)_{\mathrm{i,j-\frac{1}{2}}}, \\ \mathrm{E} = \mathrm{A}+\mathrm{B}+\mathrm{C}+\mathrm{D},\ \mathrm{F} = \frac{6\mu\omega r}{\Delta\theta}(\phi_c h)_{\mathrm{i,j}},\ \mathrm{G} = \frac{6\mu\omega r}{\Delta\theta}(\phi_c h)_{\mathrm{i-1,j}},\ \mathrm{H} = 3\mu\omega r\sigma\frac{(\phi_s)_{\mathrm{i+1,j}} - (\phi_s)_{\mathrm{i-1,j}}}{\Delta\theta} \end{array} \quad (9)$$

where $k_{\mathrm{i+\frac{1}{2},j}}$, $k_{\mathrm{i-\frac{1}{2},j}}$, $k_{\mathrm{i,j+\frac{1}{2}}}$, and $k_{\mathrm{i,j-\frac{1}{2}}}$ are node diffusion coefficients.

Due to the significant increase in the film thickness gradient near the texture boundary, the node diffusion coefficients are introduced and discretized in an arithmetic average method to improve the calculation accuracy [35].

The liquid film thicknesses (h) of diffusion coefficients are discretized using an arithmetic average method, given as

$$\begin{array}{l} h_{\mathrm{i-1/2,j}} = (h_{\mathrm{i,j}} + h_{\mathrm{i-1,j}})/2,\ h_{\mathrm{i+1/2,j}} = (h_{\mathrm{i,j}} + h_{\mathrm{i+1,j}})/2 \\ h_{\mathrm{i,j-1/2}} = (h_{\mathrm{i,j}} + h_{\mathrm{i,j-1}})/2,\ h_{\mathrm{i,j+1/2}} = (h_{\mathrm{i,j}} + h_{\mathrm{i,j+1}})/2 \end{array} \quad (10)$$

Furthermore, the liquid pressure (p) and the liquid film density ratio (Φ) can be expressed as

$$\begin{array}{l} p_{\mathrm{i,j}} = \frac{\mathrm{A}p_{\mathrm{i+1,j}} + \mathrm{B}p_{\mathrm{i-1,j}} + \mathrm{C}p_{\mathrm{i,j+1}} + \mathrm{D}p_{\mathrm{i,j-1}} - \mathrm{F}\Phi_{\mathrm{i,j}} + \mathrm{G}\Phi_{\mathrm{i-1,j}} - \mathrm{H}}{\mathrm{E}} \\ \Phi_{\mathrm{i,j}} = \frac{\mathrm{A}p_{\mathrm{i+1,j}} + \mathrm{B}p_{\mathrm{i-1,j}} + \mathrm{C}p_{\mathrm{i,j+1}} + \mathrm{D}p_{\mathrm{i,j-1}} - \mathrm{E}p_{\mathrm{i,j}} + \mathrm{G}\Phi_{\mathrm{i-1,j}} - \mathrm{H}}{\mathrm{F}} \end{array} \quad (11)$$

2.2.2. Mechanical Analysis

When the metal seals of the roller cone bit work stably, the opening force (F_{open}) and closing force (F_{close}) between the end faces of the stator and rotor are in balance, and a stable liquid film forms in the sealing gap. Since the metal seals are contact seals, the opening force (F_{open}) comprises the above bearing force of liquid film (F_{p}) and the contact force of the asperities (F_{c}). The contact force (F_{c}) can be characterized using the plastic contact model [36,37] as follows:

$$F_{\mathrm{c}} = \int_0^{2\pi}\int_{r_{\mathrm{i}}}^{r_{\mathrm{o}}} p_{\mathrm{c}} r \mathrm{d}r \mathrm{d}\theta = H2\pi\int_{r_{\mathrm{i}}}^{r_{\mathrm{o}}}\left[\int_h^{\infty}\frac{1}{\sigma\sqrt{2\pi}}\mathrm{e}^{\left(-z^2/2\sigma^2\right)}\mathrm{d}z\right] r\mathrm{d}r \quad (12)$$

where p_{c} is the contact pressure, H is the flow stress (yield stress or compressive strength) of the softer face material, and z is the axial coordinate.

The closing force (F_{close}) is produced through the support of the O-ring and backup ring to the floating stator, as well as the influence of the inner and outer pressure. It can be determined by integrating the static contact pressure (p_{sc}) on the end faces of metal seals. Considering the significant nonlinearity and complex structural arrangement of the rubber O-ring and backup ring, a two-dimensional axisymmetric finite element numerical model of metal seals is established via ANSYS 15.0 software to conduct the mechanical analysis, as shown in Figure 4.

Both the metal rotor and stator are made of alloy steel 20CrNiMo with the elastic modulus E = 2.03 × 10^5 MPa and Poisson's ratio ν = 0.3 [38]. The rubber O-ring and backup ring use the typical hyperelastic material fluororubbers (FKM). The two-parameter Mooney–Rivlin hyperelastic model is adopted to describe the strain energy function of the rubber, given as [39,40]

$$W = C_{10}(\bar{I}_1 - 3) + C_{01}(\bar{I}_2 - 3) \quad (13)$$

where $\bar{I}_1$ and $\bar{I}_2$ are the first and second invariants of stress tensors, respectively, and C_{10} and C_{01} are the parameters, with C_{10} = 1.444 and C_{01} = 0.0165 for the O-ring and C_{10} = 1.833 and C_{01} = −0.003 for the backup ring.

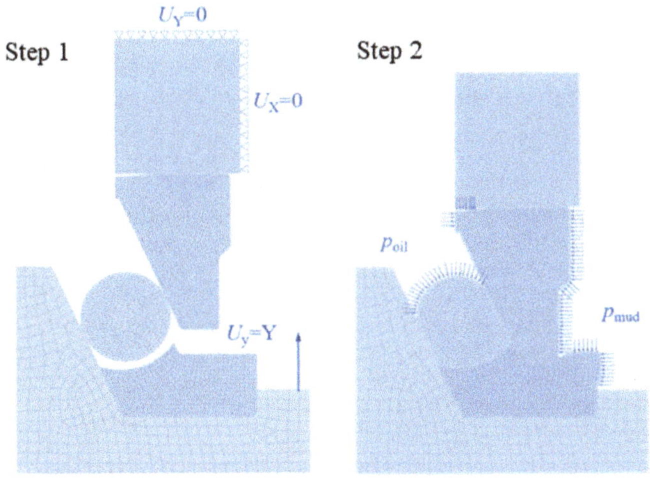

Figure 4. Finite element model and boundary conditions of metal seals.

The PLANE182 axisymmetric solid element is adopted in the analysis, and the CONTA172 element and TARGE169 element are adopted for low hardness contact surfaces and high hardness target surfaces, respectively. The finite element analysis procedure consists of the following steps (as shown in Figure 4): Step 1 entails pre-compressing the metal seals. The axial displacement constraint (U_y = Y) is applied to the journal, and the radial and axial fixed constraint boundary conditions (U_y = 0, U_X = 0) are applied to the outer diameter side and back of the rotor, respectively. Step 2 entails applying the lubricant oil pressure and mud pressure on the inner and outer sides of the metal seals. After the numerical solution, the static contact pressure (p_{sc}), close force (F_{close}), and initial deformation of the end faces on the metal seals can be obtained under environmental pressure.

2.2.3. Thermal Analysis

The following assumptions are made in the thermal analysis: the thermal conductivities of the sealing rings are constant, and the heat losses carried away by the leakage medium and caused by radiation are unaffected.

The heat generation rate (q) between the end faces of metal seals can be determined via the viscous shear heat of liquid film and the frictional heat of asperity contact, given as [41]

$$q = \int_0^{2\pi} \int_{r_i}^{r_o} \mu \frac{r^2\omega^2}{h} r\mathrm{d}r\mathrm{d}\theta + \int_0^{2\pi} \int_{r_i}^{r_o} f_0 p_c r\omega r\mathrm{d}r\mathrm{d}\theta \quad (14)$$

where ω is the rotational angular velocity, and f_0 is the dry friction coefficient [42,43].

The rotor is convectively heated with the lubricant oil on the inner side, and the contact zones between the top and outer sides of the rotor and the cone are regarded as constant temperature boundaries. Similarly, the stator is convectively heated with the drilling mud on the outer side and the lubricant oil on the inner side, and the bottom and lower part of the inner side of the stator are regarded as adiabatic boundaries.

The heat conduction equation of the sealing rings is

$$\frac{\partial}{\partial r}\left(k\frac{\partial T}{\partial r}\right) + \frac{\partial}{\partial z}\left(k\frac{\partial T}{\partial z}\right) = 0 \quad (15)$$

where T is the temperature of the sealing ring, and k is the thermal conductivity of the ring material.

The boundary conditions of heat flux and convective heat transfer are

$$k\frac{\partial T}{\partial n_{\mathrm{c}}}\bigg|_{S_1} - q = 0 \tag{16}$$

$$k\frac{\partial T}{\partial n_{\mathrm{c}}}\bigg|_{S_2} + h_{\mathrm{c}}(T - T_0) = 0 \tag{17}$$

where S_1 and S_2 are the corresponding surfaces, h_{c} is the convective heat transfer coefficient, T_0 is the environmental temperature, and n_{c} is the exterior normal direction.

2.2.4. Deformation Analysis

The liquid film thickness of the metal seals is directly related to the deformations of the end faces. The total deformation of the sealing end faces (δ) is composed of mechanical deformation (δ_{m}) and thermal deformation (δ_{t}). The microscopic deformation of the stator and rotor can be solved using the influence factor method [41].

The total deformation of a radial node i on end faces is

$$\delta_{\mathrm{i}} = DI_{\mathrm{i}} + DR_{\mathrm{i}} + DS_{\mathrm{i}} \tag{18}$$

where DI_{i} is the initial deformation at radial node i, and DR_{i} and DS_{i} are the deformations of the rotor and stator at radial node i, respectively, expressed as

$$DR_{\mathrm{i}} = \sum_{j=1}^{n} MR_{\mathrm{ij}}F_{\mathrm{j}} + \sum_{j=1}^{n} TR_{\mathrm{ij}}q_{\mathrm{j}} \tag{19}$$

$$DS_{\mathrm{i}} = \sum_{j=1}^{n} MS_{\mathrm{ij}}F_{\mathrm{j}} + \sum_{j=1}^{n} TS_{\mathrm{ij}}q_{\mathrm{j}} \tag{20}$$

where F_{j} is the force acting on the node j, q_{j} is the heat generation rate, MR_{ij} and MS_{ij} are the influence factor matrices of the mechanical deformation of the rotor and stator, respectively, and TR_{ij} and TS_{ij} are the influence factor matrices of the thermal deformation of the rotor and stator, respectively.

Based on the finite element analysis of metal seals in Section 2.2.2, the influence factor matrices can be obtained by applying the unit load to the nodes one by one in the mechanical analysis and thermodynamic analysis modules of the ANSYS 15.0 software, as shown in Figure 5.

2.3. *Performance Parameters*

The key performance parameters of metal seals, including the liquid film bearing coefficient K_{f}, leakage rate Q, and frictional force F_{f}, can be expressed as [44]

$$K_{\mathrm{f}} = \frac{F_{\mathrm{p}}}{F_{\mathrm{open}}} = \frac{\int_0^{2\pi}\int_{r_{\mathrm{i}}}^{r_{\mathrm{o}}} pr\mathrm{d}r\mathrm{d}\theta}{\int_0^{2\pi}\int_{r_{\mathrm{i}}}^{r_{\mathrm{o}}} pr\mathrm{d}r\mathrm{d}\theta + \int_0^{2\pi}\int_{r_{\mathrm{i}}}^{r_{\mathrm{o}}} p_{\mathrm{c}}r\mathrm{d}r\mathrm{d}\theta} \tag{21}$$

$$Q = \int_0^{2\pi} -\frac{rh^3}{12\mu}\frac{\partial p}{\partial r}\mathrm{d}\theta \tag{22}$$

$$F_{\mathrm{f}} = \frac{1}{r_{\mathrm{m}}}\int_0^{2\pi}\int_{r_{\mathrm{i}}}^{r_{\mathrm{o}}} \tau r^2\mathrm{d}r\mathrm{d}\theta \tag{23}$$

where r_{m} is the average radius of the sealing zone.

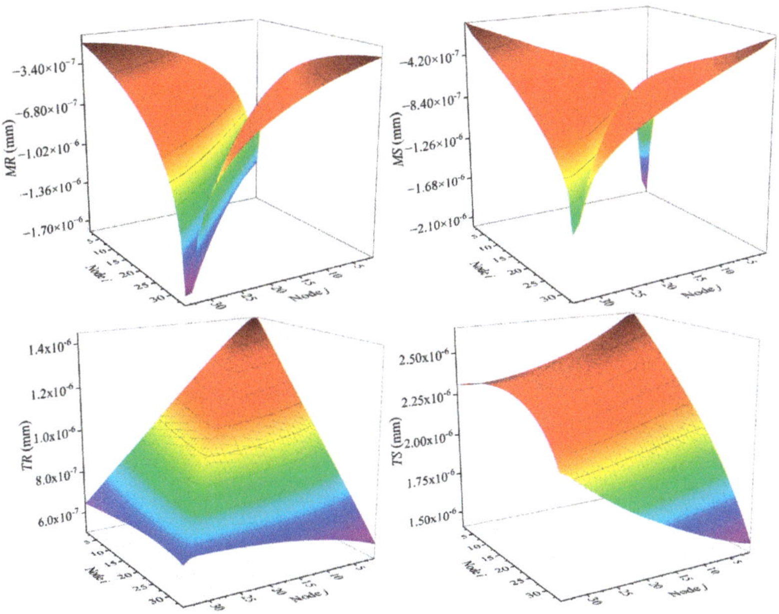

Figure 5. Influence factor matrices under initial conditions.

3. Calculation Procedure

According to the above analysis, a complex thermal–fluid–solid coupling relationship exists between the sealing rings, the surface texture, and the liquid film on the end faces of the roller cone bits. The mechanical deformations on the end faces of the rotor and stator are influenced by the liquid film pressure, viscous shear force obtained via fluid lubrication analysis, and the contact pressure obtained via mechanical analysis. Consequently, the micro-interface deformations subsequently impact the state of the liquid film and contact mechanics. The interface temperature obtained via thermal analysis induces alterations in both the viscosity of the liquid film and the thermal deformations of the end faces, thus affecting the results of fluid lubrication analysis and mechanical analysis.

Figure 6 is a flowchart of the multi-field coupling numerical model of surface-textured metal seals. The present study utilizes a combined approach involving finite element analysis and programming computation to facilitate the thermal–fluid–solid coupling of metal seals. The static contact pressure between the sealing faces, the initial deformation of the rotor and stator, and the influence factor matrices are first solved and substituted into the program. The numerical iterative program and the finite difference method deal with the strong coupling between the fluid field, mechanical field, and thermal field of the metal seals. A total of three loops are nested in the whole coupling process. The inner cycle solves the liquid film thickness (h). The middle cycle adjusts the average liquid film thickness (h_m) to satisfy the force balance between the end faces of the rotor and stator. The outer cycle adjusts the interface temperature (T) on the end faces. Finally, the end face deformation, film thickness, and interface temperature are converged. Then, the fluid film characteristics and sealing performance parameters of surface-textured metal seals can be obtained.

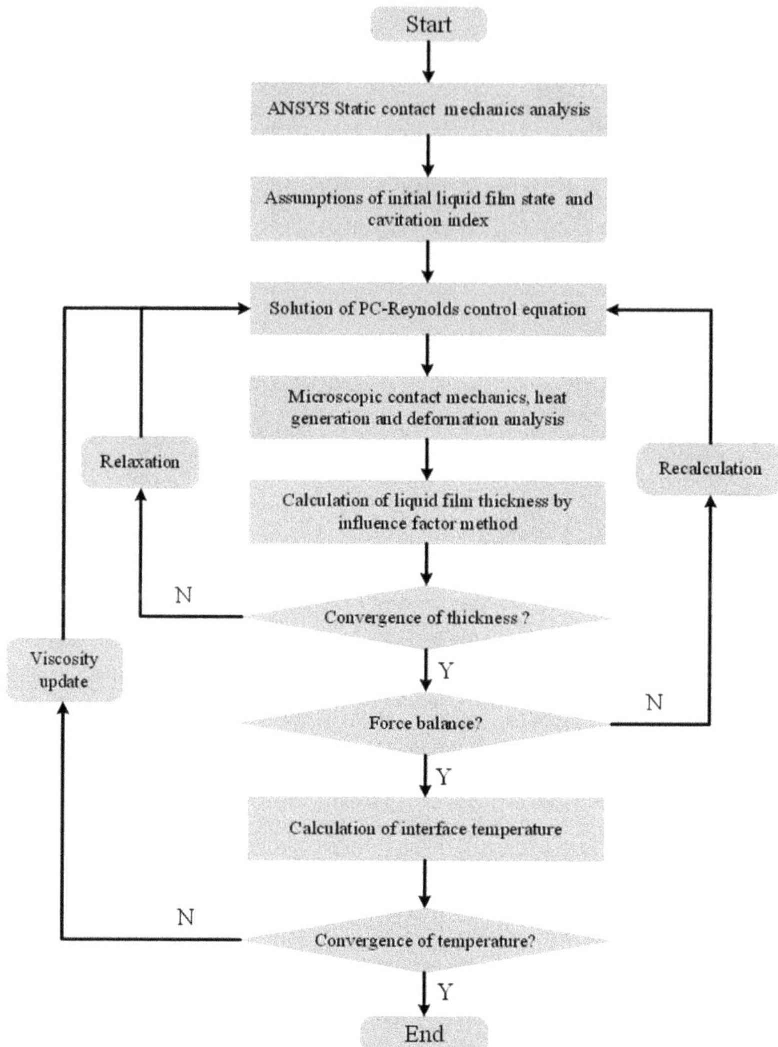

Figure 6. Flowchart of multi-field coupled numerical model.

4. Results and Discussions

4.1. Liquid Film Pressure on End Faces

Figure 7 shows the distribution of liquid film pressure (p) in a single column of the end faces of metal seals with different shape textures under the initial conditions of environmental pressure p_{oi} = 3 MPa, pressure difference Δp_i = 0.5 MPa, rotational speed n_i = 200 rpm, the texture depth h_p = 3 μm, and the area ratio S_p = 0.2. Different from the gradual decline in the liquid film pressure on the untextured flat end face, the presence of micro-textures on metal seals results in the formation of textured convergent and divergent zones along the rotation direction. This strong hydrodynamic effect leads to higher peak values of liquid film pressure (p_{max}) in the textured convergent zone, reaching approximately 3.9 to 5.62 times the environmental pressure. This magnitude far exceeds that observed in untextured metal seals. Hence, the textured end face can provide a specific bearing capacity of liquid film. As a consequence of the rotational shear exerted on the end faces, the fluid density and pressure in the divergent zone of the textured liquid film significantly decrease, leading to the occurrence of cavitation at the textured zones. The extent of cavitation, which varies depending on the shape of the textures, accounts for approximately 5% to 27% of the overall texture area.

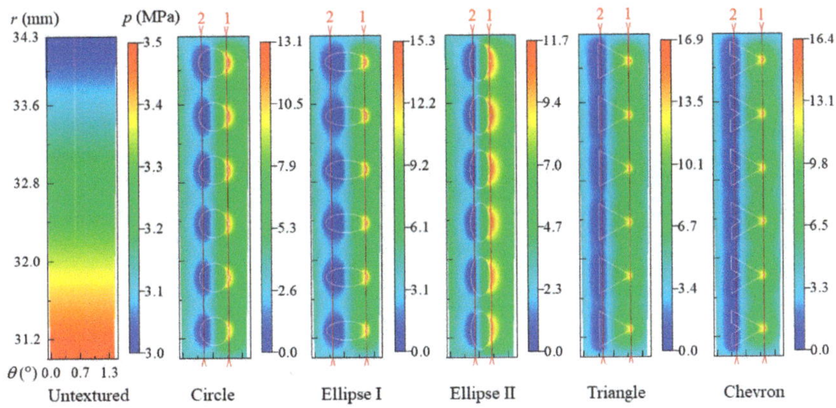

Figure 7. Distribution contours of fluid film pressure with different shape textures.

The middle intercepting lines 1 and 2 of the textured convergent zone and divergent zone of liquid film on the end faces are extracted, respectively, as shown in Figure 7. The distributions of liquid film pressure along the two intercepting lines with radial position (r) are shown in Figure 8. For line 1, the radial liquid film pressures (p) in the convergent zone of the triangle- and chevron-shaped textures are sharper and higher than those of the circle-, ellipse I-, and ellipse II-shaped textures. Notably, the triangle-shaped textures obtain the highest peak pressure of the liquid film ($p_{\max}$) among the five shape textures. In comparison to the circle- and ellipse-shaped textures, the triangle- and chevron-shaped textures exhibit a pronounced angularity in the direction of the fluid flow, thereby facilitating a more significant hydrodynamic effect in the convergent zone of the liquid film. For line 2, the liquid film pressure (p) at the bottom of these textures diminishes to zero. The distributions of liquid film pressure in the divergent zone of the circle-, ellipse I- and ellipse II-shaped textures are comparable, with their film pressure values surpassing those of the triangle and chevron textures. The sharp-angle texture on the end faces of metal seals promotes the concentration and divergence of the liquid film pressure.

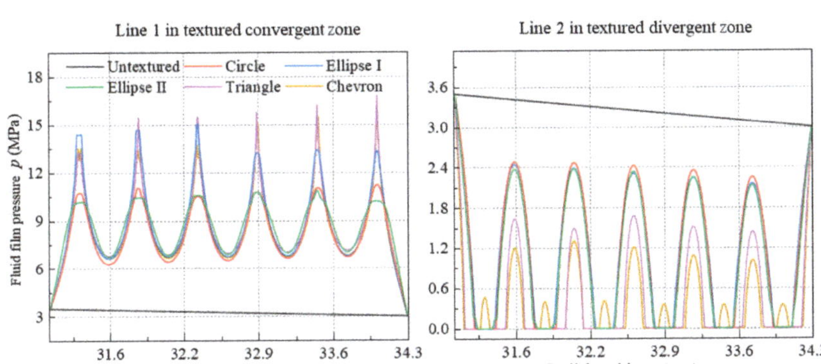

Figure 8. Distributions of film pressure in textured zone.

4.2. Influence of Operating Conditions on Sealing Performance

Figures 9 and 10 show the key sealing parameters of metal seals with different shape textures, such as the liquid film bearing coefficient (K_f), maximum interface temperature rise (ΔT_{max}), leakage rate (Q), and frictional force (F_f), respectively, under different rotational speeds (n = 100~500 rpm) and environmental pressures (p_o = 3~69 MPa), corresponding to a formation depth of l = 260~6000 m.

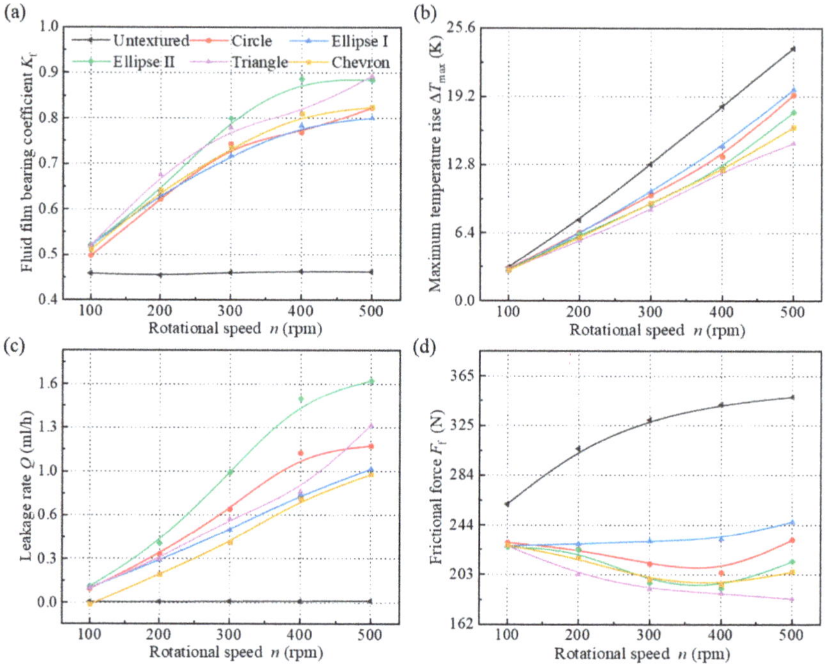

Figure 9. Sealing performance of metal seals under different rotational speeds. (**a**) Fluid film bearing coefficient; (**b**) Maximum temperature rise; (**c**) Leakage rate; (**d**) Frictional force.

As the rotational speed (n) increases from 100 rpm to 500 rpm, the hydrodynamic pressure effects and liquid film bearing coefficients (K_f) of the textured end faces gradually increase, and the leakage rates (Q) increase significantly due to the increase in film thickness. In contrast, the liquid film bearing coefficient (K_f) and leakage rate (Q) of the untextured end face undergo minimal alterations. Furthermore, the liquid film bearing capacities of the triangle- and ellipse II-textured end faces are greater than those of the other three textured end faces. However, the leakage rate (Q) of the ellipse II-textured end face exhibits the highest value among the five different textured end faces. The maximum temperature rise (ΔT_{max}) of both the untextured and textured end faces increase approximately linearly with the rotational speed (n), while the variation in frictional force (F_f) on the end faces differs greatly with the rotational speed (n). The untextured end face generates substantial contact pressure and frictional heat due to its minimal film thickness, resulting in significantly larger values of maximum temperature rise and frictional force (ΔT_{max}, F_f) compared to those of the textured end faces. The order of the texture shapes, namely, ellipse I, circle, ellipse II, chevron, and triangle, correspond to the decreasing maximum temperature rise and frictional force (ΔT_{max}, F_f). Specifically, when the rotational speed (n) is 500 rpm, the frictional force (F_f) of the triangle-textured end face decreases to 52.7% and to 74.5% regarding the untextured and ellipse I-textured end faces, respectively. This indicates that the triangle texture enhances the bearing capacity of the liquid film, making it easier to open the sealing end face and resulting in reduced contact frictional heat.

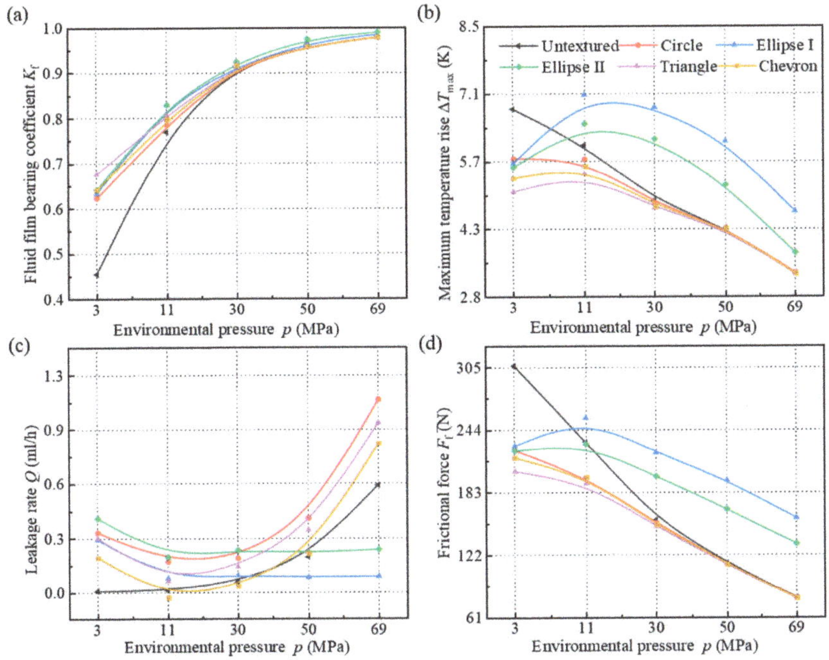

Figure 10. Sealing performance of metal seals under different environmental pressures. (**a**) Fluid film bearing coefficient; (**b**) Maximum temperature rise; (**c**) Leakage rate; (**d**) Frictional force.

With there is an increase in environmental pressure (p) from 3 MPa to 69 MPa, the liquid film bearing coefficient (K_f) and leakage rates (Q) of the untextured end face increase as a result of the radial redistribution of the film thickness, leading to a notable reduction in frictional heat between the sealing rings. The liquid film bearing coefficients (K_f) of the textured end faces are larger than that of the untextured end face and gradually increase up to 98.9%. It is noteworthy that the hydrodynamic effect of the textured end faces diminishes gradually when the environmental pressure is $p \geq 30$ MPa, resulting in liquid film bearing coefficients that are no longer superior to that of the untextured end face. This can be attributed to the substantial alterations in film thickness and contact pressure, leading to a fluctuation in the leakage rates (Q) of the circle-, triangle- and chevron-textured end faces. The leakage rates (Q) initially decrease and then increase with the environmental pressure (p), ultimately reaching the minimum values at $p = 11$ MPa. The maximum temperature rise and frictional force (ΔT_{max}, F_f) of these three textured end faces decrease with the environmental pressure (p), consistently remaining lower than those of the untextured end face. It can effectively weaken the wear of the sealing end faces. The leakage rates (Q) of the ellipse I- and ellipse II-textured end faces gradually decrease to a stable value with the environmental pressure (p). However, except for the low-pressure condition, the ellipse I- and ellipse II-textured end faces exhibit higher maximum temperature rise and frictional force (ΔT_{max}, F_f) compared to the untextured, circle-, triangle-, and chevron-textured end faces. This can be attributed to the significant shear effect of the liquid film on the ellipse-textured end faces, resulting in an inferior sealing performance under high-pressure conditions. Therefore, the influence of surface textures on hydrodynamic pressure is crucial for enhancing the sealing efficacy in low-pressure conditions. As the environmental pressure increases, the hydrodynamic pressure effect induced by the textures weakens, while the circle-, triangle- and chevron-shaped textures still help to reduce the contact pressure and frictional heat of the sealing rings.

4.3. Influence of Texture Parameters on Sealing Performance

Based on the above comparisons, the triangle-textured end face has the best comprehensive sealing performance among the five textured end faces. Considering that the hydrodynamic effect is closely related to the texture parameters, the influences of texture depth (h_p) and area ratio (S_p) are further investigated.

Figures 11 and 12 show the fluid film pressure and sealing parameters of metal seals under different texture depths (h_p = 2.6~8.6 μm) and area ratios (S_p = 0.1∼0.4), respectively. With the increase in the texture depth (h_p), the hydrodynamic pressure effect resulting from the fluid extrusion in the textured convergent zone and the cavitation area caused by the fluid shear in the textured divergent zone of the liquid film both weaken. The pressure distribution of the fluid film on the triangle-textured end face becomes more uniform with the increase in h_p. Additionally, the peak value of liquid film pressure (p_{max}) at h_p = 8.6 μm is 71.6% lower compared to h_p = 2.6 μm. Consequently, the liquid film bearing coefficient (K_f) and the opening degree of the sealing end face decrease significantly. In particular, the leakage rate (Q) of metal seals reduces rapidly to less than 0.1 mL·h^{-1}. However, the maximum temperature rise and frictional force (ΔT_{max}, F_f) of the triangle-textured end face increase and gradually stabilize after $h_p \geq$ 4.3 μm.

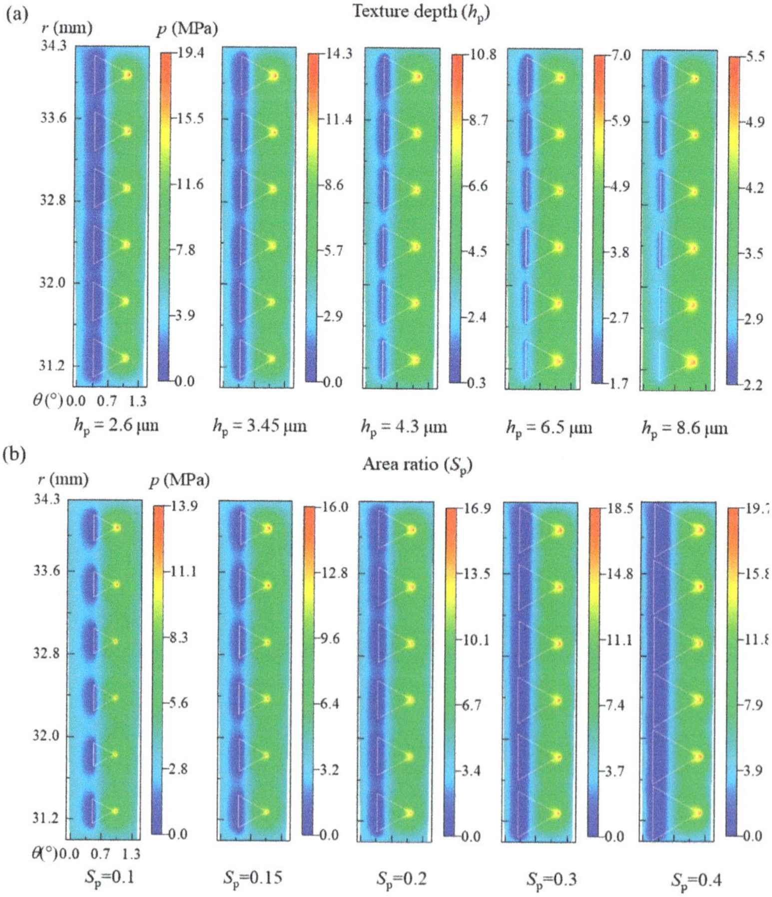

Figure 11. Contours of fluid film pressure under different texture parameters. (**a**) Texture depth; (**b**) Area ratio.

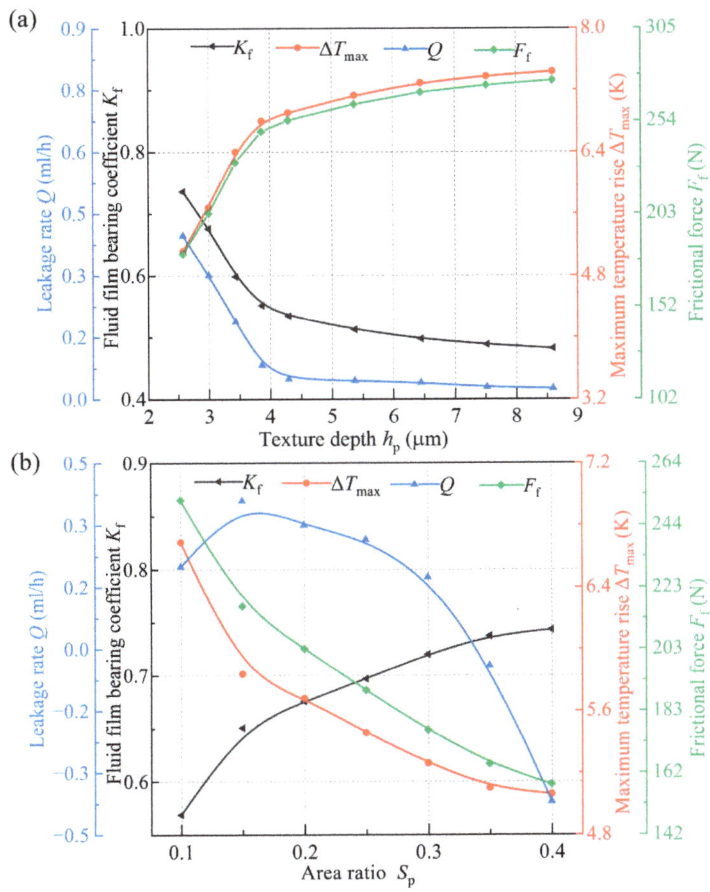

Figure 12. Sealing performance under different texture parameters. (**a**) Texture depth; (**b**) Area ratio.

As the area ratio (S_p) increases from 0.1 to 0.4, the textured convergent zone and divergent zone of the liquid film continue to expand. This expansion is accompanied by a heightened hydrodynamic effect and shear effect, resulting in a 41.9% increase in the peak pressure (p_{max}) and a 17.4% increase in the bearing coefficient of the liquid film (K_f) on the triangle-textured end face. Accordingly, the maximum temperature rise and frictional force (ΔT_{max}, F_f) decrease with the attenuated contact area. It is noteworthy that the leakage rate (Q) of the metal seals increases first and then decreases with the area ratio (S_p). Eventually, a phenomenon known as 'negative leakage' emerges. Specifically, when S_p increases from 0.3 to 0.35, the leakage rate (Q) changes from positive to negative, and the 'zero leakage' of metal seals can be obtained theoretically at $S_p = 0.34$ and $h_p = 3$ μm. Increasing the texture area ratio makes the textures gradually approach the outer diameter side of the end faces, resulting in a certain degree of directional pumping effects. This leads to the invasion of the external drilling mud and accelerates the wear failure on the end faces of metal seals.

4.4. Experimental Validations

A performance test bench was constructed to validate the accuracy of the simulation results for surface-textured metal seals in roller cone bits, as shown in Figure 13. The test bench mainly consists of the power transmission, pressure supply, sealing test, and data acquisition systems. The power transmission system controls the rotating speed of the motor shaft by adjusting the frequency of the inverter cabinet. The pressure supply system provides the sealing medium for the test cavities. The sealing test system includes a metal sealing cavity and an auxiliary sealing cavity. The two sealing cavities collaborate with the sleeve and the end cover to form the inner and outer cavities, characterized by a specific pressure differential. The data acquisition system can dynamically collect the sealing performance parameters including sealing pressure, end face temperature, and other relevant signals during the test.

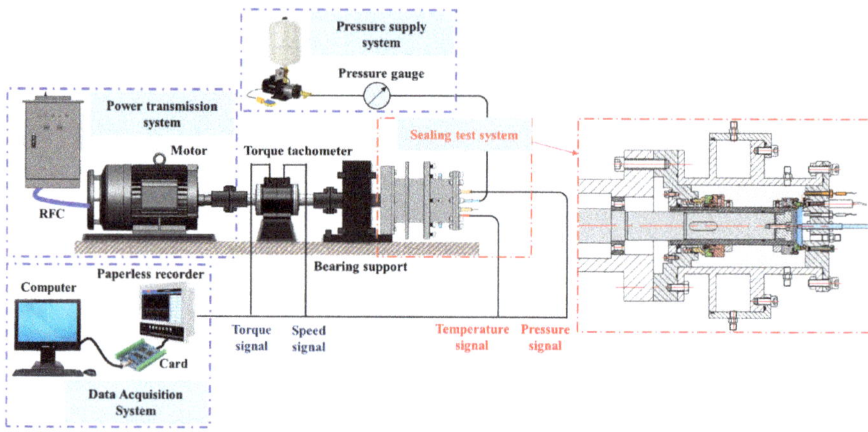

Figure 13. Performance test bench of metal seals.

Laser processing technology, one of the most widely used surface texture processing technologies, was used to process the surface textures of various shapes on the stator of metal seals. A flowchart of the surface texture processing of metal seals is shown in Figure 14. The materials of the rotor and stator in the experiment are the same as those in the numerical calculation, that is, alloy steel 20CrNiMo. The texture shapes of circle, ellipse I, ellipse II, triangle, and chevron with an area ratio S_p = 0.2 are adopted and compared with the untextured flat end face of metal seals. The circumferential and radial arrangements of the textures in the experiment are consistent with those in the numerical calculation.

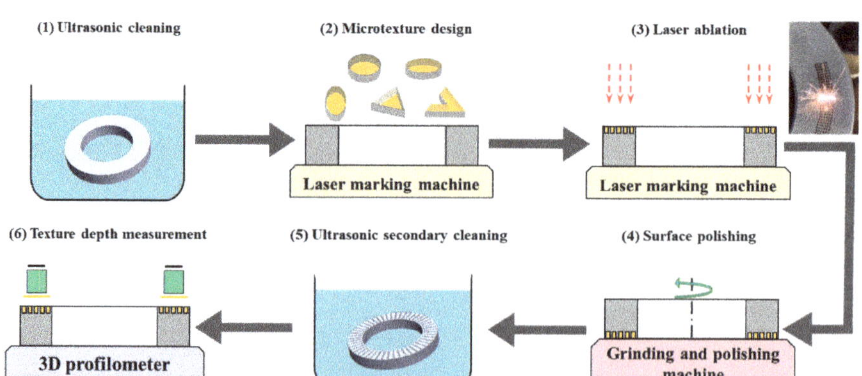

Figure 14. Surface texture processing.

Considering the safety and feasibility of the test, a comprehensive performance test of metal seals with different shape textures was carried out under the pressure difference of Δp = 0.5 MPa, the sealing medium of water, and the rotational speed range of n = 100~500 rpm. During the experiment, we adjusted the rotational speed n to 100, 200, 300, 400, and 500 rpm sequentially after the temperature on the end face stabilized at each stage. The test time took about 30 min. After the experiment, the surface wear morphology of the sealing rings was measured using a 3D laser microscope to verify the accuracy of the multi-field coupling simulation of surface-textured metal seals.

Figure 15 shows the time-varying curves of the temperature rise of the untextured and textured end faces of metal seals under different rotational speeds (n). Figure 16 compares the interface temperature rise of the untextured and textured end faces of metal seals obtained via numerical simulation and experiments under the same experimental conditions. With an increase in rotational speed (n) from 100 rpm to 500 rpm, the contact frictional heat and viscous shear heat between the end faces of the rotor and stator increase. Both the numerical and experimental average values of the interface temperature rise (ΔT) increase. Notably, the values and change gradient of the temperature rise for the untextured end face are significantly higher than those of the textured end faces, and it requires a lengthier duration for the temperature rise to reach stability. The experimental results prove that the textures on the end faces can indeed reduce the asperity friction and temperature rise of the metal sealing rings. For the textured end faces, the experimental temperature rise of the ellipse I- and ellipse II-textured end faces are comparable to those of the circle- and chevron-textured end faces, respectively. Moreover, when the rotational speed (n) reaches 500 rpm, the triangle-textured end face has the lowest temperature rise among these textured end faces. The observed phenomenon aligns with the predominant dynamic pressure impact of triangle textures as determined through numerical simulation. In contrast, the experimental values of interface temperature rise (ΔT) are lower than the numerical values. This discrepancy can be attributed to the steady-state assumptions employed in the numerical simulations of metal seals, whereas actual operational tests are subject to diverse external factors including shaft vibration, fluctuations in film thickness, and interference in heat transfer. In general, the multi-field coupling numerical method of metal seals in this study is suitable for reflecting the sealing performance and its variation law with different parameters.

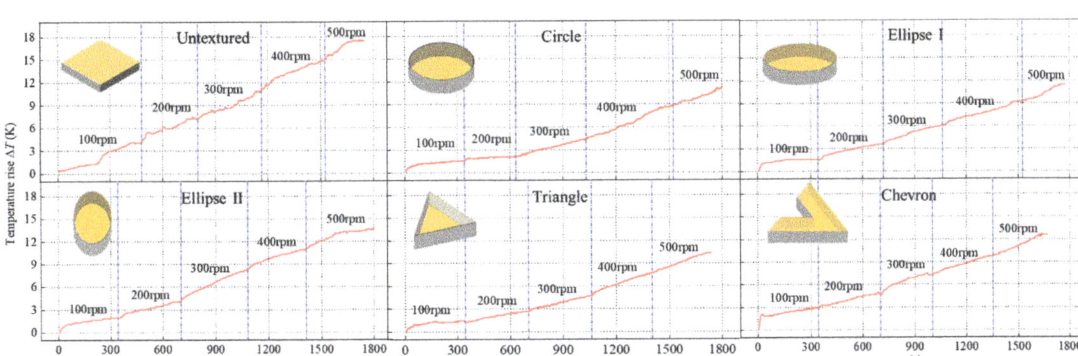

Figure 15. Time-varying curves of temperature rise with different textures.

The worn surface morphologies of the inner diameter (ID) and outer diameter (OD) side of the rotor and stator with the untextured and textured end surfaces are shown in Figure 17. The rotor and stator with an untextured end face experience significant wear due to the insufficient bearing capacity of the liquid film. The direct contact between the end faces exacerbates the semi-dry friction state of the untextured metal seals, and the seals are close to failure after the friction pairs run for a certain time. Affected by the convergent sealing gap at low environmental pressure, the wear on the OD of the rotor and stator is generally more serious than that on the ID of the rotor and stator. Similar to some contact mechanical seals [45,46], obvious wear grooves parallel to the sliding direction and material losses occur on the ID and OD of untextured end faces. Corresponding to the changes in temperature rise with the rotational speed, clear indications of adhesive wear are evident on the untextured surfaces of the rotor and stator. A portion of the wear debris adheres to the untextured end face, exacerbating the wear of sealing rings. The existence of textures makes a thicker lubricating liquid film form between the end faces so that the wear depth and range of the textured end faces of the rotor and stator significantly weaken in a mixed friction state. Different degrees of wear debris, grooves, and material shedding can be observed on the five textured end faces. Take the circle-textured end faces as an example. Deep grooves and pits are generated on the OD of the stator during the frictional process, and the wear debris is transferred and accumulated on the OD of the rotor. The grooves on the ID are shallower than those on the OD of the rotor and stator. Due to the ability to produce a greater liquid film bearing capacity, the triangle-textured end faces have a greater opening degree between the rotor and stator. This improvement significantly enhances the wear resistance of the friction pairs. Black abrasive grooves and material attachment appear on the OD of the rotor, and the wear scars and material shedding on the ID of the rotor are greatly reduced. Among the five different shapes of textured end faces, the triangle-textured end faces demonstrate the smallest wear area and quantity. Therefore, the texture design of metal seals is conducive to reducing the temperature rise, friction, and wear of the end faces and has the effect of accommodating abrasive particles and external intrusion mud.

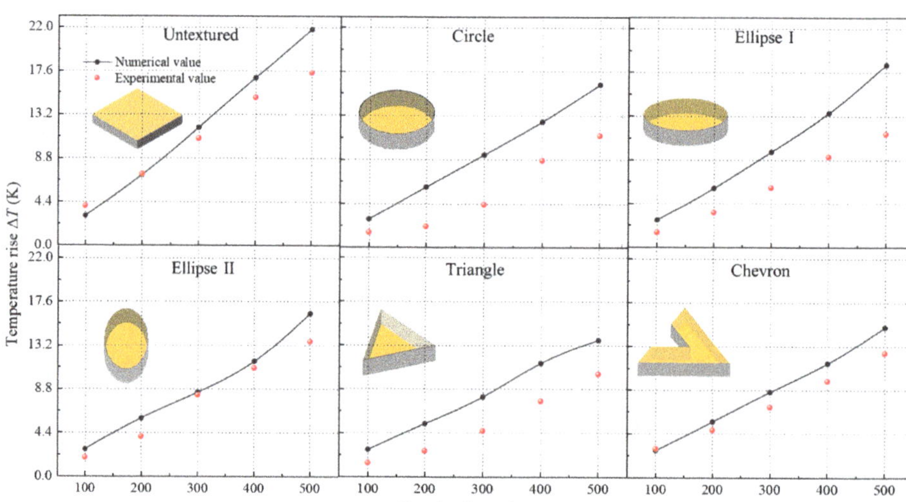

Figure 16. Comparison of temperature rises with different textures.

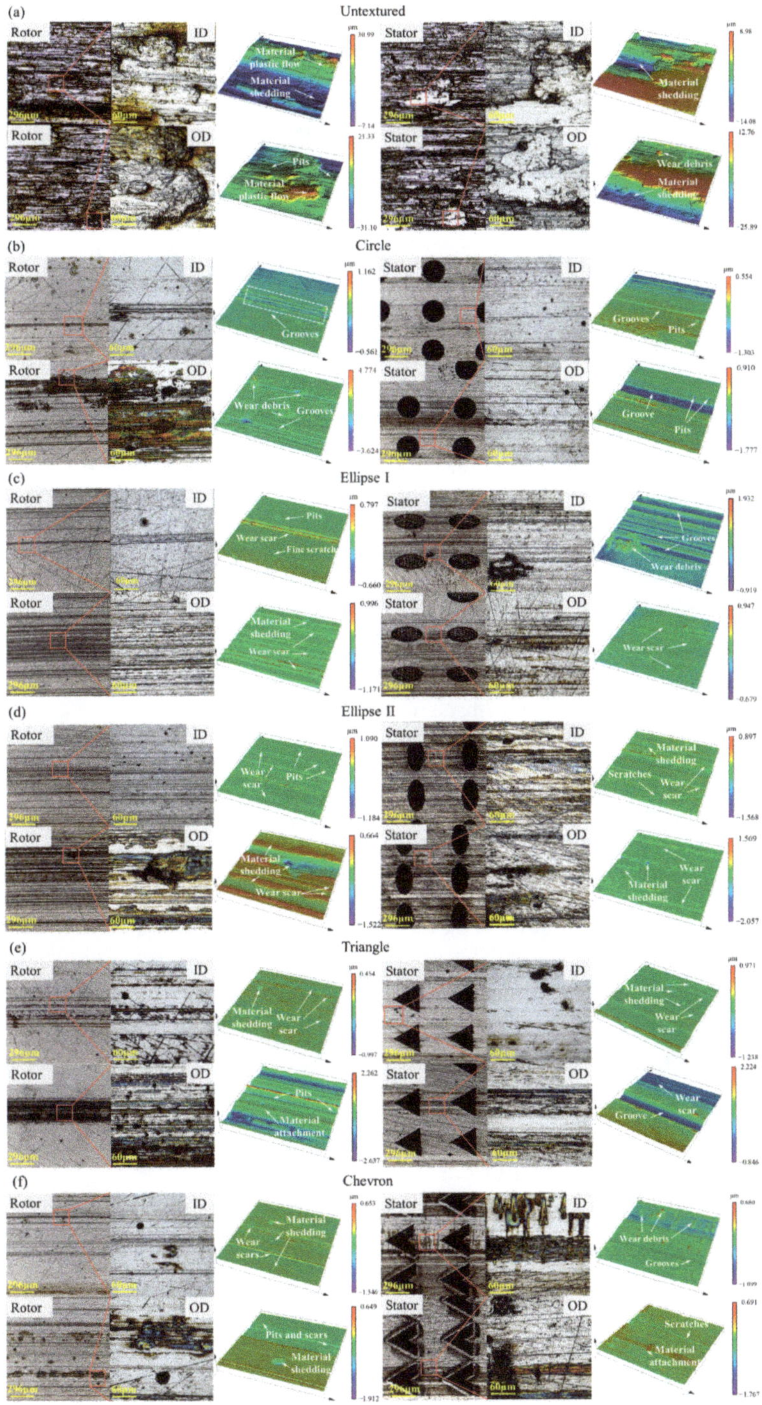

Figure 17. Worn surface morphology of rotor and stator with different shape textures. (**a**) Untextured; (**b**) Circle; (**c**) ellipse I; (**d**) ellipse II; (**e**) Triangle; (**f**) Chevron.

5. Conclusions

This study employs numerical simulation and experimental validation to examine the effectiveness of surface textures applied to the metal seals of roller cone bits. The liquid film state, sealing performance, and wear characteristics of surface-textured metal seals are studied, considering the various texture shapes of circle, ellipse I (horizontal), ellipse II (vertical), triangle, and chevron and the different texture parameters of depth and area ratio. With the discussions above, some meaningful results can be achieved.

- A comprehensive multi-field coupling numerical model for the surface-textured metal seals of roller bits is established, taking into account the influence of surface roughness, asperity contact, micro-deformation, and cavitation. Under the initial condition of p_{oi} = 3 MPa and n_i = 200 rpm, the textured convergent zones and divergent zones of the liquid film are formed along the rotational direction on end faces with different shape textures. The sharp-angle textures, including triangle and chevron textures, can produce a stronger hydrodynamic pressure effect on the end faces of metal seals. The cavitation area of different shape textures accounts for about 5% to 27% of the total texture area.
- The liquid film bearing coefficient, leakage rate, and maximum temperature rise of the metal seals with five shape textures increase with an increase in the rotational speed (n) from 100 rpm to 500 rpm. The temperature rise and frictional force of the textured end faces under different rotational speeds are significantly improved when compared to the untextured flat end face, particularly in the case of the triangle-textured end face. As the environmental pressure increases, the hydrodynamic pressure effect induced by the textures weakens, although the textures continue to contribute to a reduction in the contact pressure and frictional heat of the sealing rings. However, when the environmental pressure is $p \geq 30$ MPa, the textured end face gradually loses its advantage in the sealing performance of metal seals.
- The liquid film characteristics and sealing performance of triangle-textured end faces are significantly affected by texture parameters, specifically the depth and area ratio. The hydrodynamic pressure effect and cavitation phenomenon intensify with a decrease in the depth (h_p) from 8.6 μm to 2.6 μm and with an increase in the area ratio (S_p) from 0.1 to 0.4. Consequently, the maximum temperature rise and frictional force (ΔT_{max}, F_f) decrease with the area ratio (S_p). The effect of depth (h_p) on the sealing parameters is significantly weakened when $h_p \geq 4.3$ μm. The leakage rate (Q) changes from positive to negative in the range of S_p = 0.3~0.35, and the optimal texture parameters of S_p = 0.34 and h_p = 3 μm can achieve an ideal state of 'zero leakage'.
- A sealing performance test bench of surface-textured metal seals is constructed to validate the accuracy of multi-field coupling simulation. The application of surface textures to metal seals offers advantages in terms of reducing temperature rise, friction and wear on the end faces. The untextured and textured end faces are in a semi-dry friction and mixed friction state during the test, respectively, resulting in the occurrence of abrasive wear and adhesive wear on the end faces of the rotor and stator. The triangle texture has proven to have the best hydrodynamic lubrication and wear resistance among the five shape textures.

In further studies, the test device of the metal seals will continue to be improved to better simulate the downhole drilling conditions and systematically study the influence of surface texture parameters on the sealing performance.

Author Contributions: Conceptualization, Y.M., X.M. and X.P.; methodology, Z.Y.; software, Z.L.; validation, Z.L. and J.J.; formal analysis, Y.M. and X.M.; investigation, Y.M. and Z.Y.; data curation, Y.M. and Z.Y.; writing—original draft preparation, Y.M. and Z.L.; writing—review and editing, X.P.; supervision, X.P.; funding acquisition, Y.M. All authors have read and agreed to the published version of the manuscript.

Funding: This research was funded by the National Natural Science Foundation of China, grant number 52375211 and 51975527.

Data Availability Statement: All relevant data are presented in the paper.

Conflicts of Interest: The authors declare no conflicts of interest.

References

1. Wang, C.; Wang, X.; Li, S.; Jiao, Y.; Yang, R.; Li, G. Evaluation method for tooth wear of roller bits based on the fractal dimension of the rock surface. *J. Petrol. Sci. Eng.* **2022**, *210*, 110039. [CrossRef]
2. Naganawa, S. Feasibility study on roller-cone bit wear detection from axial bit vibration. *J. Petrol. Sci. Eng.* **2012**, *82–83*, 140–150. [CrossRef]
3. Li, B.; Liu, J.; Lu, D.; Wu, M.; Chen, H. Optimization of single metal seal structure of cone bit. *Lubr. Eng.* **2019**, *44*, 98–102.
4. Aminfar, O. Innovations in Rotary Drill Bit Design to Reduce Vibration and Improve Durability. Master's Thesis, University of Waterloo, Waterloo, ON, Canada, 2008.
5. Elmgerbi, A.; Thonhauser, G. Holistic autonomous model for early detection of downhole drilling problems in real-time. *Process Saf. Environ.* **2022**, *164*, 418–434. [CrossRef]
6. Keller, A.M.; Feng, T.H.; Demirer, N.; Darbe, R.; Chen, D.M. Rate of penetration estimation downhole with machine learning for drilling position control. *Geoenergy Sci. Eng.* **2022**, *224*, 211593. [CrossRef]
7. Al-Wahedi, T.; Hay, M.A.; Waheed, T.; Grimes, T. New slim hole technology maximizes productivity in middle east horizontal drilling programs. In Proceedings of the SPE/IADC Drilling Conference, SPE-92376-MS, Amsterdam, The Netherlands, 23–25 February 2005.
8. Zhou, Y.; Hu, J.; Tan, B.; Jiang, Y.; Tang, Y. Wear of spiral seal cone bit in complex formations. *SPE J.* **2021**, *26*, 3575–3590. [CrossRef]
9. Etsion, I. State of the art in laser surface texturing. *J. Tribol.* **2005**, *127*, 248–253. [CrossRef]
10. Gropper, D.; Wang, L.; Harvey, T.J. Hydrodynamic lubrication of textured surfaces: A review of modeling techniques and key findings. *Tribol. Int.* **2016**, *94*, 509–529. [CrossRef]
11. Chen, T.; Ji, J.; Fu, Y.; Yang, X.; Fu, H.; Fang, L. Tribological performance of UV picosecond laser multi-scale composite textures for C/SiC mechanical seals: Theoretical analysis and experimental verification. *Ceram. Int.* **2021**, *47*, 23162–23180. [CrossRef]
12. Xiong, S.; Salant, R.F. A numerical model of a rock bit bearing seal. *Tribol. Trans.* **2000**, *43*, 542–548. [CrossRef]
13. Xiong, S.; Salant, R.F. A dynamic model of a contacting mechanical seal for down-hole tools. *J. Tribol.* **2003**, *125*, 469–479. [CrossRef]
14. Ma, Y.; Yuan, Z.; Ni, Y.; Meng, X.; Peng, X. Performance prediction and multi-objective optimization of metal seals in roller cone bits. *J. Petrol. Sci. Eng.* **2022**, *208*, 109316. [CrossRef]
15. Ma, Y.; Xu, Y.; Chen, Y.; Meng, X.; Peng, X. Dynamic characteristics of metal seals in roller cone bits. *Lubricants* **2022**, *10*, 280. [CrossRef]
16. Etsion, I.; Kligerman, K.; Halperin, G. Analytical and experimental investigation of laser-textured mechanical seal faces. *Tribol. Trans.* **1999**, *42*, 511–516. [CrossRef]
17. Etsion, I.; Halperin, G. A laser surface textured hydrostatic mechanical seal. *Tribol. Trans.* **2002**, *45*, 430–434. [CrossRef]
18. Brunetiere, N.; Tournerie, B. Numerical analysis of a surface-textured mechanical seal operating in mixed lubrication regime. *Tribol. Int.* **2012**, *49*, 80–89. [CrossRef]
19. Yang, X.; Peng, X.; Meng, X.; Jiang, J.; Wang, Y. Thermo-elasto-hydrodynamic analysis of triangular textured mechanical face seals. *J. Zhejiang Univ.-Sci. A* **2019**, *20*, 864–881. [CrossRef]
20. Qiu, Y.; Khonsari, M.M. Experimental investigation of tribological performance of laser textured stainless steel rings. *Tribol. Int.* **2011**, *44*, 635–644. [CrossRef]
21. Dingui, K.; Brunetière, N.; Bouyer, J.; Adjemout, M. Surface texturing to reduce temperature in mechanical seals. *Tribol. Online* **2020**, *15*, 222–229. [CrossRef]
22. Grimes, B.; Kirkpatrick, B. Step change in performance: Upgraded bit technology significantly improves drilling economics in GOM motor applications. In Proceedings of the SPE Annual Technical Conference and Exhibition, San Antonio, TX, USA, 24–27 September 2006.
23. Lin, C. Oil and Gas Drilling Bit Tribology. In *Encyclopedia of Tribology*; Springer: New York, NY, USA, 2013; pp. 2467–2475.
24. Liu, Z.; Kuang, Y.; Wang, Y. Research on working principle of a roller bit's seal testing machine. *Mach. Des. Manuf.* **2009**, *47*, 191–193.
25. Qiu, M.; Delic, A.; Raeymaekers, B. The effect of texture shape on the load-carrying capacity of gas-lubricated parallel slider bearings. *Tribol. Lett.* **2012**, *48*, 315–327. [CrossRef]
26. Adjemout, M.; Andrieux, A.; Bouyer, J.; Brunetière, N.; Marcos, G.; Czerwiec, T. Influence of the real dimple shape on the performance of a surface textured mechanical seal. *Tribol. Int.* **2017**, *115*, 409–416. [CrossRef]
27. Patir, N.; Cheng, H.S. An average flow model for determining effects of three-dimensional roughness on partial hydrodynamic lubrication. *J. Lubri. Technol.* **1978**, *100*, 12–17. [CrossRef]
28. Patir, N.; Cheng, H.S. Application of average flow model to lubrication between rough sliding surfaces. *J. Lubri. Technol.* **1979**, *101*, 220–229. [CrossRef]
29. Meng, X.; Bai, S.; Peng, X. Lubrication film flow control by oriented dimples for liquid lubricated mechanical seals. *Tribol. Int.* **2014**, *77*, 132–141. [CrossRef]

30. Yu, H.; Wang, X.; Zhou, F. Geometric shape effects of surface texture on the generation of hydrodynamic pressure between conformal contacting surfaces. *Tribol. Lett.* **2010**, *37*, 123–130. [CrossRef]
31. Jakobsson, B.; Floberg, L. The finite journal bearing considering vaporization. *Trans. Chalmers Univ. Technol.* **1957**, *190*, 1–116. Corpus ID: 136822854.
32. Olsson, K. Cavitation in dynamically loaded bearings. *Trans. Chalmers Univ. Technol.* **1965**, *308*, 1–60. Corpus ID: 135904611.
33. Meng, X.; Tu, Z.; Ma, Y.; Jiang, J.; Peng, X. Topology optimization of liquid lubricating zero-leakage mechanical face seals. *Tribol. Int.* **2022**, *169*, 107490. [CrossRef]
34. Shen, C.; Khonsari, M.M. Texture shape optimization for seal-like parallel surfaces: Theory and experiment. *Tribol. Trans.* **2016**, *59*, 698–706. [CrossRef]
35. Xiong, S.; Wang, Q. Steady-state hydrodynamic lubrication modeled with the Payvar-Salant mass conservation model. *J. Tribol.* **2012**, *134*, 031703. [CrossRef]
36. Abbott, E.J.; Firestone, F.A. Specifying surface quality-A method based on accurate measurement and comparison. *J. Mech. Eng.* **1933**, *55*, 569–572.
37. Harp, S.R.; Salant, R.F. Analysis of mechanical seal behavior during transient operation. *J. Tribol.* **1998**, *120*, 191–197. [CrossRef]
38. Liu, J. Research on Sealing Structure of New Roller Bit Bearing. Master's Thesis, Degree-Southwest Petroleum University, Chengdu, China, 2018.
39. Qu, J.; Chen, G.; Yang, Y.W. Finite element analysis of rubber sealing ring resilience behavior. *Adv. Mater. Res.* **2013**, *705*, 410–414. [CrossRef]
40. Kim, B.; Lee, S.B.; Lee, J.; Cho, S.; Park, H.; Yeom, S.; Park, S.H. A comparison among Neo-Hookean model, Mooney-Rivlin model, and Ogden model for chloroprene rubber. *Int. J. Precis. Eng. Man.* **2012**, *13*, 759–764. [CrossRef]
41. Ruan, B.; Salant, R.F.; Green, I. A mixed lubrication model of liquid/gas mechanical face seals. *Tribol. Trans.* **1997**, *40*, 647–657. [CrossRef]
42. Li, Y.; Pei, X. Investigation on tribological behaviors of surface micro-forming 20CrNiMo steel under laser shock processing. *Laser Technol.* **2012**, *36*, 814–817.
43. Zhong, L.; Wei, G.; Wang, G.; He, X.; Liao, Y.; Xie, N. Effects of femtosecond laser texture on the tribological properties of 20CrNiMo/beryllium bronze tribo-pairs of rock bit sliding bearings under non-Newtonian lubrication. *Proc. Inst. Mech. Eng. Part J J. Eng. Tribol.* **2019**, *233*, 1293–1305. [CrossRef]
44. Wang, Y.; Wu, J.; Li, Z.; Xu, L. Effect of slip position on the hydrodynamic performance of liquid film seal. *Ind. Lubri. Tribol.* **2021**, *73*, 405–413. [CrossRef]
45. Li, X.; Yang, Z.; Xu, J.; Chen, R.; Yang, H. The fractal leakage model of contact mechanical seals considering wear and thermal deformation. *J. Braz. Soc. Mech. Sci.* **2019**, *41*, 521. [CrossRef]
46. Ni, X.; Sun, J.; Ma, C.; Zhang, Y. Wear model of a mechanical seal based on piecewise Fractal Theory. *Fractal Fract.* **2023**, *7*, 251. [CrossRef]

lubricants

MDPI

Article

Stability Analysis of the Secondary Motion of a Textured Piston

Igal Cohen *, Shachar Tresser and Roman Goltsberg

Department of Mechanical Engineering, Technion, Haifa 32000, Israel; shachar.tresser@gmail.com (S.T.); groman@me.technion.ac.il (R.G.)

* Correspondence: igalcohen@campus.technion.ac.il

Abstract: Piston–cylinder pairs are very common in industrial mechanisms. While a piston is primarily designed for axial reciprocating motion, the occurrence of secondary motions—lateral and rotational—due to the small clearance between the piston and the cylinder may lead to frictional contact, energy loss, wear and an increase in unwanted leakage. This study focuses on mitigating the secondary motion of a ringless piston. The influence of a Rayleigh step bearing and partial surface texturing with numerous micro-dimples on the dynamic stability of the secondary motion is studied. A linear model was used to obtain the trajectory of the secondary motion and Floquet theory was applied to analyze the stability and draw stability maps. The influence of various texturing and step parameters, including the dimple area density and aspect ratio for partial texturing, as well as the length and depth of treatment for both partial texturing and step profiles, on the stability of the secondary motion was examined. The normalization method is presented, enabling the expansion of the results for various operating conditions and geometries. Conclusions and guidelines regarding the optimal parameters, in terms of a wider stability range and higher decay rate, are formulated.

Keywords: ringless piston; laser surface texturing; partial texturing; Navier–Stokes; hydrodynamic lubrication; secondary motion

1. Introduction

Piston–cylinder systems are widely used in various engineering applications. These include internal combustion engines, spool valves, various actuators, capsule-pipeline transportation devices, viscometers, timing devices, piston compressors, plunger pumps and free pistons in some versions of the Stirling engine [1]. In addition to the primary reciprocating axial motion of the piston, the small clearance between the piston and the cylinder liner allows lateral and rotational motion in the direction of the clearance during the reciprocating motion, also known as secondary motion [2]. For hydraulic pistons, unlike those in internal combustion engines, the lateral forces due to the axial motion mechanism are small; hence, the main contributors to the radial forces are the hydrodynamic forces, which may lead to instability of the secondary motion [3].

The secondary motion of an untreated hydraulic piston (smooth and cylindrical in shape) is unstable, as was observed both analytically through the perturbation method [4] and experimentally [5] in piston hydraulic machines for various fluid properties, geometric properties and operation conditions. Unstable motion results in collisions of the radially oscillating piston with the cylinder [1], which is detrimental to the piston–cylinder system and may also lead to a 50%–100% increase in leakage [6]. The secondary motion of the piston should be stabilized to reduce its detrimental effect. In other words, the piston should be centered during its reciprocating axial motion. Several methods for centering a piston have been presented in the literature. For instance, Xie et al. [7] introduced a method for centering a piston using a closed-loop active magnetic control system. In Refs. [8,9], it was shown for a stepped-profile piston (also known as a Rayleigh step bearing), assuming small eccentricities (without rotation), that centering of the piston can be achieved due to a radial returning force, and the optimum step design was discussed. In [10], Kligerman et al.

showed that the secondary motion of the piston (lateral and rotational) during reciprocating motion can be stabilized by utilizing axial step profiling of the piston at the end of the high-pressure region.

However, while a Rayleigh step bearing is easy to model mathematically, it may not be applicable at the micron scale. In [11], Tonder suggested that an effective step or slope profile could be produced by removing material through texturing. Various texturing techniques can be used for micro-surface texturing, including machining, ion beam texturing and etching techniques. Nevertheless, it seems that laser surface texturing (LST) offers the most promising concept of all practical micro-surface texturing methods [12]. This is due to its high flexibility and accuracy. The implementation of LST is quick, cheap and environmentally clean [13]. In Ref. [14], it was shown that LST in the form of spherical dimples can improve tribological performance. In Ref. [15], the topography of dimples was measured, and it showed good agreement with a spherical shape. LST produces a very large number of micro-dimples on the surface. The dimples' diameter and height can be precisely controlled; the diameter typically ranges from 20 to 200 microns, and the height ranges from 0 to 100 microns [16]. In addition to generating a separating force in the case of full lubrication, each dimple may serve as a micro-reservoir or a micro-trap for wear debris in the case of starved lubrication [17].

Similar to a Rayleigh step bearing, LST can be implemented on part of the piston at the end of the high-pressure region, a treatment that is also known as "partial texturing" [18]. In Ref. [19] for thrust bearings and in Ref. [20] for piston rings, the pressure distribution for partial texturing was shown to be similar to the pressure distribution for Rayleigh step bearings. These studies also investigated the impact of various texturing parameters, including dimple depth, textured portion and dimple area density. The results indicated that the dimple area density, which is the ratio of the total area of the dimples to the area of the textured zone, should be maximized for optimal performance. The optimal values for dimple depth and other parameters were found to be dependent on specific operational conditions. In Ref. [21], a computational fluid dynamics (CFD)-based investigation was performed to compare the pressure distribution of a partially textured surface with that of an equivalent-volume step. The influence of different texturing parameters on the applicability of the equivalent-volume step to model partial texturing was examined, and an analytic tool to predict the pressure distribution for surface texturing was proposed. The authors noted that while a CFD-based model for surface texturing showed good agreement with the experimental results, the agreement with the Reynolds equation was poor. This conclusion is compatible with Ref. [22], in which it was stated that the Reynolds equation cannot handle the surface discontinuities added by texturing. Therefore, the mathematical model that should be used is the more general mass and momentum conservation equations (Navier–Stokes equations) rather than the Reynolds equation.

In the following, several methods that were used in the literature for analyzing the stability of secondary motion are presented. Fagotti et al. [23] developed a numerical model to calculate the secondary motion of a ringless piston in a small reciprocating compressor. The piston trajectory of the secondary motion during the reciprocating motion was obtained through simultaneous solution of the Reynolds equation and the piston's equation of motion (EOM). The authors showed the effect of different parameters on the secondary motion of the piston, which significantly affects the power consumption and oil leakage. They suggested using their model for piston design optimization. Tan et al. [24] established an experimental technique for measuring the secondary motion of a piston using laser sensors. In their later study [25], the measurement technique was used to validate the numerical model of pistons' secondary motion. The numerical model assumed linear dynamics, and a nonlinear term was added for cases of contact between the piston and the cylinder. The agreement between the numerical model and the experiment was good, especially for low frequencies of reciprocating motion. It was found that the vibration amplitude decreased as the damping and stiffness coefficients increased. In [10], Kligerman et al. used simultaneous solution of the Reynolds equation and the EOM for analyzing the stability of the secondary motion of a stepped-profile piston. In their later study [26], the solution acquired from the Reynolds equation was validated with a more accurate model, which is based on simultaneous solution of the full Navier–Stokes equations and the dynamic equations. The validity limits of the Reynolds equation in terms of normalized parameters were discussed. Kumar and Bergada [27] investigated the effect of groove cuts along the piston's surface on the secondary motion stability of pump pistons. By solving the Reynolds equation for various eccentric displacements and angular positions of the piston during the reciprocating motion, the authors showed that axial groove cuts can increase the stability of the secondary motion by increasing the stiffness of the system. In [28], a simultaneous solution of the Reynolds equation, the EOM and the deformation equations was applied for a grooved piston in an axial piston pump. It was found that grooves that are located near the displacement chamber (a chamber where suction and pumping of the liquid occur) can prevent contact between the piston and cylinder and reduce viscous friction power loss. However, as stated in [29], the analytical models should be further improved to better design piston profiles for enhanced tribological performance.

While it was shown in some of the abovementioned studies that the secondary motion of a piston can be stabilized, most of the studies did not provide mathematical proof of the stability of the secondary motion. The Lyapunov method [30] can potentially be employed to test the stability of the EOM. However, due to the coupling of the EOM with the Reynolds equation, this approach becomes cumbersome. Nevertheless, when a linear model is employed, as demonstrated in [25], for small oscillations around the equilibrium point, which is usually a concentric piston, additional methods can be utilized to test the stability. For linear time–invariant (LTI) systems, the classic modal analysis (see [31]) can be used. By representing the LTI system in state space, the stability can be proven if all the eigenvalues of the state matrix have negative real parts. Nevertheless, the secondary motion of the piston has time-varying behavior due to reciprocating motion, and therefore, other methods for testing stability should be used. Among the suitable methods for time-varying systems are perturbation methods (see [32]), such as the averaging and multiple-scale methods. However, in perturbation methods, small damping forces are assumed, and therefore, these methods might not be suitable. Another method that can be used for time-varying systems in which small damping forces are not assumed is Floquet theory (see [33]). Because of the applicability of this method, it can be used for analyzing the stability of the secondary motion.

Research Objectives

The motivation of the present work is to reduce energy loss and wear by stabilizing the secondary motion of a piston. The stability of the secondary motion of a piston with partial laser surface texturing and of a stepped-profile piston during axial reciprocating motion inside a cylinder liner is studied. The trajectory of the piston's secondary motion during the reciprocating motion is obtained by simultaneously solving both the Navier–Stokes equation and the EOM and by a more convenient linear model based on mass, stiffness and damping matrices. A unique approach for analyzing the stability of the secondary motion based on Floquet theory due to the time-varying stiffness and damping terms, which arise due to the axial motion, is presented. The use of Floquet theory with a matrix-based model allows the performance of stability analysis and the generation of stability maps for various operation conditions and different parameters of dimple texturing and step bearings with relatively low computational costs. The main goals of the current study are to examine the stability range for various texturing and step parameters and to draw guidelines for the optimal texturing parameters.

2. Theoretical Model

In the following, the physical model of the problem and the models for the step bearing and partial laser surface texturing are presented.

In Figure 1, a schematic cross section of the investigated ringless piston–cylinder system is presented. The piston, surrounded by a viscous fluid, performs an axial reciprocating motion within the cylinder according to a predetermined law in time. The piston separates the high and low pressure regions of the cylinder. The parameters that describes the model are as follows:

- $U(t)$—the instantaneous axial velocity of the piston;
- P_h—the pressure at the high-pressure region;
- P_l—the pressure at the low-pressure region;
- $\Delta P(P_h - P_l)$—the pressure drop;
- L—the length of the piston;
- R—the radius of the piston;
- C—the clearance between the piston and the cylinder;
- z and r—the axial and radial coordinates, respectively.

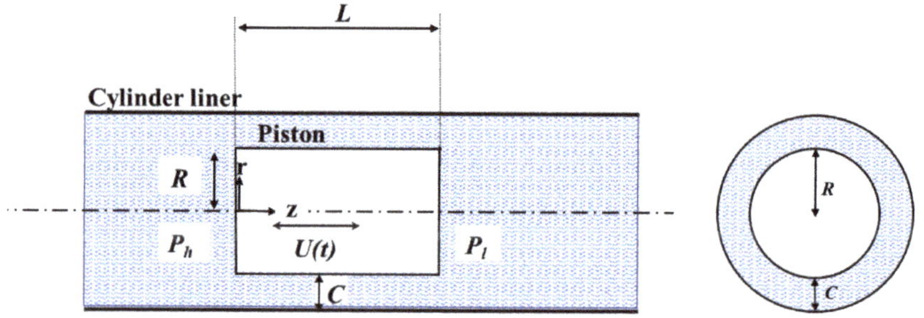

Figure 1. Schematics of a ringless piston–cylinder system: (**a**) cross section (r-z plane) and (**b**) side view (r-θ plane).

In the present study, it is assumed that the secondary motion of the piston (translation and rotation) occurs in one plane only (the r-z plane). This assumption is reasonable because for small oscillations, translation and rotation in one plane do not affect motion in the perpendicular plane. Nevertheless, the geometry is not axisymmetric; therefore, the fluid flow is modeled by 3-D geometry. In addition, it is assumed that the axial motion mechanism does not enable the piston to rotate relative to its axis (about the z-axis) and that the axial motion of the piston is not affected by the piston's secondary motion.

Based on these assumptions, the piston has two degrees of freedom, namely, radial translation and rotation around the center of mass, as presented in Figure 2. The translation and rotation are determined by the hydrodynamic forces acting on the piston. ε is the instantaneous eccentricity during the piston radial translation, and γ is the instantaneous angle during its rotation. Thus, $\varepsilon^{\cdot}$ and $\gamma^{\cdot}$ are the instantaneous radial and angular velocities, respectively, of the piston.

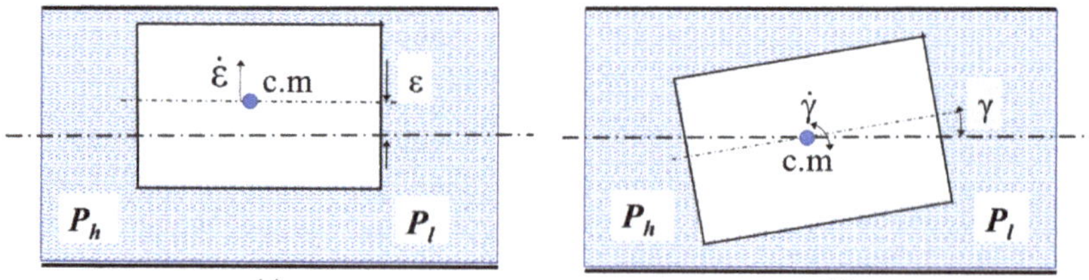

Figure 2. The degrees of freedom of the secondary motion of the piston: (**a**) translation and (**b**) rotation around the center of mass.

2.1. Step Profiling and Surface Texturing

In the following, the geometric properties of the step bearing and of the partial surface texturing are presented. In Figure 3, an axisymmetric section of a concentric stepped-profile piston and a cyclic section of a concentric textured piston are presented. The step is characterized by its length and height, which are denoted as follows:

- $\alpha \times L$—the length of the step;
- h_s—the height of the step.

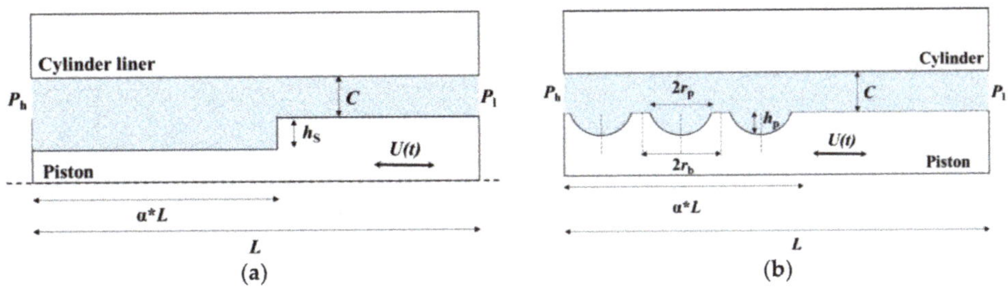

Figure 3. Concentric piston. (**a**) Axisymmetric section of a stepped-profile piston and (**b**) cyclic section of a textured piston.

The spherical-segment-shaped dimples of the partial surface texturing are characterized by 4 parameters:

- $\alpha \times L$—the length of the textured zone;
- r_p—the radius of the dimples;
- $2r_b$—the imaginary square cell length;
- h_p—the dimple depth.

The imaginary square cell length and the dimple base radius determine the dimple area density, $S_p = \pi r_p^2 / 4r_b^2$, which was shown to have a significant effect on the tribological performance [34]; hence, this parameter was examined in the present study.

2.2. Governing Equation

The solution process of the presented problem involves simultaneously solving two sets of coupled equations: the hydrodynamic equations for the pressure and velocity fields and the EOM for the relationship between the forces acting on the piston (due to shear and pressure differences in the fluid) and its secondary motion.

The main assumptions of the model are as follows:

1. The fluid flow is laminar and fully developed.
2. The fluid is Newtonian and incompressible with constant viscosity (isothermal—the energy equation is not being solved).
3. The body and gravitational forces are negligible.
4. The operating pressure is high, and therefore, no cavitation occurs.
5. No slip and no penetration conditions are applied to the surfaces.
6. The piston and the cylinder are rigid bodies.

Based on the abovementioned assumptions, the mass and momentum conservation equations (Navier–Stokes equations) in vector notation are given by [35]:

$$\nabla \cdot \mathbf{v} = 0 \tag{1}$$

$$\rho[\frac{\partial \mathbf{v}}{\partial t} + (\mathbf{v} \cdot \nabla)\mathbf{v}] = -\nabla p + \mu \nabla^2 \mathbf{v} \tag{2}$$

where the variables are defined as follows:

- μ—the dynamic viscosity;

- ρ—the density of the fluid;
- p—the local pressure;
- **v**—the velocity vector;
- t—the time.

The EOM describing the piston radial translational and rotation are given by:

$$\begin{aligned} m\ddot{\varepsilon} &= F \\ I\ddot{\gamma} &= M \\ U(t) &= U_{\text{amp}} \cdot \sin(f \cdot 2\pi t) \end{aligned} \tag{3}$$

where the variables are defined as follows:

- m—the piston's mass;
- I—the piston's mass moment of inertia;
- $\ddot{\varepsilon}$ and $\ddot{\gamma}$—the radial and angular accelerations of the piston, respectively;
- F and M—the radial force and moment exerted on the piston by the fluid, respectively;
- U_{amp}—the amplitude of the axial velocity;
- f—the frequency of the axial velocity.

F and M are calculated by integration of the instantaneous pressure and shear stress fields. A more detailed explanation of the governing equations and boundary conditions can be found in Ref. [26].

3. Numerical Model

The numerical model that was used to obtain the full simultaneous dynamic model is described in Ref. [26]. The commercial ANSYS Fluent 2021 R1 software was used. For more information regarding this solver, the reader is referred to [36]. The mesh design of the fluid in the gap is presented in Figure 4.

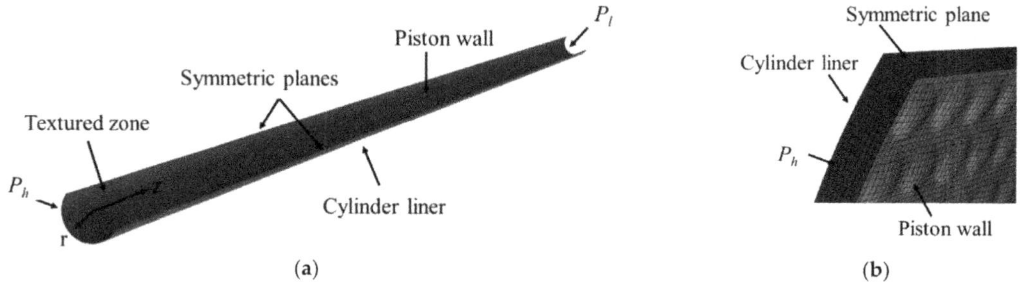

Figure 4. Mesh design of the fluid region: (**a**) image of the entire domain and (**b**) zoom-in image of the textured zone.

The outer cylindrical surface corresponds to the fluid layer adjacent to the inner surface of the stationary cylinder liner. The inner cylindrical surface is the fluid layer adjacent to the piston. The pressure inlet is denoted by P_h, and the pressure outlet is denoted by P_l. The mesh for each imaginary square cell has 10×10 elements and 8 layers in the gap, as presented in Figure 4b. The time step for the dynamic model was 1×10^{-5} [Sec]. Increasing the number of elements changed the results by less than 5%. Increasing the time step resulted in numerical instability. A smaller time step changed the results by less than 1%. The chosen numerical properties make it possible to obtain physical insights at a reasonable computational cost.

To apply the axial velocity, moving wall boundary conditions were employed. Generally, the moving wall boundary condition should be applied to the piston. However, despite the simplicity and efficiency of the moving wall boundary condition, it considers only the tangential velocity (see Ref. [36]); therefore, it may not be accurate to physical reality for textured surfaces, especially for deep dimples. In our previous study [19], we reported differences between the movement of textured and untextured surfaces based on the use of moving wall boundary conditions. However, upon further investigation, we identified that the application of these conditions in the context of piston–cylinder systems is incorrect. This new study addresses this issue by applying moving wall boundary conditions to the cylinder, which is more appropriate for the configurations involving textured surfaces.

3.1. Linear Model

To obtain the trajectory of the piston's secondary motion, the simultaneous solution of the coupled equations should be used. However, simultaneous transient solutions require discretizing the time domain and solving the Navier–Stokes equation for each time step. The complex geometry and the numerous mesh elements that are required to model the textured piston demand expensive computational resources (a single case may take more than a month). Therefore, to study the influence of different types of texturing and to enable solutions for multiple cases in a relatively short time, a more efficient model should be used.

By assuming small perturbations and neglecting time-dependent inertia effects of the fluid (common assumptions in hydrodynamic lubrication), a linear dependency between ε, $\dot{\varepsilon}$, γ, $\dot{\gamma}$ and the resulting force F and moment M can be assumed, and Equation (3) can be rewritten in matrix notation:

$$\mathbf{M}\begin{bmatrix}\ddot{\varepsilon}\\ \ddot{\gamma}\end{bmatrix}+\mathbf{K}(\mathrm{t})\begin{bmatrix}\varepsilon\\ \gamma\end{bmatrix}+\overline{\mathbf{D}}(\mathrm{t})\begin{bmatrix}\dot{\varepsilon}\\ \dot{\gamma}\end{bmatrix}=0 \tag{4}$$

where $\mathbf{M}$ is the mass matrix, which consists of the mass and mass moment of inertia of the piston ($\mathbf{M}$ is a 2×2 diagonal matrix); $\mathbf{K}$ is the stiffness matrix, which consists of the relationship between ε, γ and the resulting forces and moments; and $\overline{\mathbf{D}}$ is the damping matrix, which consists of the relationship between $\dot{\varepsilon}$, $\dot{\gamma}$ and the resulting forces and moments. Equation (4) can be integrated over time by using the Runge–Kutta method to obtain the trajectory of the piston's secondary motion without the need to repeatedly solve the Navier–Stokes equation, and from that, an efficient linearized model is proposed. The linear model is validated in the Results section.

Based on the Reynolds equation, when assuming that there is no cavitation, the relation between the axial velocities, the pressure drop and the resulting force in the gap direction is linear and commutative, as shown in Appendix A. The current numerical model shows similar results for small values of the reduced Reynolds number, $\mathrm{Re}^* < 0.01$, which are within the considered range of the current study. Therefore, the stiffness matrix, $\mathbf{K}$, can be separated into two matrices: $\Delta P\mathbf{K}^P$, which denotes the stiffness matrix for a pressure drop of ΔP without axial motion; and $\mu U_{\mathrm{amp}}\sin(2\pi f t)\mathbf{K}^U$, which denotes the time-dependent stiffness matrix for an axial velocity with an amplitude of U_{amp}, a frequency of f and a viscosity of μ without a pressure drop. In addition, the damping matrix, $\overline{\mathbf{D}}$, which is related to the squeeze effect, has a linear dependency on the viscosity and is not dependent on the pressure drop or the axial velocity; therefore, it can be expressed as $\mu\mathbf{D}$ Therefore, Equation (4) can be rewritten as:

$$\mathbf{M}\begin{bmatrix}\ddot{\varepsilon}\\ \ddot{\gamma}\end{bmatrix}+\Delta P\mathbf{K}^P\begin{bmatrix}\varepsilon\\ \gamma\end{bmatrix}+\mu U_{\mathrm{amp}}\sin(2\pi f t)\mathbf{K}^U\begin{bmatrix}\varepsilon\\ \gamma\end{bmatrix}+\mu\mathbf{D}\begin{bmatrix}\dot{\varepsilon}\\ \dot{\gamma}\end{bmatrix}=0 \tag{5}$$

For given matrices $\mathbf{M}$, $\mathbf{K}^P$, $\mathbf{K}^U$, $\mathbf{D}$ and given values of ΔP, μ, U_{amp}, the dynamic response due to initial excitation can be achieved by integrating Equation (5) over time. The coefficients of the matrices in Equation (5) can be rewritten explicitly as follows:

$$\begin{bmatrix}m & 0\\ 0 & I\end{bmatrix}\begin{bmatrix}\ddot{\varepsilon}\\ \ddot{\gamma}\end{bmatrix}+\Delta P\begin{bmatrix}k^P_{\varepsilon\varepsilon} & k^P_{\varepsilon\gamma}\\ k^P_{\gamma\varepsilon} & k^P_{\gamma\gamma}\end{bmatrix}\begin{bmatrix}\varepsilon\\ \gamma\end{bmatrix}+\mu U_{\mathrm{amp}}\sin(2\pi\omega t)\begin{bmatrix}k^U_{\varepsilon\varepsilon} & k^U_{\varepsilon\gamma}\\ k^U_{\gamma\varepsilon} & k^U_{\gamma\gamma}\end{bmatrix}\begin{bmatrix}\varepsilon\\ \gamma\end{bmatrix}+\mu\begin{bmatrix}d_{\varepsilon\varepsilon} & d_{\varepsilon\gamma}\\ d_{\gamma\varepsilon} & d_{\gamma\gamma}\end{bmatrix}\begin{bmatrix}\dot{\varepsilon}\\ \dot{\gamma}\end{bmatrix}=0 \tag{6}$$

The eigenvalues of the system (without axial velocity) are obtained via modal analysis. The state matrix **A** can be derived such that:

$$\dot{\mathbf{x}} = \mathbf{A}\mathbf{x}, \quad \mathbf{A} = \begin{bmatrix} 0 & I \\ -\mathbf{M}^{-1}\Delta P\mathbf{K}^{P} & -\mathbf{M}^{-1}\mu\mathbf{D} \end{bmatrix}, \quad \mathbf{x}(\mathrm{t}) = \begin{bmatrix} \varepsilon \\ \gamma \\ \dot{\varepsilon} \\ \dot{\gamma} \end{bmatrix} \tag{7}$$

where **x** is the state vector of the system. The eigenvalues of the system are the eigenvalues of the state matrix **A**. The stability criterion of the secondary motion (centering of the piston) is:

$$\max(\mathrm{Re}(\lambda(\mathbf{A})))<0 \tag{8}$$

where λ are the eigenvalues of **A**. The solution of Equation (7) is a linear combination of all mode shapes (eigenvectors of **A**), where the response of the ith mode shape is proportional to $e^{\lambda_i t}$. The real part of λ_i is the decay rate of the ith mode.

3.2. Derivation of the Damping and Stiffness Coefficients

In the following, two methods for deriving the matrix coefficients are presented. The first method is based on harmonic motion, and the second method is a simplified method based on finite perturbation, neglecting inertia terms.

To derive the matrix coefficients, a harmonic radial motion, $\varepsilon = \varepsilon_0 \sin(\omega t)$, $\gamma = 0$, should be applied to the piston, which leads, according to the frequency response theorem, to a harmonic radial force and moment on the piston with phase shift, $F = F_0 \sin(\omega t + \phi)$ and $M = M_0 \sin(\omega t + \phi)$. The harmonic motion can then be applied to ANSYS Fluent through a user-defined function (UDF). A similar treatment should be performed for angular harmonic motion. The forces applied by the fluid on the piston consist of inertia, damping and stiffness components; therefore, the EOM of the piston are:

$$\begin{aligned} m\ddot{\varepsilon} &= F = -d_{\varepsilon\varepsilon}\dot{\varepsilon} - k_{\varepsilon\varepsilon}\varepsilon - \widehat{\mathbf{M}}_{\varepsilon\varepsilon}\ddot{\varepsilon} \\ I\ddot{\gamma} &= M = -d_{\gamma\varepsilon}\dot{\varepsilon} - k_{\gamma\varepsilon}\varepsilon - \widehat{\mathbf{M}}_{\gamma\varepsilon}\ddot{\gamma} \end{aligned} \tag{9}$$

where $\widehat{\mathbf{M}}_{\varepsilon\varepsilon}$ and $\widehat{\mathbf{M}}_{\gamma\varepsilon}$ are the coefficients of the inertia force matrix, which are related to the inertia of the fluid being pushed by the motion of the piston and are also known as the added mass (these parameters were neglected in the previous subsection). Substituting the harmonic motions into Equation (9) yields:

$$\begin{aligned} F &= -(d_{\varepsilon\varepsilon}\omega\cos(\omega t) - k_{\varepsilon\varepsilon}\sin(\omega t) - \widehat{\mathbf{M}}_{\varepsilon\varepsilon}\omega^2\sin(\omega t))\varepsilon_0 \\ M &= -(d_{\gamma\varepsilon}\omega\cos(\omega t) - k_{\gamma\varepsilon}\sin(\omega t) - \widehat{\mathbf{M}}_{\gamma\varepsilon}\omega^2\sin(\omega t))\varepsilon_0 \end{aligned} \tag{10}$$

Similarly, applying harmonic angular motion, $\gamma = \gamma_0\sin(\omega t)$, $\varepsilon = 0$ yields:

$$\begin{aligned} F &= -(d_{\varepsilon\gamma}\omega\cos(\omega t) - k_{\varepsilon\gamma}\sin(\omega t) - \widehat{\mathbf{M}}_{\varepsilon\gamma}\omega^2\sin(\omega t))\gamma_0 \\ M &= -(d_{\gamma\gamma}\omega\cos(\omega t) - k_{\gamma\gamma}\sin(\omega t) - \widehat{\mathbf{M}}_{\gamma\gamma}\omega^2\sin(\omega t))\gamma_0 \end{aligned} \tag{11}$$

The harmonic motion should be applied in the radial and angular directions, while a pressure drop, ΔP, or axial velocity, U, should be applied at several different frequencies, since no unique solution for the inertia and stiffness parameters can be achieved from a single frequency. The resultant forces and moments should be substituted into the LHS of Equations (10) and (11) to derive the coefficients of the matrices in Equation (6).

However, the computational resources for this approach are expensive. As an alternative, a simplified approach can be used by neglecting the inertia of the fluid (a common assumption in lubrication theory). Equation (10) becomes the following (a similar treatment should be performed for angular harmonic motion):

$$\begin{aligned} F &= -(d_{\varepsilon\varepsilon}\omega\cos(\omega t) - k_{\varepsilon\varepsilon}\sin(\omega t))\varepsilon_0 \\ M &= -(d_{\gamma\varepsilon}\omega\cos(\omega t) - k_{\gamma\varepsilon}\sin(\omega t))\varepsilon_0 \end{aligned} \quad (12)$$

For the simplified approach, the stiffness and damping coefficients can be found by the instantaneous forces of the harmonic motion at two time frames: maximum movement and zero velocity, and vice versa. These time frames can be simulated with the finite perturbation method without using harmonic motion. The dynamic mesh approach was used for applying the perturbations ($\varepsilon, \dot{\varepsilon}, \gamma, \dot{\gamma}$) to the piston. The abovementioned CFD model was used. The matrix coefficients are found by dividing the resultant force or moment by the disturbance. A similar approach was used in [37]. The differences in the values of the coefficients between the first approach and the simplified approach for the present study were found to be less than 1% (see [38]), while the computation time was significantly lower (less than an hour for the simplified approach and more than a week to apply harmonic motion). These results also indicate that the neglect of the time-dependent inertia effects in the present work is justified.

4. Results

In the following, the normalization method and the linear model, which are used to obtain multiple results and to generalize the results, are validated. In addition, by using the linear model and the normalization method, the stability of the secondary motion of the piston, for different geometries and operating conditions for various texturing parameters and step profiling is analyzed. The ranges of the examined parameters for the current study are as follows:

- $20\,[\mu\text{m}] < C < 500\,[\mu\text{m}]$;
- $0.01\,[\text{m}] < R < 0.1\,[\text{m}]$;
- $0.1\,[\text{m}] < L < 1\,[\text{m}]$;
- $5\,[\mu\text{m}] < h_p < 20\,[\mu\text{m}]$;
- $70\,[\mu\text{m}] < 2r_p < 300\,[\mu\text{m}]$.

The dimples' parameters are in accordance with [16].

4.1. Validation of the Normalization Approach and Linear Model

In the following, the normalization method that enables generalization of the results is presented and validated. The normalization is based on the characteristic parameters which are as follows:

- U_0—the reference axial velocity;
- T—the reference piston vibration cycle time;
- ΔP^*—the reference pressure drop.

The characteristic parameters are used to normalize the dimensional parameters of the velocity, time and pressure, and from that, a universal solution in terms of U_{amp}/U_0, $f \cdot T$ and $\Delta P/\Delta P^*$ can be obtained. In addition, the dimensionless parameter Re^*, which is the reduced Reynolds number, is presented. These parameters were developed in [26].

The reference axial velocity, U_0, and the characteristic value of the piston's radial vibration cycle time, T, are given by [26]:

$$U_0 = \frac{\Delta p C^2}{L\mu},\ T = \frac{24L^2\pi\mu}{7\Delta p C^2} \quad (13)$$

Extensive simulations have shown that the characteristic pressure, ΔP^*, can be obtained as follows:

$$\Delta P^* = L^3 R^3 \mu^2 m^{-1} C^{-5} \quad (14)$$

The meaning of the characteristic pressure is that the same ratio of $\Delta P/\Delta P^*$ yields the same normalized dynamic response.

The reduced Reynolds number, Re*, for the present study is given by [26]:

$$\text{Re}^* = \frac{\rho U_0 C^2}{L\mu} = \frac{\rho \Delta p C^4}{(L\,\mu)^2} \tag{15}$$

Numerical simulations showed (see [38]) that values of $\text{Re}^* < 0.01$ do not affect the results; therefore, the current study is limited to this range.

The normalization method for a stepped-profile piston was validated in [26]. To demonstrate the validity of the suggested normalization method for a partially textured piston, solutions for different fluid parameters and geometries (with the same normalized parameters) were calculated using the linearized model. The fluid and geometry parameters for three different example cases are presented in Table 1. All the cases have the same normalized parameters, $\Delta P/\Delta P^*$, $f * T$ and U/U_0. The parameters of the texturing (relative to the other geometric parameters) were kept constant. The variations in time of the normalized radial eccentricity and angle for an initial radial eccentricity of 0.3 for the different cases are presented in Figure 5. The time scale was normalized according to the cycle time of the axial velocity of the piston. The radial eccentricity was normalized according to the clearance and the angle was normalized according to the maximum geometrical angle—$2C/L$. The axial velocity was normalized according to U_0 and is presented by the dashed line related to the secondary vertical axis. Cases 1, 2 and 3 are represented by the white circle markers, solid line and black circle markers, respectively. Despite the different dimensions of the geometric features and operating conditions, the difference between the normalized solutions is less than 1%; therefore, it can be concluded that the normalization approach is valid. Further numerical analysis showed (see also [26,38]) that while the normalized results are not sensitive to m, μ, ΔP or U (it can also be shown from the linearity of Equation (6)), they are valid only for high L/R ratios (see [26]). Further numerical results indicate that for step bearings, the normalization method is valid for $L/R > 10$, while for partial texturing, it is valid for $L/R > 20$ (see [38]).

Table 1. Parameters used for the validation of the results normalization.

	Parameter	Case 1	Case 2	Case 3
Geometry of the piston	Piston length, L [m]	0.2	0.1	0.2
	Piston radius, R [m]	0.02	0.01	0.02
	Radial clearance, C [μm]	100	50	100
	Piston mass, m [kg]	5	2.5	500
Operating conditions	Pressure drop, ΔP [MPa]	1	1	100
	Velocity amplitude, U_{amp} [m/s]	2	1	2
	Viscosity, μ [Pa·s]	0.0022	0.0022	0.22
	Velocity frequency, f [Hz]	5	5	5
Texturing parameters	Imaginary square cell length, $2r_b$ [μm]	174	87	174
	Dimple base diameter, $2r_p$ [μm]	153	75	153
	Dimple depth, h_p [μm]	10	5	10
	Textured portion, α	0.045	0.045	0.045
	Dimple area density [%]	60	60	60
	Number of dimples	50	50	50

Table 1. *Cont.*

	Parameter	Case 1	Case 2	Case 3
Normalized parameters	U/U_0	0.088	0.088	0.088
	$f \times T$	0.5	0.5	0.5
	$\Delta P/\Delta P^*$	0.16	0.16	0.16

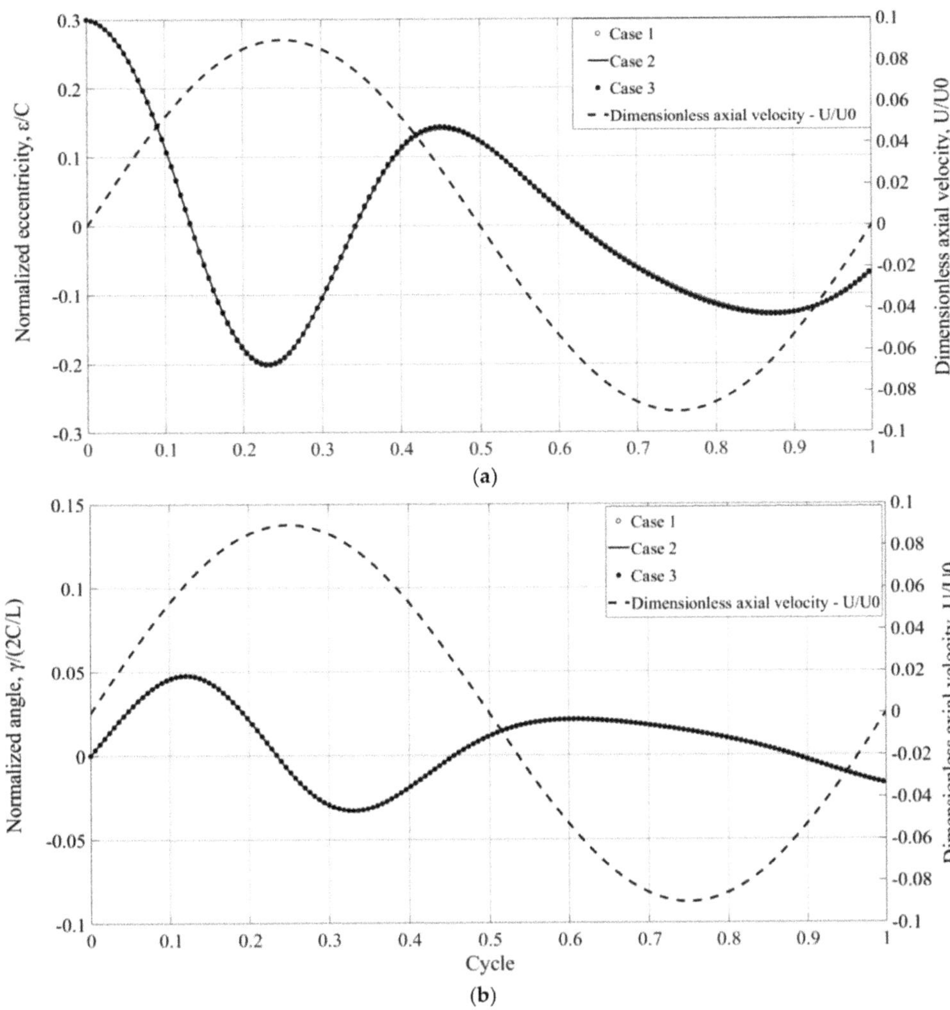

Figure 5. A typical solution for (**a**) dimensionless eccentricity and (**b**) dimensionless tilt for an initial dimensionless eccentricity of 0.3 for different cases.

Since the normalization method is valid, a smaller radius can be used. Based on Equation (14), decreasing the radius by a factor of 10 and the mass by a factor of 1000 will yield the same normalized results. Therefore, to save computation time (fewer dimples and, hence, fewer mesh elements in the circumferential direction), R = 0.002 [m] was used in all the subsequent results of the present work.

To validate the linearized model for secondary motion, the full simultaneous model was used. A comparison between the models is presented in Figure 6. The results are presented in terms of the minimum piston–cylinder gap, which varies over time. This gap represents the closest distance between any point on the piston and the cylinder wall. Since the minimum gap depends on both radial eccentricity and rotational motion, it is not fixed at a single point on the piston. The gap can range from 0 (indicating contact) to 1 (indicating a concentric position). The full simultaneous solution and the linear solution are presented by the solid and dot-and-dash lines, respectively. The axial velocity is presented by the dashed line related to the secondary vertical axis. The differences are small (less than 5% difference in the time period of the oscillations and less than 1% difference in the peak value of the oscillation); therefore, the linearized model is reasonable and can be used for the stability analysis. As for the computation time, a single case of simultaneous solution takes more than two weeks, while the computation time of the linear model is a few seconds.

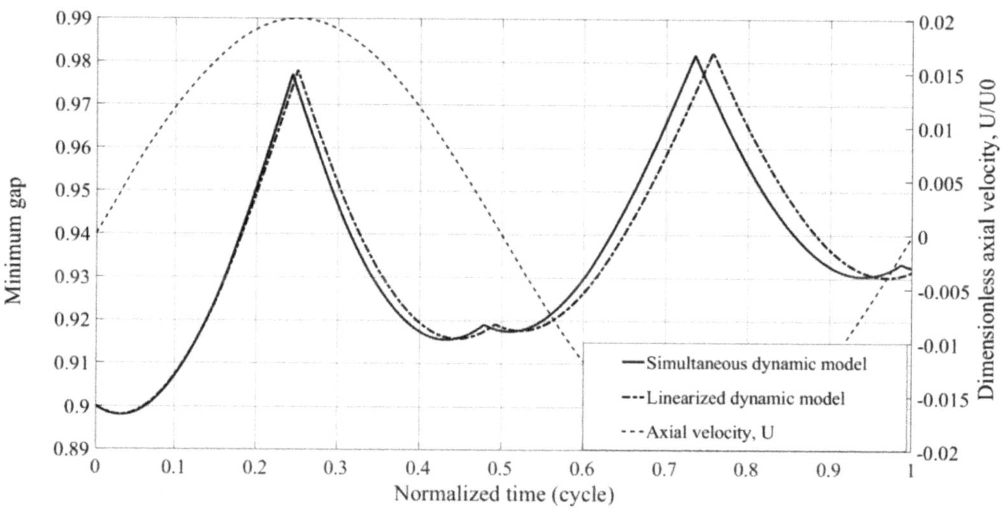

Figure 6. Minimum gap varying in time for the full simultaneous and linearized models.

4.2. The Form of the Stiffness Matrix

In the following, a numeric example for $\mathbf{K}^P$, $\mathbf{K}^U$ and $\mathbf{D}$ is presented. In addition, the general form of the stiffness matrices is discussed and an example for the root-locus map is presented.

For the parameters that are presented in case 1 (see Table 1), the stiffness and damping matrices obtained from the CFD model are:

$$\mathbf{K}^P = \begin{bmatrix} 28000 & -3 \times 10^5 \\ 2400 & 330 \end{bmatrix}, \mathbf{K}^U = \begin{bmatrix} 1.9 \times 10^6 & -11 \times 10^6 \\ 1.5 \times 10^5 & 19000 \end{bmatrix}, \mathbf{D} = \begin{bmatrix} 23 \times 10^6 & -10000 \\ -10000 & 50000 \end{bmatrix} \quad (16)$$

All the units are in MKS (they are different for each term; see Equation (6)). The matrices of Equation (16) can be used to obtain the trajectory of the secondary motion for any ΔP, μ, U_{amp} (see Equation (5)) for a given initial perturbation and to generate stability maps. The values of the matrix terms in Equation (16) are affected by the geometry of the piston and by the texturing or step parameters. However, since the results in this study are presented in dimensionless form, the values of the matrix terms for the other cases are not presented. Generally, $k_{\varepsilon\gamma}$ (see Equation (6)) and the damping matrix are quite insensitive to the texturing and the step, unlike the other terms, which are significantly more sensitive. For an untreated piston, $k_{\varepsilon\varepsilon} = k_{\gamma\varepsilon} = 0$, and the secondary motion is unstable regardless of ΔP, μ, U_{amp} (see [10]).

From Equation (16), it can be noticed that the cross-coupled stiffness coefficients (the terms on the secondary diagonal) are high relative to the direct stiffness coefficients. In addition, the cross-coupled stiffness coefficients have opposite signs. While, for most mechanical systems, the stiffness matrix is symmetric, non-symmetric stiffness matrices can be found, for instance, in journal bearings (see [39]) and robotic grasps [40]. To explain the opposite signs of the cross-coupled stiffness coefficients, an illustration of the perturbations and the resultant forces and moments is presented in Figure 7. For radial eccentricity (see Figure 7a), the center of pressure (the average location of all of the pressure acting on the piston) is close to the end of the texturing zone. This is because the local pressure increase due to eccentricity at the end of the texturing zone is maximal. Hence, when the textured portion is below 50%, the moment's direction is positive, e.g., $k^U_{\gamma\varepsilon}$ is positive. On the other hand, a positive angle leads to converging clearance (in the direction of the axial velocity, from left to right) on one side and diverging clearance on the other side; therefore, the direction of the force is negative, e.g., $k^U_{\varepsilon\gamma}$ is negative. In addition to the non-symmetry of the stiffness matrix, it is also time dependent (due to the time-dependent axial velocity). To the best of the authors' knowledge, a non-symmetric time-dependent stiffness matrix is unique to this study. It yields unique stability maps, which are presented in the upcoming subsection.

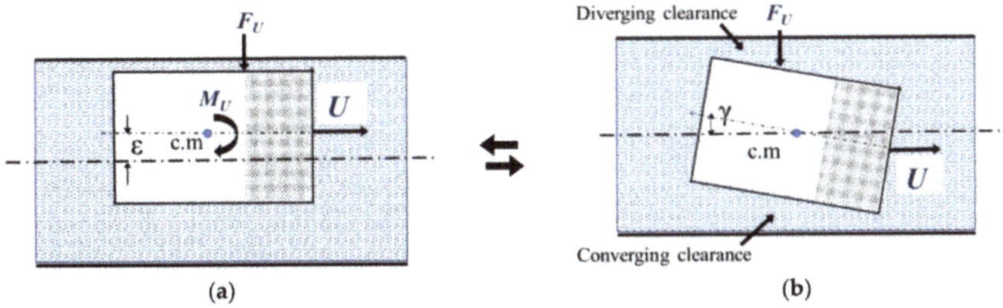

Figure 7. Illustration of the derivation of the stiffness matrix. (**a**) Moment due to eccentricity ($k^U_{\gamma\varepsilon}$) and (**b**) force due to tilt angle ($k^U_{\varepsilon\gamma}$).

4.3. Modal Solution

In Figure 8, a root-locus map for the considered parameters with varying ΔP is presented (without axial motion). For convenience, the map is divided into two parts. Each pole is presented by different color. It seems that there are two poles (eigenvalues of the state matrix **A**) with high negative real parts, which means that their corresponding eigenvectors are damped very quickly, and, hence, are outside the scope of our interest in the current study. On the other hand, there is a composite pole with a real part that is close to zero, and it becomes unstable (positive real part) for $\Delta P > 13.5$ [Mpa]. The instability of the system is due to relatively low damping (which is unaffected by pressure). Since the left poles dampen very quickly, the natural frequency of the piston is determined only by the imaginary part of the right poles (the upcoming results that discuss the decay rate and the natural frequency refer to the right pole)s. The corresponding mode shape (the two first terms) of the right poles for $\Delta P = 10$ [MPa] is $[-2.5 \times 10^{-5} \pm 2.7 \times 10^{-6}i; 1 \times 10^{-4} \pm 0.0032i]$.

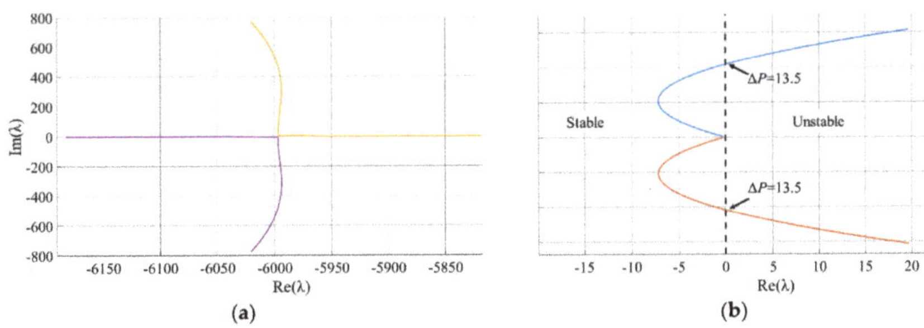

Figure 8. Root-locus map for the considered parameters with varying ΔP; (**a**) zoomed-in graph for the left poles and (**b**) zoomed-in graph for the right poles.

4.4. Influence of Step Parameters

In the following, the influence of different step parameters (length and depth) are examined. For convenience, the step depth is presented in dimensionless form as follows: $\beta = (h_S + C)/C$. The parameters were compared in terms of the stability range for low axial velocities that are negligible in comparison to the pressure drop ($U_{amp} << U_0$) obtained by using modal analysis. They were also compared in terms of the stability range for high amplitudes of the velocity and the stability range for low frequencies obtained by Floquet theory.

The ranges of the examined parameters for the current study are as follows:

- $0.1 < \alpha < 0.3$;
- $1.04 < \beta < 1.2$.

Figures 9 and 10 present a comparison between different step lengths ($0.1 < \alpha < 0.3$) for a dimensionless depth of $\beta = 1.1$. Figure 9a presents the maximum real part of the eigenvalues (decay rate), normalized by the radial vibration characteristic time, T, as a function of $\Delta P/\Delta P^*$, for the case without axial velocity. The solid, dashed and dotted lines indicate the different values of α, and the dashed red line indicates the stability threshold, $\max(\mathrm{Re}(\lambda(\mathbf{A}))) = 0$. Figure 9a shows that the secondary motion remains stable for a wide range of parameters at low axial velocities. For low values of $\Delta P/\Delta P^*$, lower values of α (among the considered parameters) result in faster convergence as the decay rate increases. The ratio $\Delta P/\Delta P^*$ depends on various parameters (see Equation (14)). However, the most significant parameter is the radial clearance, C, for instance, for the considered parameters in [10], in which $C = 25$ [μm] and $\Delta P/\Delta P^* \approx 10^{-5}$, and for the considered parameters in [26], in which $C = 100$ [μm] and $\Delta P/\Delta P^* \approx 0.2$. Generally, for $C < 100$ [μm], $\Delta P/\Delta P^* < 10$ for a wide range of parameters. For higher values of $\Delta P/\Delta P^*$ (which can be found especially for relatively large clearances), higher values of α result in better convergence (faster convergence and a wider range of stability).

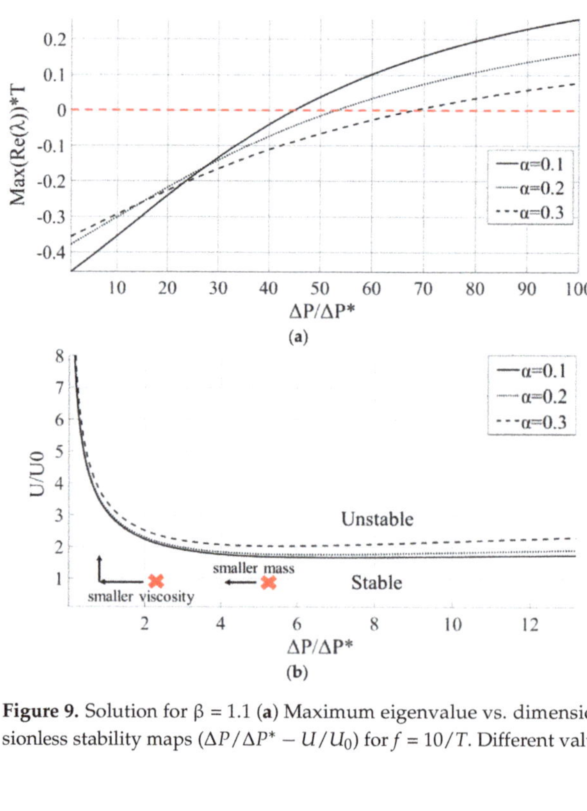

Figure 9. Solution for $\beta = 1.1$ (**a**) Maximum eigenvalue vs. dimensionless pressure and (**b**) dimensionless stability maps ($\Delta P/\Delta P^* - U/U_0$) for $f = 10/T$. Different values of α.

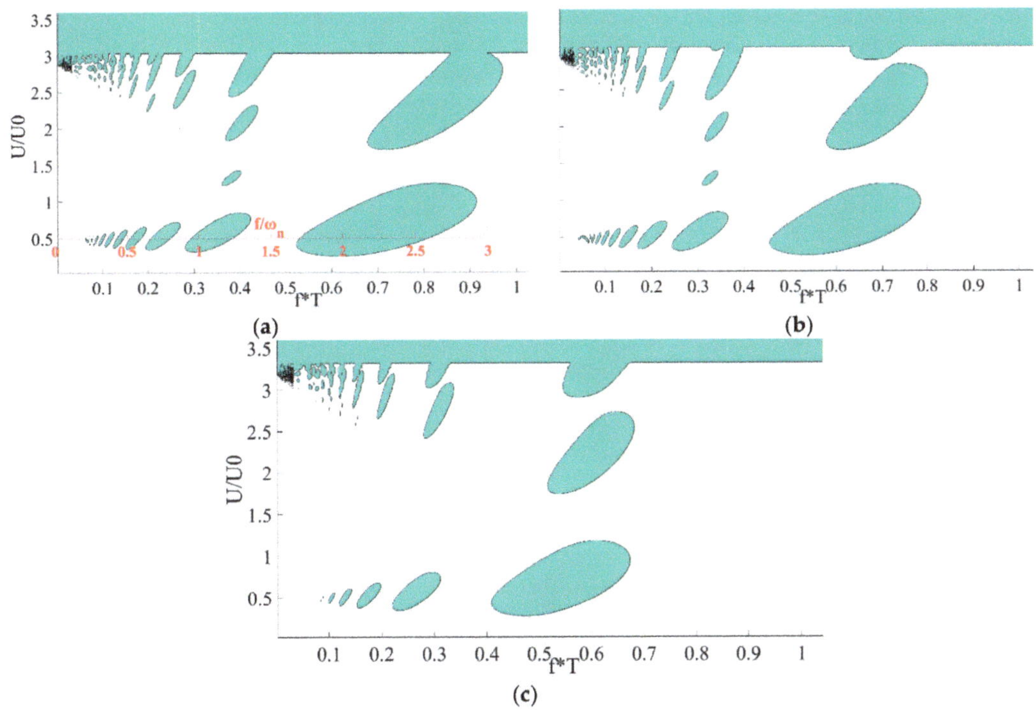

Figure 10. Dimensionless stability maps ($f \times T - U/U_0$ plane) for β = 1.1 when (**a**) α = 0.1, (**b**) α = 0.2 and (**c**) α = 0.3.

Figure 9b shows the dimensionless stability maps in the $\Delta P/\Delta P^* - U/U_0$ plane for different values of α and $f = 10/T$ (for $f \times T > 1$, the results are not sensitive to f). The solid, dashed and dotted lines indicate the threshold values of stability. Based on Figure 9b, it appears that higher values of α result in better convergence, as the secondary motion is stable for higher values of the axial velocity. In addition, in Figure 9b, it is demonstrated how a universal solution can be obtained. According to Equations (4) and (8), reducing the mass, m, leads to an increase in ΔP^*, and therefore, the normalized pressure decreases. In addition, increasing μ leads to an increase in ΔP^* and a decrease in U_0, as depicted by the arrows in Figure 9b. It can also be noted from Equation (5) that while increasing μ leads to increased damping and therefore increases the stability, it also leads to an increase in the parametric excitation, $\mu \mathbf{K}^U$, which may lead to instability. Similarly, the influence of the other parameters on the stability can be examined. For some cases, changing one of the parameters can cause a stable case to become unstable.

Figure 10 presents stability maps (Strutt diagrams) in the $f \times T - U/U_0$ plane for $\Delta P/\Delta P^* = 1$ for different values of α. The results show a unique phenomenon, derived from the non-symmetry of the stiffness matrix described above, of bubble-shaped areas of instability. Usually, Strutt diagrams are presented as the magnitude of the parametric excitation vs. the frequency, normalized by the natural frequency. However, since different values of α result in different natural frequencies and to enable generalization of the results, the frequency of the axial velocity, f, is normalized according to T, which is not dependent on the texturing or the step, and not according to the natural frequency. In our previous study (see Ref. [26]), it was shown that T is on the same order of magnitude as the natural frequency, which can be derived from the imaginary part of the right pole of matrix **A**. For all the results of the current study, the largest area of instability (the column of bubbles furthest to the right) is around twice the natural frequency (principal parametric resonance) as expected, and the second from the right column is around the natural frequency, as depicted on the secondary x-axis of Figure 10a. For smaller frequencies, multiple columns appear without any noticeable pattern, and for higher frequencies, there are no bubbles at all (a wider range than presented, up to $f \times T = 10$, was examined). These results (as well as the instability bubbles) are not compatible with other Strutt diagrams from the literature.

Further numerical results (see [38]) indicate that the frequencies of the axial velocity at which the bubble phenomenon occurs do not depend on $\Delta P/\Delta P^*$. Figure 10 shows that instability bubbles appear at relatively low frequencies ($f \times T < 1$), especially around $U = 0.5U_0$. For now, an explanation for this value was not found, and it should be addressed in the future. The results show, for the considered parameters, that the range of instability (frequencies and amplitudes of the velocity) decreases as α increases; therefore, for low frequencies and for the considered parameters, higher values of α are more stable.

Figures 11 and 12 present a comparison between different step depths ($1.04 < \beta < 1.2$) for a dimensionless length of $\alpha = 0.1$. Figure 11a shows the decay rate (for the case without axial velocity), normalized according to the radial vibration characteristic time, T, as a function of $\Delta P/\Delta P^*$. Figure 11a shows, similar to the results in Figure 9, that the secondary motion remains stable for a wide range of parameters at low axial velocities. For low values of $\Delta P/\Delta P^*$, $\beta = 1.1$ and $\beta = 1.2$ result in similar rates of convergence, while $\beta = 1.04$ has a slower rate of convergence. However, for higher values of $\Delta P/\Delta P^*$, $\beta = 1.2$ results in a faster rate of convergence and a wider range of stability. Figure 11b shows the dimensionless stability maps in the $\Delta P/\Delta P^* - U/U_0$ plane for different values of β and $f = 10/T$. Based on Figure 11b, it appears that for high axial velocities, $\beta = 1.1$ is the optimum. However, the differences are small. Figure 12 presents the Strutt diagram for $\Delta P/\Delta P^* = 1$ for different values of β. As with the results for α, there are bubbles of instability, especially around $U = 0.5U_0$. The results show, for the considered parameters, that the natural frequency and width of the bubbles increase as β increases. As for the amplitude of the axial motion, $\beta = 1.1$ is more stable, since the instability bubbles occur at higher amplitudes.

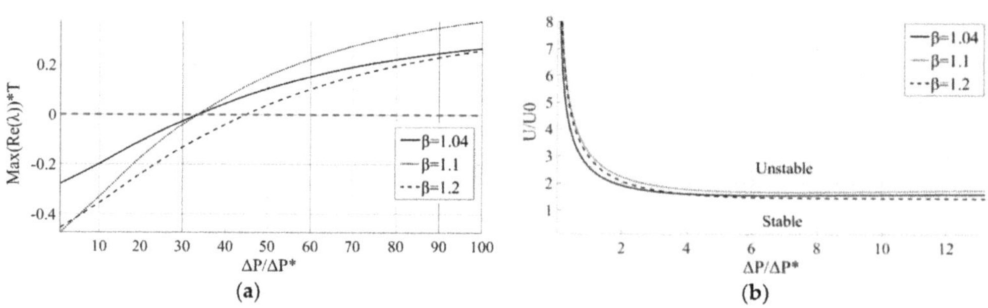

Figure 11. Solutions for $\alpha = 0.1$. (**a**) Maximum eigenvalue vs. dimensionless pressure and (**b**) dimensionless stability maps ($\Delta P/\Delta P^* - U/U_0$ plane) for $f = 10/T$. Different values of β.

4.5. Influence of Texturing Parameters (Dimples)

In this subsection, the influence of different texturing parameters on the stability of the secondary motion of a partially textured piston is examined. The following parameters are examined: aspect ratio, $2r_p/h_p$; textured portion, α; dimple depth, h_p/C; and dimple area density, S_p (see Figure 3). The reference dimensionless parameters are as follows: aspect ratio—$2r_p/h_p = 15$, textured portion—$\alpha = 0.05$, dimensionless dimple depth—$h_p/C = 0.1$ and dimple area density—$S_p = 60\%$.

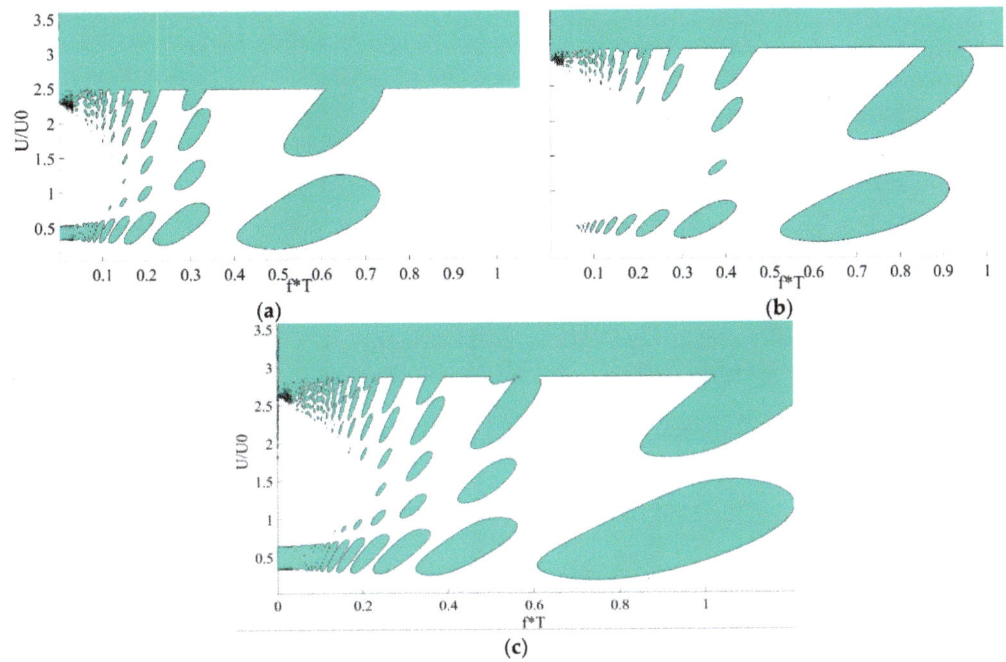

Figure 12. Dimensionless stability maps ($f \times T - U/U_0$ plane) for α = 0.1 when (**a**) β = 1.04, (**b**) β = 1.1 and (**c**) β = 1.2.

The ranges of the examined parameters for the current study are as follows:

- $0.05 < \alpha < 0.15$;
- $20\% < S_p < 60\%$;
- $0.05 < h_p/C < 0.2$;
- $7.5 < 2r_p/h_p < 30$.

In Figures 13 and 14, a comparison between different portions of textured area is presented (the other parameters are the same as in the reference case). The examined range ($0.05 < \alpha < 0.15$) is more limited than the range used for the step due to numerical limitations (higher values of α require a denser mesh). Figure 13a shows the decay rate (for the case without axial velocity), normalized according to the radial vibration characteristic time, T, as a function of $\Delta P/\Delta P^*$. The results indicate that the secondary motion remains stable for $\Delta P/\Delta P^* < 10$. The rate of convergence and the stability range are similar for the considered values of α. Figure 13b shows the dimensionless stability maps in the $\Delta P/\Delta P^* - U/U_0$ plane for different values of α and $f = 10/T$. The influence of α on the stability range is small. Figure 14 presents the Strutt diagram for $\Delta P/\Delta P^* = 1$ for different values of α. It seems that the phenomenon of "bubbles" occurs within the same range of frequencies for all the values of α. Hence, the natural frequency is not sensitive to α. Nevertheless, the instability bubbles are smaller for higher values of α. This is because $k^U_{\gamma\varepsilon}$ is smaller for higher values of α, and therefore, the non-symmetry of $\mathbf{K}^U$ is smaller. For the considered parameters, it seems that the textured piston results in a slower convergence rate (higher values of the maximum eigenvalue) and a smaller stability range than the step-profiled piston (see Figures 9–11). However, for small frequencies, the secondary motion of the textured piston is more stable, as derived from Figure 14.

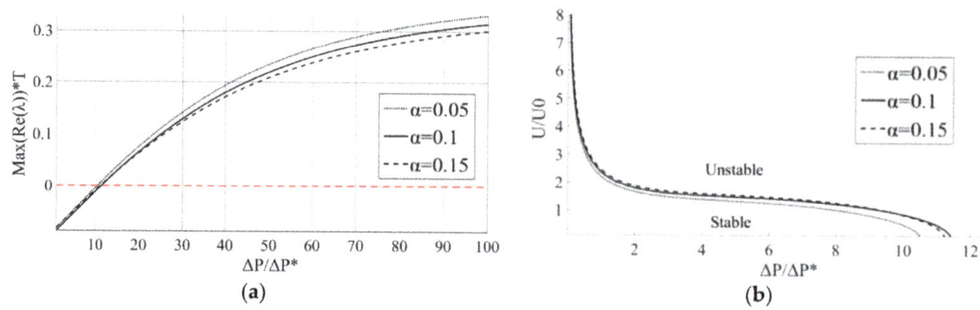

Figure 13. Solutions for different values of α: (**a**) maximum eigenvalue vs. dimensionless pressure and (**b**) dimensionless stability maps ($\Delta P/\Delta P^* - U/U_0$ plane).

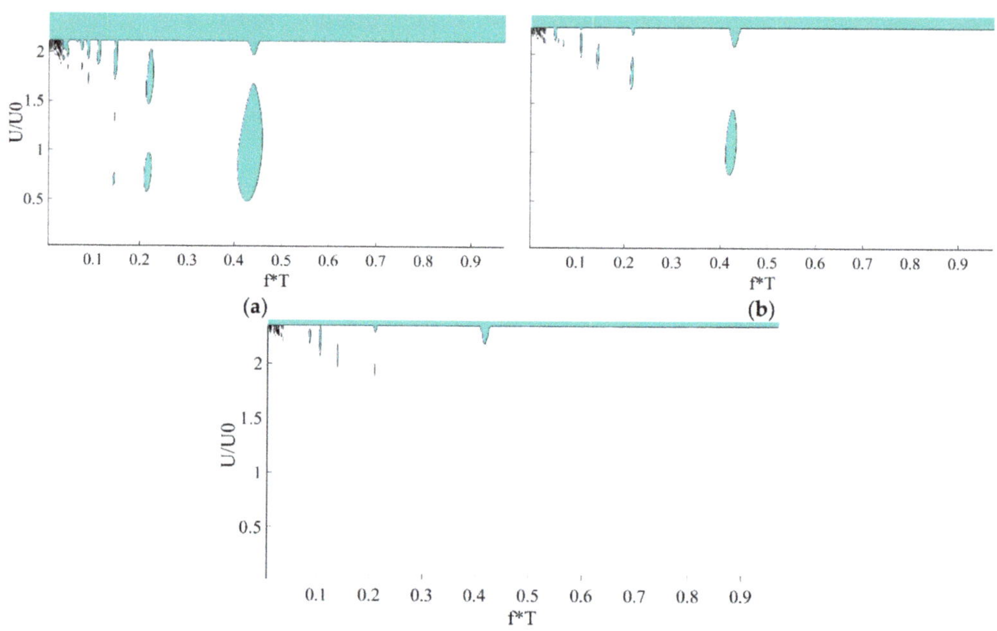

Figure 14. Dimensionless stability maps ($f \times T - U/U_0$ plane) for (**a**) α = 0.05, (**b**) α = 0.1 and (**c**) α = 0.15.

In Figures 15 and 16, a comparison between different dimple area densities (20% < S_p < 60%) is presented (the other parameters are the same as in the reference case). Figure 15a shows the decay rate (for the case without axial velocity), normalized according to the radial vibration characteristic time, T, as a function of $\Delta P/\Delta P^*$. It seems that for low velocities, S_p should be as high as possible. These results are compatible with the results in [19], in which there is no optimum for S_p. Figure 15b presents the dimensionless stability maps in the $\Delta P/\Delta P^* - U/U_0$ plane for different values of S_p and $f = 10/T$. The results indicate that for high velocities and low values of $\Delta P/\Delta P^*$, $S_p = 20\%$ results in a greater range of stability, unlike the results for low velocities (see Figure 15a). This is due to smaller values of $\mathbf{K}^U$ (smaller parametric excitation). Figure 16 presents the Strutt diagram for $\Delta P/\Delta P^* = 1$ for different values of S_p. The results indicate that for low frequencies, $S_p = 20\%$ is significantly less stable. This is because $k^U_{\gamma\varepsilon}$ is relatively small (significantly smaller than all the other cases in the current study);

therefore, the non-symmetry of the stiffness matrix, $\mathbf{K}^U$, is small, leading to a more classic Strutt diagram, which is less stable at the natural frequency and double that value (principal parametric resonance).

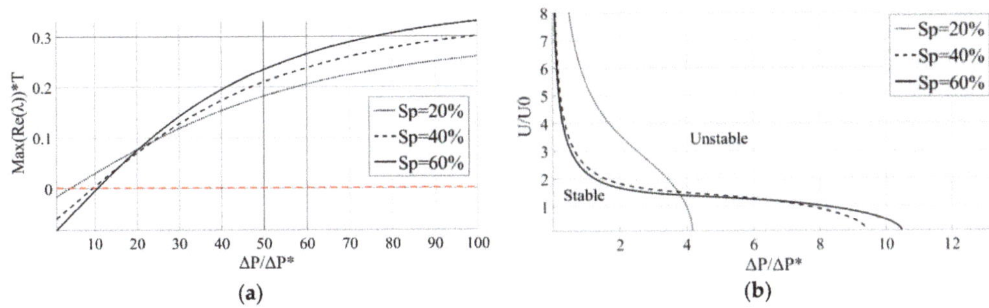

Figure 15. Solutions for different values of S_p: (**a**) maximum eigenvalue vs. dimensionless pressure and (**b**) dimensionless stability maps ($\Delta P/\Delta P^* - U/U_0$ plane).

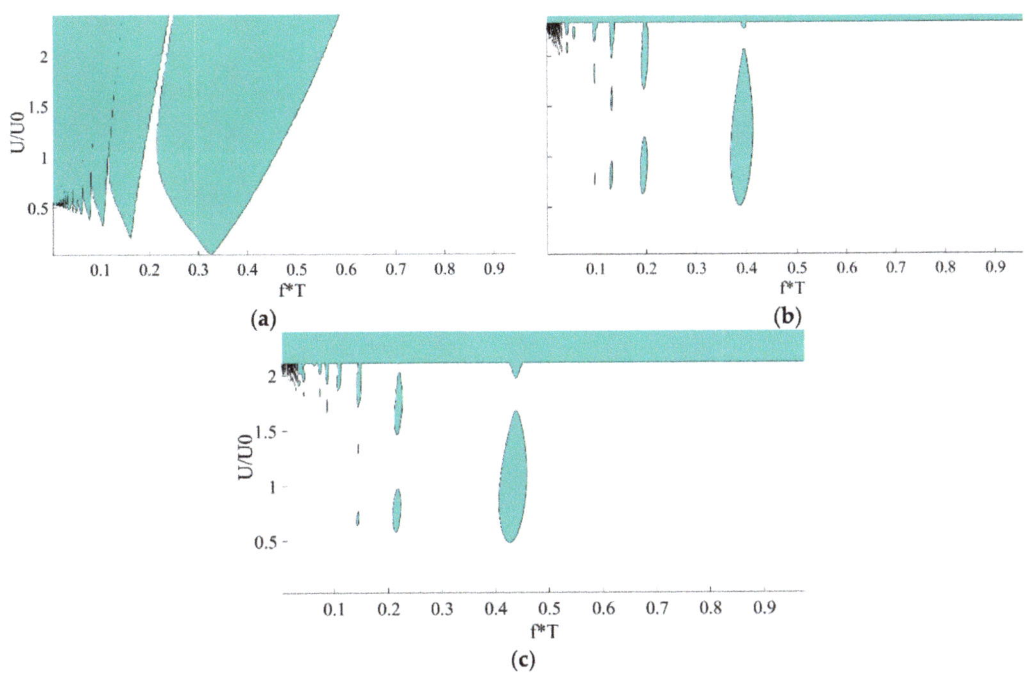

Figure 16. Dimensionless stability maps ($f \times T - U/U_0$ plane) for (**a**) $S_p = 20\%$, (**b**) $S_p = 40\%$ and (**c**) $S_p = 60\%$.

In Figures 17 and 18, a comparison between different dimensionless dimple depths ($0.05 < h_p/C < 0.2$) is presented (the other parameters are the same as in the reference case). Figure 17a shows the decay rate (for the case without axial velocity), normalized according to the radial vibration characteristic time, T, as a function of $\Delta P/\Delta P^*$. The results indicate that while $h_p/C = 0.2$ is more stable for low values of $\Delta P/\Delta P^*$, for higher values, $h_p/C = 0.05$ is more stable, since it remains stable regardless of $\Delta P/\Delta P^*$. This is because, for this case, the direct stiffness coefficients are high relative to the cross-coupled stiffness coefficients, and as a result, the eigenvalues of the state matrix, **A**, for the case without damping (μ = 0) are purely imaginary. Therefore, minimum damping is sufficient to stabilize the system. Figure 17b presents the dimensionless stability maps in the $\Delta P/\Delta P^* - U/U_0$ plane for different values of h_p/C and $f = 10/T$. It seems that the stability range for $h_p/C = 0.05$ is significantly greater, as it remains stable for high values of U/U_0 (for low values of $\Delta P/\Delta P^*$, the stability threshold is greater than the values that are presented). This is because the values of $\mathbf{K}^U$ are small relative to the values of $\mathbf{K}^P$. Figure 18 presents the Strutt diagram for $\Delta P/\Delta P^* = 1$ for different values of h_p/C. It seems that the natural frequency and the range of frequencies in which the instability bubbles occur for $h_p/C = 0.05$ are smaller.

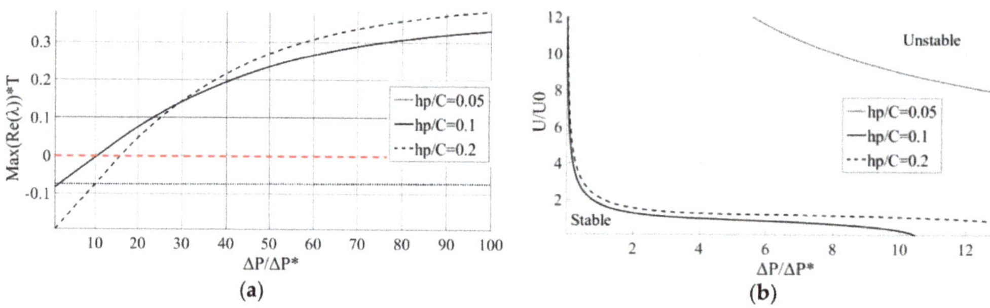

Figure 17. Solutions for different values of h_p/C: (**a**) maximum eigenvalue vs. dimensionless pressure and (**b**) dimensionless stability maps ($\Delta P/\Delta P^* - U/U_0$ plane).

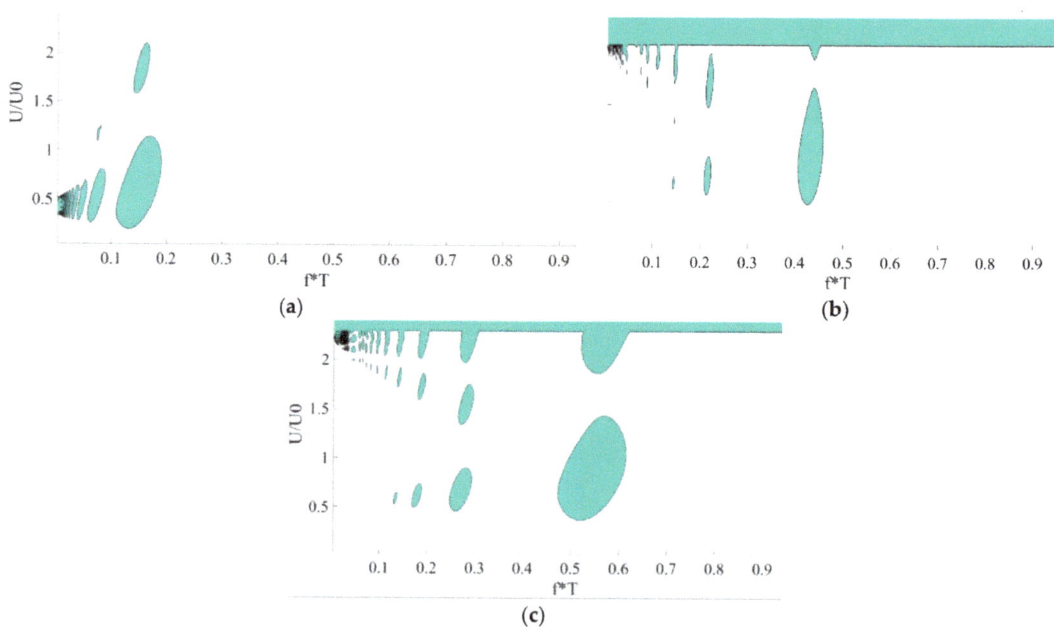

Figure 18. Dimensionless stability maps ($f \times T - U/U_0$ plane) for (**a**) $h_p/C = 0.05$, (**b**) $h_p/C = 0.1$ and (**c**) $h_p/C = 0.2$.

In Figures 19 and 20, a comparison among different aspect ratios, $7.5 < 2r_p/h_p < 30$, is presented. Figure 19a shows the decay rate (for the case without axial velocity), normalized according to the radial vibration characteristic time, T, as a function of $\Delta P/\Delta P^*$. The results show that for the considered parameters, for low axial velocities, $2r_p/h_p = 30$ is the optimum. Figure 19b presents the dimensionless stability maps in the $\Delta P/\Delta P^* - U/U_0$ plane for different values of $2r_p/h_p$ and $f = 10/T$. The results show that the stability range for $2r_p/h_p = 30$ is greater. Figure 20 presents the Strutt diagram for $\Delta P/\Delta P^* = 1$ for different values of $2r_p/h_p$. It seems that for low frequencies, $2r_p/h_p = 15$ is the optimum, since the range of the bubbles (frequency and amplitude) is smaller. In addition, for $2r_p/h_p = 30$, the density of the bubbles seems to be the highest among all the considered parameters of the current study. This is due to the high non-symmetry of $\mathbf{K}^U$.

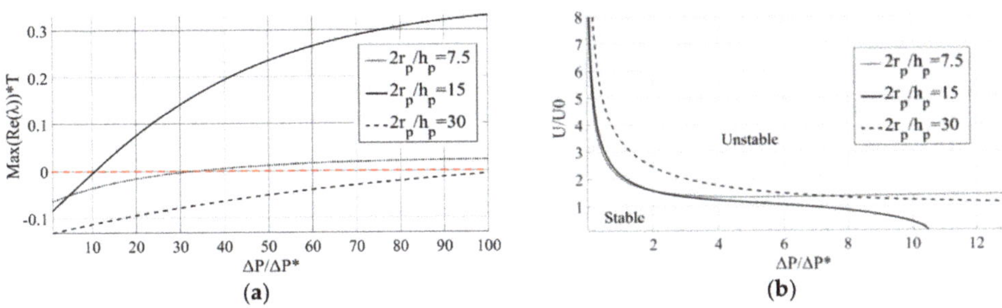

Figure 19. Solutions for different values of $2r_p/h_p$: (**a**) maximum eigenvalue vs. dimensionless pressure and (**b**) dimensionless stability maps ($\Delta P/\Delta P^* - U/U_0$ plane).

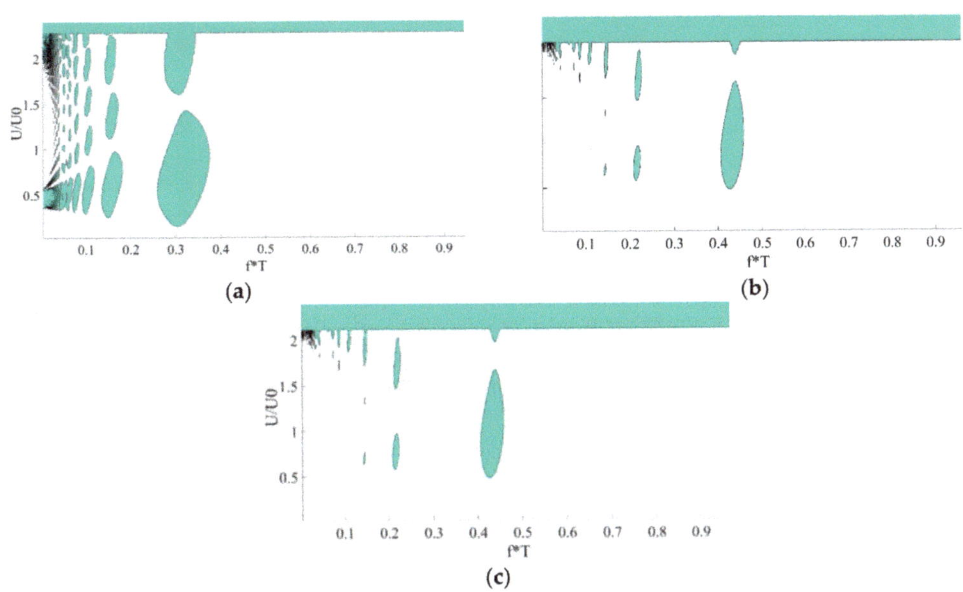

Figure 20. Dimensionless stability maps ($f * T - U/U_0$ plane) for (**a**) $2r_p/h_p = 7.5$, (**b**) $2r_p/h_p = 15$ and (**c**) $2r_p/h_p = 30$.

5. Conclusions

A linear model based on CFD simulations and Floquet theory was developed to analyze the stability of the secondary motion of partially textured and step-profiled pistons around the working point of a concentric piston. This linearized model was validated against the full simultaneous model. A normalization approach was introduced and validated, allowing the results to be generalized across various geometries and operating conditions. It was shown that the axial motion leads to parametric excitation, resulting in a time-dependent and non-symmetric stiffness matrix, which leads to a unique phenomenon of bubble-shaped areas of instability in the Strutt diagrams for low frequencies (up to about 2.5 times the natural frequency). Unlike the classic cases, in which increasing the parametric excitation at some dangerous frequencies (especially the natural frequency and double that value) leads to instability of the system, the presence of instability bubbles indicates that increasing excitation up to a certain level can stabilize the system. Furthermore, it was demonstrated that while a minor degree of non-symmetry in K^U yields a more traditional yet less stable Strutt diagram, excessive non-symmetry increases the number and size of instability bubbles. Thus, an optimal level of non-symmetry exists that enhances stability.

The influence of different step and texturing parameters on the stability of the secondary motion was examined for three different cases: low axial velocities with various pressures, high axial velocities with various pressures and low frequencies with various velocities. Some guidelines can be drawn regarding the optimal parameters based on the operating conditions.

The following was found for length:

- For a step-profiled piston, the influence of the step length, α, seems to be small. For all the considered parameters, the stability range of the secondary motion is wide. For low velocities, $\alpha = 0.1$ results in faster stabilizing for most of the cases, while for high velocities and for low frequencies, $\alpha = 0.3$ is more stable.
- For the length of the textured zone, a similar trend was observed for low axial velocities (for texturing, the range of α that was examined is $0.05 < \alpha < 0.15$). However, for high axial velocities, there is an optimum ($\alpha = 0.1$), and for low frequencies, the texturing length does not affect the stability range.

For depth, the following was found:

- For a step-profiled piston, the step depth, β, for the considered parameters and for low velocities, should be as high as possible. For high velocities, there is an optimum ($\beta = 1.1$), and for low frequencies, β should be as low as possible.
- For the texturing depth, h_p/C, the range of h_p/C that was examined is $0.05 < h_p/C < 0.2$. $h_p/C = 0.05$ seems to be the optimum texturing among all the considered dimensionless texturing and step parameters for a wide range of parameters, as it remains stable for high values of $\Delta P/\Delta P^*$ and for high amplitudes of the axial velocities. In addition, the range of the instability bubbles is the smallest. Nevertheless, for low axial velocities and low values of $\Delta P/\Delta P^*$, $h_p/C = 0.2$ is more stable.

For the other texturing parameters, the following was found:

- For the dimple area density, S_p, the range of S_p that was examined is $20\% < S_p < 60\%$. For the considered texturing parameters at low amplitudes of velocity, S_p should be as high as possible. For high velocities, low values of S_p are more stable. Therefore, for applications with high reciprocating velocities and relatively low pressure drops, S_p should be low. However, low values may be significantly less stable for low frequencies.
- For the aspect ratio $2r_p/h_p$, the range of $2r_p/h_p$ that was examined is $7.5 < 2r_p/h_p < 30$. For the considered parameters, at low velocities, the aspect ratio should be as high as possible, and at high velocities and low frequencies, there is an optimum.

For the considered texturing parameters, the results are very sensitive to the texturing depth and aspect ratios; therefore, further investigation for different depths is needed. In addition, the current model assumes that the axial motion is independent, and therefore, it does not consider the radial forces resulting from the mechanism of the axial velocity. This limitation should be addressed in future work to make the model applicable to a wider range of piston–cylinder systems such as piston engines.

Author Contributions: Conceptualization, I.C., S.T. and R.G.; methodology, I.C., S.T. and R.G.; software, I.C.; validation, I.C.; investigation, I.C.; writing—original draft preparation, I.C.; writing—review and editing, S.T. and R.G.; supervision, S.T. and R.G. All authors have read and agreed to the published version of the manuscript.

Funding: This research received no external funding.

Informed Consent Statement: This study complied with ethical standards.

Data Availability Statement: Data are contained within the article.

Conflicts of Interest: The authors declare that they have no known competing financial interests or personal relationships that could have appeared to influence the work reported in this paper.

Nomenclature

$\mathbf{A}$	State matrix
C	Nominal clearance between the piston and the cylinder
$\overline{\mathbf{D}}$	Damping matrix
$\mathbf{D}$	Modified damping matrix
f	Frequency of the axial velocity
F	Radial force acting on the piston
h_p	Dimple depth
h_s	Step depth
I	Piston's axial mass moment of inertia
$\mathbf{K}$	Stiffness matrix
$\mathbf{K}^P$	Pressure stiffness matrix
$\mathbf{K}^U$	Axial velocity stiffness matrix
L	Length of the piston
m	Mass of the piston
M	Moment acting on the piston
$\mathbf{M}$	Mass matrix
$\widehat{\mathbf{M}}$	Added mass matrix
p	Local pressure of the fluid
P_h	Pressure on the high-pressure side
P_l	Pressure on the low-pressure side
ΔP	Pressure drop $(P_h - P_l)$
ΔP^*	Characteristic pressure drop
R	Radius of the piston
r, z, θ	Coordinates
$2r_b$	Imaginary square cell length
r_p	Dimple base radius
Re*	Reduced Reynolds number, $\rho \Delta p C^4/(L\,\mu)^2$
S_p	Dimple density
T	Piston's characteristic vibration cycle time
t	Time
t_0	Reference time
$U(t)$	Axial velocity of the piston
U_{amp}	Amplitude of the axial velocity
U_0	Reference axial velocity
$\mathbf{v}$	Velocity vector of the fluid
$\mathbf{x}$	State vector
α	Textured portion
β	Dimensionless step depth
γ	Tilt angle of the piston
ε	Radial displacement of the piston
λ	Eigenvalues of the state matrix
μ	Viscosity of the fluid
ρ	Density of the fluid

Appendix A

In the following, it is shown analytically, based on the Reynolds equation, that $\mathbf{K}^P$, K^U and $\mathbf{D}$ can be separated and that the matrices have linear dependency on ΔP, μU_{amp} and μ, respectively. This is needed in order to derive Equation (5).

The two-dimensional, time-dependent form of the Reynolds equation for an incompressible Newtonian fluid in laminar flow is given by [14]:

$$\begin{array}{l}\frac{\partial}{\partial z}(h^3(z,\theta)\frac{\partial p}{\partial z}) + \frac{1}{R^2}\frac{\partial}{\partial \theta}(h^3(z,\theta)\frac{\partial p}{\partial \theta}) = 12\mu\frac{\partial h(z,\theta)}{\partial t} - 6\mu U(t)\frac{\partial h(z,\theta)}{\partial z} \\ p(z=0) = P_h p(z=L) = P_l\end{array} \quad (A1)$$

where h is the local value of the film thickness and θ is the circumferential coordinate. Due to the linearity of the Reynolds equation, it can be separated into three equations. The pressure distribution and the radial force and moment due to the pressure drop, which are used to calculate $\mathbf{K}^P$, can be obtained as follows:

$$\begin{array}{l}\frac{\partial}{\partial z}(h^3(z,\theta)\frac{\partial p}{\partial z}) + \frac{1}{R^2}\frac{\partial}{\partial \theta}(h^3(z,\theta)\frac{\partial p}{\partial \theta}) = 0 \\ p(z=0) = P_h p(z=L) = P_l\end{array} \quad (A2)$$

The pressure distribution and the radial force and moment due to the initial velocity (squeeze effect), which are used to calculate $\mathbf{D}$, can be obtained as follows:

$$\begin{array}{l}\frac{\partial}{\partial z}(h^3(z,\theta)\frac{\partial p}{\partial z}) + \frac{1}{R^2}\frac{\partial}{\partial \theta}(h^3(z,\theta)\frac{\partial p}{\partial \theta}) = 12\mu\frac{\partial h(z,\theta)}{\partial t} \\ p(z=0) = 0 p(z=L) = 0\end{array} \quad (A3)$$

The pressure distribution and the radial force and moment due to the axial velocity, which are used to calculate $\mathbf{K}^U$, can be obtained by:

$$\begin{array}{l}\frac{\partial}{\partial z}(h^3(z,\theta)\frac{\partial p}{\partial z}) + \frac{1}{R^2}\frac{\partial}{\partial \theta}(h^3(z,\theta)\frac{\partial p}{\partial \theta}) = -6\mu U(t)\frac{\partial h(z,\theta)}{\partial z} \\ p(z=0) = 0 p(z=L) = 0\end{array} \quad (A4)$$

The sum of the solutions for Equations (A2)–(A4), which are not coupled, provides the solution for Equation (A1); therefore, $\mathbf{K}^P$, $\mathbf{K}^U$ and $\mathbf{D}$ can be separated. In addition, since Equation (A2) and the boundary conditions of Equations (A3) and (A4) are homogenous, there is a linear dependency between ΔP, μU_{amp} and the resultant pressure distribution (as well as the overall forces) of Equations (A2)–(A4).

References

1. Etsion, I.; Magen, M. Piston Vibration in Piston-Cylinder Systems. *J. Fluids Eng.* **1980**, *102*, 26–33. [CrossRef]
2. Forero, J.D.; Ochoa, G.V.; Alvarado, W.P. Study of the Piston Secondary Movement on the Tribological Performance of a Single Cylinder Low-Displacement Diesel Engine. *Lubricants* **2020**, *8*, 97. [CrossRef]
3. Yamaguchi, A. Motion of the Piston in Piston Pumps and Motors: The Case of Metallic Contact. *JSME Int. J. Ser. 3 Vib. Control Eng. Eng. Ind.* **1990**, *33*, 627–633. [CrossRef]
4. Yamaguchi, A. Motion of Pistons in Piston-Type Hydraulic Machines: 3rd Report, Exponential Function-Type Piston. *Bull. JSME* **1976**, *19*, 413–419. [CrossRef]
5. Yamaguchi, A.; Tanioka, Y. Motion of Pistons in Piston-Type Hydraulic Machines: 2nd Report: Experiment. *Bull. JSME* **1976**, *19*, 408–412. [CrossRef]
6. Cheng, J.; Zeng, X.; Liu, Z.; Yu, X.; Feng, Q. Research on Dynamic Modeling and Electromagnetic Force Centering of Piston/Piston Rod System for Labyrinth Piston Compressor. *Proc. Inst. Mech. Eng. Part I J. Syst. Control Eng.* **2016**, *230*, 786–798. [CrossRef]
7. Xie, J.; Zeng, X.; Zhang, M.; Cheng, J.; Wang, H. Implementation of Active Magnetic Control System for Piston Centering in Labyrinth Piston Compressor. *Mechatronics* **2018**, *54*, 52–67. [CrossRef]
8. Etsion, I. An Optimum Step Design for Centering of Pistons. *J. Fluids Eng.* **1975**, *97*, 621–624. [CrossRef]
9. Etsion, I.; Hamrock, B.J. Optimum Step Design for Centering of Pistons Moving in an Incompressible Fluid. *J. Fluids Eng.* **1977**, *99*, 675–680. [CrossRef]
10. Kligerman, Y.; Cohen, I.; Shinkarenko, A. Centering and Friction Reduction between Parts of the Cylinder-Piston Group. In Proceedings of the 10th International Scientific Conference BALTTRIB 2019, Kaunas, Lithuania, 14–16 November 2019; pp. 94–101. [CrossRef]
11. Tønder, K. Hydrodynamic Effects of Tailored Inlet Roughnesses: Extended Theory. *Tribol. Int.* **2004**, *37*, 137–142. [CrossRef]
12. Kumar, V.; Verma, R.; Kango, S.; Sharma, V.S. Recent Progresses and Applications in Laser-Based Surface Texturing Systems. *Mater. Today Commun.* **2021**, *26*, 101736. [CrossRef]
13. Etsion, I. Modeling of Surface Texturing in Hydrodynamic Lubrication. *Friction* **2013**, *1*, 195–209. [CrossRef]

14. Ronen, A.; Etsion, I.; Kligerman, Y. Friction-Reducing Surface-Texturing in Reciprocating Automotive Components. *Tribol. Trans.* **2001**, *44*, 359–366. [CrossRef]
15. Wu, Z.; Xing, Y.; Huang, P.; Liu, L. Tribological Properties of Dimple-Textured Titanium Alloys under Dry Sliding Contact. *Surf. Coat. Technol.* **2017**, *309*, 21–28. [CrossRef]
16. Mao, B.; Siddaiah, A.; Liao, Y.; Menezes, P.L. Laser Surface Texturing and Related Techniques for Enhancing Tribological Performance of Engineering Materials: A Review. *J. Manuf. Process.* **2020**, *53*, 153–173. [CrossRef]
17. Etsion, I. State of the Art in Laser Surface Texturing. *J. Tribol.* **2005**, *127*, 248–253. [CrossRef]
18. Etsion, I.; Halperin, G. A Laser Surface Textured Hydrostatic Mechanical Seal. *Tribol. Trans.* **2002**, *45*, 430–434. [CrossRef]
19. Brizmer, V.; Kligerman, Y.; Etsion, I. A Laser Surface Textured Parallel Thrust Bearing. *Tribol. Trans.* **2003**, *46*, 397–403. [CrossRef]
20. Kligerman, Y.; Etsion, I.; Shinkarenko, A. Improving Tribological Performance of Piston Rings by Partial Surface Texturing. *J. Tribol.* **2005**, *127*, 632–638. [CrossRef]
21. Cohen, I.; Goltsberg, R. Partial Surface Texturing in Hydrodynamic Lubrication: A CFD-Based Investigation. *Lubricants* **2023**, *11*, 395. [CrossRef]
22. Gropper, D.; Wang, L.; Harvey, T.J. Hydrodynamic Lubrication of Textured Surfaces: A Review of Modeling Techniques and Key Findings. *Tribol. Int.* **2016**, *94*, 509–529. [CrossRef]
23. Fagotti, F.; Fernandes, J.R.S. Dynamic Analysis of Piston Secondary Motion for Small Reciprocating Compressors. *J. Tribol.* **2000**, *122*, 752–760. [CrossRef]
24. Tan, Y.C.; Ripin, Z.M. Technique of Measuring Piston Secondary Motion Using Laser Displacement Sensors. *Exp. Mech.* **2012**, *52*, 1447–1459. [CrossRef]
25. Tan, Y.C.; Ripin, Z.M. Analysis of Piston Secondary Motion. *J. Sound Vib.* **2013**, *332*, 5162–5176. [CrossRef]
26. Cohen, I.; Kligerman, Y.; Goltsberg, R. CFD Analysis of a Ringless Piston's Secondary Motion and the Validity of Reynolds Equation. *Meccanica* **2023**, *58*, 1347–1364. [CrossRef]
27. Kumar, S.; Bergada, J.M. The Effect of Piston Grooves Performance in an Axial Piston Pumps via CFD Analysis. *Int. J. Mech. Sci.* **2013**, *66*, 168–179. [CrossRef]
28. Jiang, J.; Wang, K.; Yan, W.; Sun, Y. A Novel Fluid Structure Interaction Model for the Grooved Piston-Cylinder Interface in Axial Piston Pump. *AIP Adv.* **2019**, *9*, 055013. [CrossRef]
29. Delprete, C.; Razavykia, A. Piston Dynamics, Lubrication and Tribological Performance Evaluation: A Review. *Int. J. Engine Res.* **2020**, *21*, 725–741. [CrossRef]
30. Haddad, W.M.; Chellaboina, V. Chapter Three. Stability Theory for Nonlinear Dynamical Systems. In *Nonlinear Dynamical Systems and Control*; Princeton University Press: Princeton, NJ, USA, 2019; pp. 135–206. [CrossRef]
31. Silva, J.M.M.; Maia, N.M.M. (Eds.) *Modal Analysis and Testing*; Springer: Dordrecht, The Netherlands, 1999.
32. Nayfeh, A.H. *Perturbation Methods*; WILEY-VCH Verlag GmbH & Co. KGaA: Hoboken, NJ, USA, 2000.
33. Richards, J.A. *Analysis of Periodically Time-Varying Systems*; Springer: Berlin/Heidelberg, Germany, 1983.
34. Kligerman, Y.; Etsion, I. Analysis of the Hydrodynamic Effects in a Surface Textured Circumferential Gas Seal. *Tribol. Trans.* **2001**, *44*, 472–478. [CrossRef]
35. Szeri, A.Z. *Fluid Film Lubrication*, 2nd ed.; Cambridge University Press: Cambridge, UK, 2010; ISBN 978-0-511-78202-2.
36. Cohen, I.; Goltsberg, R.; Kligerman, Y.; Tresser, S. *Dynamic Analysis for Optimizing Tribological Performance of Ringless Piston Cylinder Systems/Igal Cohen*; Supervision: Roman Goltsberg, Shachar Tresser, Yuri Kligerman; Technion—Israel Institute of Technology: Haifa, Israel, 2023.
37. *Ansys Fluent Theory Guide*; ANSYS: Canonsburg, PA, USA, 2021; Volume 15317.
38. Li, Q.; Zhang, S.; Ma, L.; Xu, W.; Zheng, S. Stiffness and Damping Coefficients for Journal Bearing Using the 3D Transient Flow Calculation. *J. Mech. Sci. Technol.* **2017**, *31*, 2083–2091. [CrossRef]
39. Dimond, T.W.; Younan, A.A.; Allaire, P. Comparison of Tilting-Pad Journal Bearing Dynamic Full Coefficient and Reduced Order Models Using Modal Analysis (GT2009-60269). *J. Vib. Acoust.* **2010**, *132*, 0510091–05100910. [CrossRef]
40. Li, J.; Kao, I. Grasp Stiffness Matrix-fundamental Properties in Analysis of Grasping and Manipulation. In Proceedings of the 1995 IEEE/RSJ International Conference on Intelligent Robots and Systems. Human Robot Interaction and Cooperative Robots, Pittsburgh, PA, USA; 1995; Volume 2, pp. 381–386. [CrossRef]

MDPI AG
Grosspeteranlage 5
4052 Basel
Switzerland
Tel.: +41 61 683 77 34

Lubricants Editorial Office
E-mail: lubricants@mdpi.com
www.mdpi.com/journal/lubricants

www.ingramcontent.com/pod-product-compliance
Lightning Source LLC
Chambersburg PA
CBHW061836210926
51850CB00006B/904
9783725851034